MODELLING OF METAL FORMING PROCESSES

Of Related Interest:

J. Caldwell & R. Bradley (eds.) Industrial Vibration Modelling, 1987,
    ISBN 90–247–3423–1
O. D. D. Soares & Perez-Amor (eds.) Applied Laser Tooling, 1987,
    ISBN 90–247–3486–x
R. J. Moreau (ed.) Measurement and Control in Liquid Metal Processing, 1987,
    ISBN 90–247–3510–6
T. J. Smith (ed.) Modelling the Flow and Solidification of Metals, 1987,
    ISBN 90–247–3526–2
G. N. Pande & J. Middleton (eds.) Numerical Techniques for Engineering Analysis
    and Design, NUMETA 87, Volume 1, 1987, ISBN 90–247–3564–5
G. N. Pande & J. Middleton (eds.) Transient/Dynamical Analysis and Constitutive
    Laws for Engineering Materials, NUMETA 87, Volume 2, 1987,
    ISBN 90–247–3565–3

Modelling of Metal Forming Processes

Proceedings of the Euromech 233 Colloquium, Sophia Antipolis,
France, August 29–31, 1988

edited by

J.L. CHENOT

Centre de Mise en Forme des Matériaux,
Ecole Nationale Supérieure des Mines de Paris,
Sophia Antipolis, Valbonne, France

and

E. OÑATE

Escuela Técnica Superior
de Ingenieros de Caminos y Puertos,
Barcelona, Spain

KLUWER ACADEMIC PUBLISHERS
DORDRECHT / BOSTON / LONDON

Library of Congress Cataloging in Publication Data

Euromech Colloquium (233 : 1988 : Sophia Antipolis,
    France)
    Modelling of metal forming processes.

    1. Metal-work--Computer simulation--Congresses.
2. Finite element method--Congresses. I. Chenot,
J. L. II. Oñate, E. III. Title.
TS213.E97 1988   671.3   88-13242

ISBN-13: 978-94-010-7131-4   e-ISBN-13: 978-94-009-1411-7
DOI: 10.1007/978-94-009-1411-7

Published by Kluwer Academic Publishers,
P.O. Box 17, 3300 AA Dordrecht, The Netherlands.

Kluwer Academic Publishers incorporates
the publishing programmes of
D. Reidel, Martinus Nijhoff, Dr W. Junk and MTP Press.

Sold and distributed in the U.S.A. and Canada
by Kluwer Academic Publishers,
101 Philip Drive, Norwell, MA 02061, U.S.A.

In all other countries, sold and distributed
by Kluwer Academic Publishers Group,
P.O. Box 322, 3300 AH Dordrecht, The Netherlands.

All Rights Reserved
© 1988 by Kluwer Academic Publishers
Softcover reprint of the hardcover 1st edition 1988
No part of the material protected by this copyright notice may be reproduced or
utilized in any form or by any means, electronic or mechanical
including photocopying, recording or by any information storage and
retrieval system, without written permission from the copyright owner

PREFACE

The physical modelling of metal forming processes has been widely used both in University and in Industry for many years.

Relatively simple numerical models, such as the Slab Method and the Upper Bound Method, were first used and many such models are implemented in the industry for practical design or regulation of forming processes. These models are also under investigation in the University, mainly for treatments which require low cost calculations or very fast answers for on-line integration.

More recently, sophisticated numerical methods have been used for the simulation of metal flow during forming operations. Since the early works in 1973 and 1974, mainly in U.K. and U.S.A., the applications of the finite element method to metal processing have been developed in many laboratories all over the world. Now the numerical approach seems to be widely recognized as a powerful tool for comprehension oriented studies, for predicting the main technological parameters, and for the design and the optimization of new forming sequences. There is also a very recent trend for the introduction of physical laws in the thermo-mechanical models, in order to predict the local evolution of internal variable representing the microstructure of the metal.

To day more and more practicians of the Industry are asking for computer models for design of their forming processes. This volume represents an effort to assess the present state of the art, in the development of computer codes, in the validation of the models by comparison with experimental data, and to stimulate exchanges between material scientists and engineers from industry.

This volume contains contributions to :
- the analysis of the constitutive equations of material, the friction law, the prediction of defects, and the physical modelling,
- the new developments in formulation of the finite element method for metal flow including error estimation,
- application of real forming operations : cold and hot rolling, forging, drawing, sheet forming and casting.

It is a pleasure for the editors to thank Pr J. SALENÇON, who suggested the idea of the colloquium, and assisted us for its organization. They finally wish to congratulate the authors for their excellent work, and thank them for making this conference successful.

The editors

J.L. CHENOT
E. OÑATE

v

Scientific Committee

J.L. CHENOT	Ecole des Mines de Paris
G. DUVAUT	Université de Paris VI
E. OÑATE	Escuela Téchnica Superior de Ingenieros de Caminos Canales y Puertos – Barcelone
M. PREDELEANU	E.N.S.E.T. Cachan
J. SALENÇON	L.M.S. – Ecole Polytechnique

Organising Committee

M. BELLET	Ecole des Mines de Paris
C. LEVAILLANT	Ecole des Mines de Paris
E. MASSONI	Ecole des Mines de Paris
P. MONTMITONNET	Ecole des Mines de Paris

Colloquium Sponsors

ARMINES
ASSOCIATION UNIVERSITAIRE DE MECANIQUE
DIGITAL EQUIPEMENT CORPORATION
ECOLE DES MINES DE PARIS
I.R.S.I.D.
MICHELIN
PECHINEY
REGIE NATIONALE DES USINES RENAULT
TEKSID

TABLE OF CONTENTS

PART 4 : FORGING AND DRAWING

PART 5 : HOT AND COLD ROLLING

PART 1
MATERIAL BEHAVIOUR

THE DESCRIPTION OF YIELD SURFACES FOR COLD PRESTRESSED METALS WITH THE HYPOTHESIS OF THE ISOTROPY CENTERS TRANSLATION

P. MAZILU and A. SKIADAS

Institut für Umformtechnik
Technische Hochschule Darmstadt

1. INTRODUCTION

It is well known that the prestressing drastically changes the elasto-plastic characteristics of a material. An isotropic material in particular becomes strongly anisotropic after preloading. The problem of the mathematical description of such an induced anisotropy consists in finding the analytical expression of the new yield surface as well as in the derivation of the flow rule that is able to describe the anisotropic strength hardening of the metal. In the following, the problem of yield loci description as a consequence of a proportional preloading will be considered.

The simplest form of a yield surface of a prestressed material was suggested by Edelman and Drucker [1] who proposed the anisotropic yield function

$$f = 1/2 \ C_{ijkl}(\sigma_{ij} - m\varepsilon_{ij}^{p}) \ (\sigma_{kl} - m\varepsilon_{kl}^{p}) = k^{2} \ ,$$

where ε_{ij}^{p} are the prestrains and m is a constant. The applicability of this equation was investigated by Shahabi and Sheldon [2] on prestrained En 24 steel. The experiments by Shiratory, Ikegami and Kaneko [3,4] on brass, by Haythornwaite [5] and other authors however show a great deviation from the translated ellipses predicted by the Edelman and Drucker theory.

The importance of yield surfaces for the description of material behaviour under plastic deformation has grown in the time since Tresca's work. A lot of experimental and theoretical investigations have been made in order to achieve a better description of the yield loci. For the determination of yield loci one has to deal with many difficulties, concerning not only the hysteresis of loading and unloading behaviour and rate dependence but also the variety of yield stress definitions. The yield locus' shape for predeformed metals has a strong dependence on the definition of the allowed plastic flow before the establishing of a yield stress. Measurements with high sensitivity for instance 5$\mu\varepsilon$ ($\equiv$ 5*10^{-6} % ε) plastic strain have an anisotropic

3

J. L. Chenot and E. Oñate (eds.), Modelling of Metal Forming Processes, 3–10.
© *1988 by Kluwer Academic Publishers.*

form in contrast to yield loci measured with the engineering plastic strain definition of 0.2 %.

The shape of the subsequent yield loci depends on the loading and unloading paths, on induced texture as well as on the temperature and ageing effects. A review of the experimental determination of yield loci has been given by Ikegami [4] and Hecker [6].

As a consequence of the experimental works, more refined theories for description of the subsequent yield loci have been suggested. We mention here the theory based on the isotropic and kinematic hardening rule

$$f(J_2, J_3) = k^2 \quad ,$$

where J_2, J_3 denote the modified second and third invariants :

$$J_2 = 1/2 \ (\sigma'_{ij} - \alpha_{ij}) \ (\sigma'_{ij} - \alpha_{ij}) \quad ,$$

$$J_3 = 1/2 \ (\sigma'_{ij} - \alpha_{ij}) \ (\sigma'_{jk} - \alpha_{ij}) \ (\sigma'_{ki} - \alpha_{ki}) \quad ,$$

as well a generalization due to Rees [7].
A more recent approach is that of Helling and Miller [8] . They have used a unified constitutive model (multiaxial MATMOD-4V model) in order to predict several deformation phenomena including rotation, transla- tion and "flattening" of yield surfaces.
Kurtyka and Zyczkowski [9] suggested an interesting geometrical description of elliptical yield loci able to describe translations rotations and distortions of the yield surfaces. Since these two models are rather new , there are no fittings of experimental yield loci available.
Close to experimental checking another procedure for the prediction of yield surfaces is based on the Taylor-Bishop-Hill microscopical approach. This theory considers the slip of single crystals in a polycrystalline material. The macroscopic strain is assumed to be a superposition of the shears in the individual crystallites. Since there are many glide systems in each crystallite, there are many possibili- ties of summation over the single-crystal shears. According to the Taylor theory the correct superposition is corresponding to the minimum deformation work of all crystallites. Bishop and Hill use the same assumptions with a different deformation work assumption. In order to calculate the yield stress in all directions one has to know the orientations of all crystallites. Such an orientation distribution function has been suggested by Bunge [10]. Bunge's distribution function is analogous to the orientation texture and can be determined by X-ray diffraction, neutron scattering, or calculated by using a first step of the Taylor-Bishop-Hill theory.

2. THERMODYNAMICAL CONSIDERATIONS

2.1. Isotropic case

In the plastic process where the deformation is described only in terms of the stress deviator, the general yield surface has the form

$$f(\underset{\sim}{\sigma}) \; = \; k^2 \; = \; 0 \qquad . \tag{2.1}$$

In the particular case of isotropy (2.1) reduces to

$$f(II,III) \; = \; k^2 \qquad , \tag{2.2}$$

where II und III are the second and third deviatoric invariants

$$II \; = \; \sigma'_{ij}\,\sigma'_{ij} \quad ; \quad III \; = \; det(\sigma'_{ij}) \qquad . \tag{2.3}$$

The following two questions are arising :
1. How does the third invariant intervene in the yield loci equation
2. How are the yield loci of an initial isotropic material transformed after a predeformation.

In order to answer these questions we need a thermomechanical approach of the plastic flow phenomenon. The principal problem in such a thermomechanical model consists in the description of the energy conversion during a stress cycle. The simplest assumption is, that the mechanical work is split into recoverable mechanical energy and heat. Such an assumption corresponds to an elastic-plastic rheological model made of a spring and a friction element. The description of a real elasto-plastic phenomenon requires however a more sophisticated model. A possible model is represented in Fig. 1 (see [11], [12]).

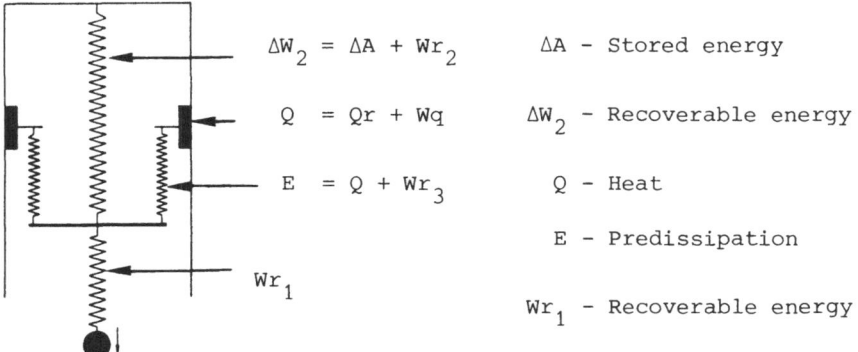

$$\Delta W_2 \; = \; \Delta A \; + \; Wr_2 \qquad \Delta A \; - \; \text{Stored energy}$$

$$Q \; = \; Qr \; + \; Wq \qquad \Delta W_2 \; - \; \text{Recoverable energy}$$

$$E \; = \; Q \; + \; Wr_3 \qquad Q \; - \; \text{Heat}$$

$$E \; - \; \text{Predissipation}$$

$$Wr_1 \qquad Wr_1 \; - \; \text{Recoverable energy}$$

Figure 1. Rheological model

The balance equation corresponding to such a rheological model is

$$E + Wr_1 + W_2 = Q + Wr_3 + Wr_1 + Wr_2 + \Delta A =$$

$$= Wr + Wq + Qr + \Delta A \quad . \tag{2.4}$$

Let $\underset{\sim}{\sigma}_\theta$ be a stress tensor having the deviator $\underset{\sim}{\sigma}'_\theta$ of the unit norm

$$\|\underset{\sim}{\sigma}'_\theta\|^2 = \Sigma\, \sigma'_{\theta ij}\, \sigma'_{\theta ij} = 1 \quad . \tag{2.5}$$

Each $\underset{\sim}{\sigma}_\theta$ will define a certain direction $\underset{\sim}{\theta}$ in the space of deviators, characterized by the generalized angle

$$\theta_{ij} = \arccos(\sigma'_{\theta ij}) \qquad i,j = 1, 2, 3 \quad , \tag{2.6}$$

$$\underset{\sim}{\sigma} = \lambda\, \underset{\sim}{\sigma}_\theta \qquad\qquad \lambda \in R \quad . \tag{2.7}$$

Let us assume the material in its virgin state and denote by $F(\underset{\sim}{\theta})$ the maximal value of the predissipation energy E for which no plastic flow occurs

$$F(\underset{\sim}{\theta}) = \max_{\underset{\sim}{\sigma} = \lambda \underset{\sim}{\sigma}_\theta} (E/\text{no plastic flow}) \quad .$$

According to these notations for a virgin material the yield surface equation is

$$E = F(\underset{\sim}{\theta}) \tag{2.8}$$

where E defines the predissipation energy. Because no predeformation is assumed and the predissipation is a part of the elastic energy, E depends only on the stress; more preci sely on the stress deviator $\underset{\sim}{\sigma}'$. In this case the yield equation has the following form

$$E(\underset{\sim}{\sigma}) = F(\underset{\sim}{\theta}) \quad . \tag{2.9}$$

For isotropic materials (2.9) reduces to (see [12])

$$E(II,III) = F(\theta) \tag{2.10}$$

where θ is the angle

$$\theta = \arccos(\sqrt{54}\ III/II^{3/2}) \tag{2.11}$$

(see [12]). By redenoting the function $F(\arccos \sqrt{54}\ III/II^{3/2})$ again by $F(III/II^{3/2})$ (2.10) becomes :

$$E(II,III) = F(III/II^{3/2}) \tag{2.12}$$

In the following special attention will be payed to the case, when the yield function $F(\theta)$ makes no difference between tension and compression, i. e. $F(\theta)$ is a symmetrical function in the principal deviatoric stresses. In that case (2.12) can be written as :

$$E(II,III) = F(III^2/II^3) \quad . \quad\quad (2.13)$$

In the section 3 a linear expression of the type (2.13) will be used in order to interpolate some experimental or numerical simulated data. For convenience this linear expression will be put into the form :

$$II + a\ III = k^2\ [1 - \alpha(1 - 54\ III^2/II^3)] \quad\quad (2.14)$$

with a, k and α constants to be determined.

2.2. Induced anisotropy

Let us consider an anisotropic material subjected to a preloading process, wich can be expressed in the deviatoric space by :

$$\underset{\sim}{\sigma}'(t) = t\ \underset{\sim}{\sigma}'^{\,o} \quad , \quad\quad t \in [0,1] \quad . \quad\quad (2.15)$$

As a consequence of the predeformation the material becomes anisotropic and the parameters II, III and θ are no longer proper to express the yield loci. Some new parameters must be determined. These parameters must reflect in some way the isotropy of the virgin material as well as the preloading process. Such parameters are well defined by the following hypothesis:

For predeformed materials the predissipation energy is a function of the translated invariants:

$$\overset{*}{II} = (\sigma'_{ij} - \overset{*}{\sigma}'_{ij})\ (\sigma'_{ij} - \overset{*}{\sigma}'_{ij}) \quad , \quad\quad (2.16)$$

$$\overset{**}{III} = \det (\sigma'_{ij} - \overset{**}{\sigma}_{ij}') \quad ; \quad\quad (2.17)$$

the flow predissipation depends on the translated angle

$$\hat{\theta} = \arccos (\sqrt{54}\ \widehat{III}/\widehat{II}^{\,3/2}) \quad\quad (2.18)$$

where

$$\widehat{II} = (\sigma'_{ij} - \hat{\sigma}'_{ij})\ (\sigma'_{ij} - \hat{\sigma}'_{ij}) \quad\quad (2.19)$$

$$\widehat{III} = \det (\sigma'_{ij} - \hat{\sigma}'_{ij}) \quad . \quad\quad (2.20)$$

According to this hypothesis the yield equation for preloaded materials must have the form (see [12])

$$E(\overset{*}{II},\overset{**}{III}) = F(\widehat{III}/\widehat{II}^{\,3/2}) \quad . \quad\quad (2.21)$$

Remark : Naturally $\overset{*}{II}$, $\overset{**}{III}$ and $\hat{\theta}$ are no longer invariant to a rigid rotation of the coordinate system. Consequently these functions are not isotropic. We note, however, that $\overset{*}{II}$, $\overset{**}{III}$ and $\hat{\theta}$ belong to a class of functions that generalizes the class of isotropic functions. An isotropic function satisfies the following requirement of invariance

$$\Phi \ (Q \underset{\sim}{\sigma} Q^T) \ = \ \Phi(\underset{\sim}{\sigma})$$

for all orthogonal tensors Q. Let us consider a scalar function $\Phi(\underset{\sim}{\sigma})$ for which there exists a $\underset{\sim}{\sigma}^o$, so that

$$\Phi \ (Q(\underset{\sim}{\sigma} - \underset{\sim}{\sigma}^o) \ Q^T + \underset{\sim}{\sigma}^o) \ = \ \Phi(\underset{\sim}{\sigma})$$

holds for all orthogonal tensors Q. In this case Φ has a translated isotropy center. It is easy to verify that II^*, III^{**} and $\hat{\theta}$ have such translated centers.

Under the assumption that the flow function F admits $\hat{\underset{\sim}{\sigma}}$ as a symmetry center, equation (2.21) can be written in the following form

$$E(\overset{*}{II}, \overset{**}{III}) \ = \ F \ (\hat{III}^2/\hat{II}^3) \ . \qquad (2.22)$$

In the next section a linear approach of (2.22) namely

$$\overset{*}{II} + a \ \overset{**}{III} \ = \ k^2 \ [1 - \alpha \ (1 - 54 \ \hat{III}^2/\hat{II}^3)] \qquad (2.23)$$

will be used to express analytically yield loci for some type of induced anisotropy. This form is a generalization of the yield loci equation proposed in [13].

3. APPLICATIONS OF THE THEORY

In the application of equation (2.23) in the two dimensional case 9 parameters have to be determined. The three translation tensors σ^*, σ^{**} and $\hat{\underset{\sim}{\sigma}}$ each with two scalar parameters and the three scalars $\underset{\sim}{k}^2, \underset{\sim}{\alpha}$ and $a.$ For the approximation some experimental and some numerical simulated yield loci are fitted. The code used to fit the data was a nonlinear least squares (NLSQ) program with Gauss-Newton or steepest descent algorithms.

First some numerically simulated data were interpolated by the theory of isotropy center translation. The data were calculated by the Taylor-Bishop-Hill theory in the GPM2 laboratory in Grenoble with the help of a computer code elaborated by Van Houtte [14].
Fig. 2 shows the fitting of two isotropic yield loci, where a) is a bcc-metal and b) is an fcc-metal. Fig. 3 shows the fitting of a rolled bcc-metal with 100% reduction calculated by the Taylor theory with the assumption of different critical resolved shear stresses in tension and compression.

Figures 4, 5 and 6 supply some interpolation examples of experimentally determined yield loci. Figure 4 shows the fitting of subsequent

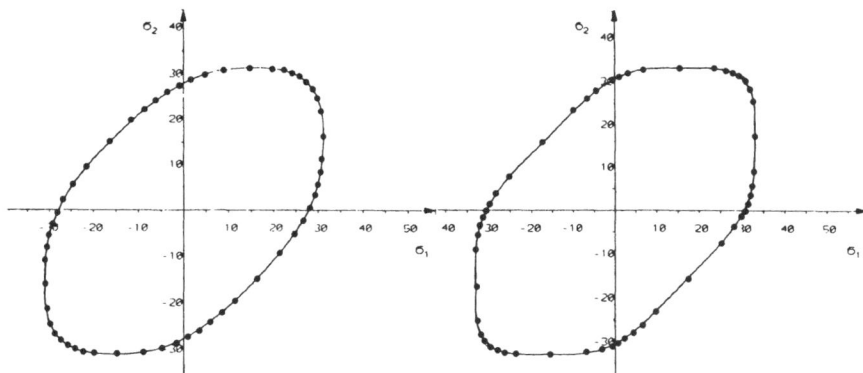

Fig. 2a. Isotropic yield surface of a bcc metal provided by the Taylor theory.

Fig. 2b Isotropic yield surface of a fcc metal provided by the Taylor theory.

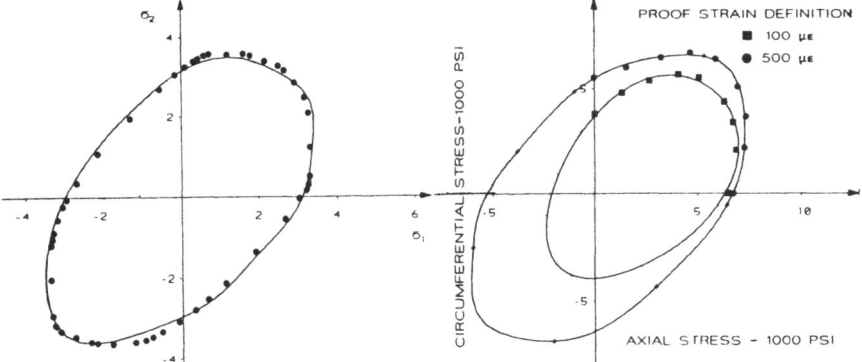

Fig. 3. Anisotropic yield surface of a bcc-metal with different CRSS in tensile and compression direction after 100% rolling. Provided by the Taylor theory.

Fig. 4. Subsequent yield loci for axially prestrained 1100-0 aluminium with different proof strain definitions. Data by Hecker [15].

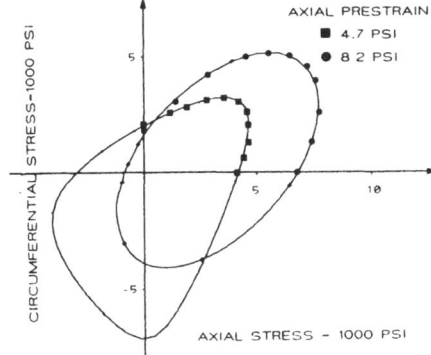

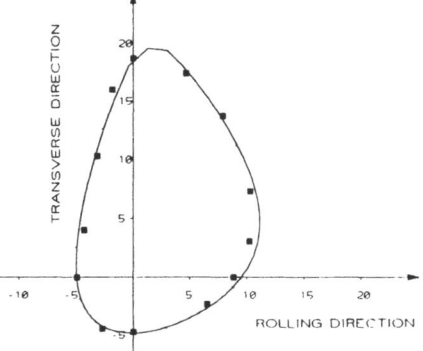

Fig. 5. Subsequent yield loci for different magnitudes of axial prestrain in 1100-0 aluminium. Data by Hecker [15]

Fig. 6. Subsequent yield locus for a magnesium probe with 1% proof strain definition. Data by Kelly and Hosfold [16].

yield loci of the same specimen with different proof strain definitions
(data by Hecker [15]). Fig. 5 shows the fitting of subsequent yield
loci data for a probe under different axial prestrains . The data in
both figures are supplied only in the positive stress region, but the
calculated fittings are drawn in the entire stress plane. In Fig. 6 the
fitting of an anisotropic Magnesium block probe yield locus is pictured
(data by Kelly and Hosford [16]). The evaluation of the dependence of
the theory parameters from prestrain and yield locus shape could be a
first step in a systematic approach of the induced anisotropy (see [17])

Acknowledgement : The autors owe many thanks to J. Bormann for the
nummerical fitting of the yield loci.

References :
1. Edelman, F.; Drucker, D.C.: Some extensions to elementary plasticity theory. J. Franklin Inst. 251 (1951) 581-605
2. Shahabi, S.N.; Sheldon, A.: The anisotropic yield flow and creep behaviour of prestrained En24-steel. J. Mech. Eng. Sci. 17 (1975) 82-92
3. Shiratori, E.; Ikegami, K.; Kaneko, K.: Subsequent yield surface determined in consideration of the Bauschinger effect. In Foundation of Plasticity. Leiden: Nordhoff Int. Publ. 1973
4. Ikegami, K.: Experimental plasticity on the anisotropy of metals. Colloques internationaux du CNRS No. 295- Comport. mécanique des solid. anisotropes, Coll. 115 Villard de Lans 1979, (Ed. J. P. Boeler) Ed. CNRS Paris (1982) 201-242
5. Haythornwaite, R. M.: A more rational approach to strain hardening data. In Eng. Plasticity Ed. J. Heyman and F. A. Leckie. Cambridge at the University Press (1968)
6. Hecker, S. S.: Experimental studies of yield phenomena in biaxially loaded metals. Constitutive equations in viscoplasticity; Computational and engineering aspects, ASME (Dez. 1976) 1-33
7. Rees, D. W. A.: The theory of scalar plastic deformation functions. ZAMM 63 (1983) 217-288
8. Helling, D. E.; Miller, A. K.: The incorporation of yield surface distortion into a unified constitutive model, Part 1: Equation development. Acta Mechanica 69 (1987) 9-23
9. Kurtyka, T.; Zyczkowski M.: A geometrical description of distortional plastic hardening of deviatoric materials. Arch. Mech. 37 Warszawa (1985) 383-395
10. Bunge, H. J.: Mathematische Methoden der Texturanalyse. Akademie Verlag Berlin 1969
11. Mazilu, P.: Variationsprinzipe der Thermoplastizität II. Mitteilungen aus dem Institut für Mechanik, Ruhr Universität Bochum, 37, August 1983
12. Mazilu, P.: Thermomechanics of the induced anisotropy. (to be published)
13. Mazilu, P.; Meyers, A.: Yield surface description of isotropic materials after cold prestrain. Ingenieur-Archiv 55 (1985) 213-220
14. van Houtte, P.; Aernoudt, E.: Solution of the generalized Taylor theory of plastic flow. Z. Metallkde. 66 (1975) 202-209, 303-306
15. Hecker, S. S.: Experimental observations on aluminium and copper subjected to biaxial load. Plasticity Workshop of The Office of Naval Research, Aerospace Eng. Depart., Texas A&M University (June 1975) ,(Ed. K. J. Saczalski, J. A. Stricklin) 190-201)
16. Kelly, E.W.; Hosford, W.F.: The determination characteristics of textured magnesium. ibid. 242 (1968) 654-661
17. Skiadas, A.: Some approximations of anisotropic yield loci by means of the theory of translation of the isotropy centers. (to be published)

CONSTITUTIVE DESCRIPTION AND NUMERICAL APPROACH IN MODELLING FOR METAL FORMING PROCESSES

Ryszard B. Pęcherski
Institute of Fundamental Technological Research,
Polish Academy of Sciences, Warsaw, Świętokrzyska 21

1. Introduction

Modelling for metal forming, strain localization and ductile fracture produces an increasing demand for the adequate constitutive description of inelastic material behaviour and the efficient numerical strategies. The constitutive equations for small elastic and finite plastic deformations with combined isotropic-kinematic hardening have been implemented in finite element programs (cf. e.g.[1]). In[2-4] the Ziegler model of kinematic hardening with the Zaremba–Jaumann rate was applied for the numerical analysis of the effect of yield surface curvature on localization in porous plastic solids. The effect of deformation induced anisotropy in plasticity of damaged solids was also studied in [5].

The introduction of the kinematic hardening is related with the formulation of objective rate-type constitutive equations. The application of the Zaremba–Jaumann rate can lead to the non-adequate prediction of the material reaction while the finite shearing with pertinent large rotations of the principal axes of the back stress tensor plays the dominant role (cf.[6-8], where the critical discussion of the related studies and references are provided).

The aim of the paper is to present the constitutive description of finite plastic deformations of porous materials with the combined isotropic-kinematic hardening. The pertinent numerical algorithm is also briefly discussed. An important aspect of gross inelastic behaviour of crystalline materials, i.e. the movement of the continuum relative to the underlying substructure and the emerging concept of plastic spin has been taken into consideration by the formulation of the objective rates (cf. [8, 21, 25-27]). The computational approach, developed in[9] and corroborated in [10], is based on the convected coordinates formulation (cf. e.g.[11]) and operator splitting methodology,[12]. In this context the return mapping algorithm was applied (cf. e.g.[1,12, 14]). As an example some results of the numerical calculations of localization in the cylindrical bar under tension with rigid grips, presented in[10], are discussed.

J.L. Chenot and E. Oñate (eds.), Modelling of Metal Forming Processes, 11–18.
© *1988 by Kluwer Academic Publishers.*

2. Physical motivation

According to the recent investigations reported in [15-18] localization
begins by a structural instability in the accumulated dislocation substructure and later becomes organized into macroscopic shear bands due
to the local concentrations of microstresses. The one of possible sequences of events can be displayed in the following way [18]. An initial
stage of the development of localization is associated with the change
from a fine to a coarse slip mode of deformation within separate grains
while stress increases. At higher stress, crystallographic coarse slip
bands, which could be associated with an avalanche-like crystallographic
glide, become powerful enough to penetrate the neighbouring grains and
to propagate in the form of micro-shear bands throughout the crossection
of the sample. The micro-shear bands are non-crystallographic in nature
and may result from a selfinduced change in deformation path.
      This can be interpreted by means of some cooperative forms of
highly concentrated dislocation motions and the onset of collective degrees of freedom that produce locally the relative rotations of the
material microvolumes with underlying substructure, [19, 20]. The propagation of micro-shear bands throughout the crossection of the sample
give rise to the geometric defect (local change of the sample crossection) which helps to nucleate micro-shear bands in other suitably oriented grains of the polycrystalline aggregate in the area affected by the
former band. The development of macroscopic band result from the syner-
getic coupling between the both mechanisms. The reported observations
lead to the general conclusion that the microscopic mechanisms produce
the internal rotational instabilities. This effect disturbs the plastic
spin and the deformation path on the macrolevel and leads to the development of necking and macroscopic shear bands.

3. Kinematics of solids with substructure

In plasticity of single crystals it is usually assumed that the dislocations traversing a volume element produce a change of its shape but
they do not change its lattice orientation. The macroscopic counterpart
of such a situation in finite deformation plasticity of polycrystals is
the Mandel's concept of the isoclinic configuration, [21, 25]. In our
case the situation is different, for the advanced plastic deformations,
associated with the formation of subboundaries and micro-shear bands,
produce large relative rotations of material microvolumes with respect
to the neighbouring material.
      Application of the concept of the isoclinic configuration and the
procedure proposed originally by Lardner [22] in the context of nonlinear
theory of continuously distributed disclinations and applied in [20, 23]
for the description of the kinematics of finite plastic deformations of
crystalline materials with lattice misorientation results in the following decomposition of the deformation gradient F:

$$F = E R_m P_i = E P \quad , \tag{1}$$

where

E – elastic transformation

R_m – orthogonal tensor responsible for the local relative rotations of material microvolumes (misorientation)

P_i – plastic transformation related with the isoclinic configuration

P – the perturbed plastic transformation related with the intermediate configuration rotated with respect to the isoclinic one with the tensor R_m .

The velocity gradient $L = \dot{F}\, F^{-1}$ takes the form:

$$L = \dot{E}\, E^{-1} + E \dot{R}_m R_m^{-1} E^{-1} + E R_m \dot{P}_i P_i^{-1} R_m^{-1} E^{-1} = \dot{E}\, E^{-1} + E \dot{P}\, P^{-1} E^{-1} . \qquad (2)$$

Due to the polar decomposition of the elastic transformation $E = V_e R_e$ the deformation gradient reads $F = V_e R_e P = V_e F_p$ and (2) leads to

$$L = \omega + \overset{\circ}{V}_e V_e^{-1} + V_e \overset{\circ}{F}_p F_p^{-1} V_e^{-1} , \qquad (3)$$

where

$$\omega = \dot{R}_e R_e^{-1} , \quad \overset{\circ}{V}_e = \dot{V}_e - \omega V_e + V_e \omega , \quad \overset{\circ}{F}_p = \dot{F}_p - \omega F_p . \qquad (4)$$

The following basic kinematical relations hold:

$$D_e = (\dot{E}\, \bar{E}^{-1})_s = (\overset{\circ}{V}_e V_e^{-1})_s , \quad W_e = (\dot{E}\, \bar{E}^{-1})_a = (\overset{\circ}{V}_e V_e^{-1})_a + \omega , \qquad (5)$$

$$D_p = (E \dot{P}\, \bar{P}^{-1} E^{-1})_s = (V_e \overset{\circ}{F}_p F_p^{-1} V_e^{-1})_s , \qquad (6)$$

$$W_p = (E \dot{P}\, P^{-1} E^{-1})_a = (V_e \overset{\circ}{F}_p F_p^{-1} V_e^{-1})_a , \qquad (7)$$

$$L = D + W , \quad D = D_e + D_p , \quad W = W_e + W_p , \qquad (8)$$

where the indicies e and p refer to elastic and plastic and the symbols $(t)_s$ and $(t)_a$ correspond to the symmetric and anti-symmetric parts of the second order tensor t.

The material substructure, crystalline lattice, is generally sub-jected to the rigid body rotations due to boundary constraints or com-patibility conditions,[24]. On the other hand, some cooperative forms of the motion of large dislocation ansambles and the onset of collec-tive degrees of freedom produce locally the selfinduced rotational ins-tabilities and contribute to the relative rotations of microvolumes with underlying substructure ([19]and[23]). Our attempt is to distin-guish between the both modes of rotation. The former one is expressed by means of the spin ω and the latter one corresponds to the misorien-tation tensor R_m . For $R_m = 1$ equations (1-8) transform into the rela-tions resulting from the theory presented in [26, 27].

4. Constitutive equations

For the sake of brevity the isothermal processes will be considered only.

Assume the mechanical state variables (π, A, S) corresponding to the isoclinic configuration, where π is the second Piola-Kirchoff stress, A represents the structural tensor variables and S denotes the scalar structural variables. The tensor variable may represent in the case of kinematic hardening the back stress α . The scalar variable can be specified as S = (f,ζ , k), where f denotes the void volume fraction,ζ is the density of local intrinsic disturbances (internal rotational instabilities) and k corresponds to the isotropic hardening parameter. The elastic Green strain $\Delta^e = 1/2(E^T E - 1)$ can be calculated from the free enthalpy function H per unit mass and the rate of elastic strain transformed to the current configuration is given by ([21, 25]):

$$D_e = M : \overset{\lor}{\sigma} \qquad , \qquad (9)$$

where M is the elastic compliance tensor calculated from the free enthalpy function and transformed to the current configuration,

$$\overset{\lor}{\sigma} = \overset{\cdot}{\sigma} - \dot{E} \bar{E}^{-1} \sigma - \sigma \bar{E}^{-T} \dot{E} + \sigma \, tr(\dot{E} \bar{E}^{-1}) \qquad (10)$$

or for the Kirchoff stress $\tau = det(E) \, \sigma$

$$\overset{\lor}{\tau} = \overset{\cdot}{\tau} - W_e \tau + \tau W_e - D_e \tau - \tau D_e \qquad . \qquad (11)$$

It is typical for the most of deformed metallic solids that their distortional elastic strains remain small under arbitrary loading conditions, whereas they can undergo large elastic dilatational changes in the shape under very high pressure. According to [28] the tensor of elastic moduli in Eulerian description can be represented as simply as in the case of infinitesimal strains, provided the logarithmic elastic strain $e^e = lnV_e$, is adopted as a state variable and that the values of the ratios of principal elastic stretches U_e, from the polar decomposition $E = V_e R_e = R_e U_e$, belong to the interval [5/6, 7/6] . In such a case theZaremba-Jaumann type rate of e^e ,corotational with W_e , can be approximated sufficiently close by D_e. Assuming that the deviatoric part of the elastic strain, $\bar{e}^e$, is infinitesimal

$$\overset{\circ}{\tau} = L : D_e + O(|\bar{e}^e|^2), \quad \overset{\lor}{\tau} = \overset{\circ}{\tau} + O(|\bar{e}^e|), \qquad (12)$$

where $\overset{\circ}{\tau}$ is the Zaremba-Jaumann type rate corotational with W_e:

$$\overset{\circ}{\tau} = \overset{\cdot}{\tau} - W_e \tau + \tau W_e = \overset{\cdot}{\tau} - (W - W_p)\tau + \tau (W - W_p) . \qquad (13)$$

Observe that (12) holds for arbitrary large dilatational part of the elastic strain e^e.

The following general rate equations are assumed:

$$D_p = < \lambda > N \, (\tau, a, s) \qquad (14)$$

$$W_p = < \lambda > W \, (\tau, a, s) \qquad (15)$$

$$\overset{\circ}{a} = < \lambda > A \, (\tau, a, s) \qquad (16)$$

$$\overset{\bullet}{s} = <\lambda> S\ (\tau,\ a,\ s) \tag{17}$$

where λ is the properly defined loading index for a state satisfying the yield condition $\Phi(\tau,\ a,\ s) = 0$ and the state variables $(\tau,\ a,\ s)$ are taken at the current configuration. The brackets $<\cdot>$ give $<\lambda> = \lambda$ if $\lambda > 0$ and $<\lambda> = 0$ if $\lambda \leq 0$. Due to the invariance requirements for a superimposed rigid body motion the constitutive functions in (14-17) are isotropic. The substructure corotational rate of a is expressed as in (13). The crucial point is to specify the new equation for the perturbed plastic spin. An approximation to the representation of the general isotropic function W for plastic spin, discussed in [25, 26], was considered in [8]:

$$W_p = (3\psi_1 - \tfrac{1}{2}\ \psi_3\ tr(\alpha^2))\ (\alpha D_p - D_p\alpha) + (\psi_2\alpha + \psi_3\alpha^2)\ (\alpha D_p - D_p\alpha)$$

$$- (\alpha D_p - D_p\alpha)\ (\psi_2\alpha + \psi_3\alpha^2)\ , \tag{18}$$

where the coefficients ψ_i, $i = 1,2,3$, are functions of invariants of α and D_p as well as of the scalar variables. The relation (18) was obtained with the application of the representation derived in [29], where different concept of objective rate of stress and kinematic hardening was studied. One of practical specifications of (18) can take the form, [7]:

$$\psi_1 = \sqrt{\tfrac{3}{2}}\ \frac{\eta(\zeta)\ \overset{\bullet}{\varepsilon}\ \varepsilon^2}{(1 + 3\varepsilon^2)((\alpha D_p - D_p\alpha):(\alpha D_p - D_p\alpha))^{1/2}}\ ,\ \psi_2 = \psi_3 = 0. \tag{19}$$

where ε and $\overset{\bullet}{\varepsilon}$ denote the equivalent plastic strain and the equivalent rate of plastic deformation, respectively. The specification (19) was applied in [23] for the study of localization induced by the perturbed plastic spin in the case of simple shear traction of rigid-plastic material with kinematic hardening and certain simple form of the function $\eta(\zeta)$. This problem was studied in [7] for $\eta(\zeta) = 1$, where the verification with the experimental results of SWIFT [30] was provided. As it has been discussed in [31] the relation (19)$_1$ leads to difficulties in the case when $\alpha = 0$ and gives inadequate prediction of material behaviour for reverse load. Therefore, the equivalent specification has been derived (cf. [31]):

$$\psi_1 = \sqrt{\tfrac{3}{2}}\ \frac{2\ \eta(\zeta)\ \alpha_{eq}}{h_\alpha^2 + 3\ \alpha_{eq}^2}\ ,\ \psi_2 = \psi_3 = 0\ , \tag{20}$$

where h_α is the kinematic hardening modulus and $\alpha_{eq} = \sqrt{\tfrac{3}{2}\ \alpha:\alpha}$.

The constitutive equations (14-17) specified for elastic-plastic material with developing voids and combined isotropic-kinematic hardening can be written concisely (cf. [10]):

$$\Phi = \tfrac{3}{2}\ \frac{(\tau' - \alpha):(\tau' - \alpha)}{\sigma_M^2} + 2\ f\ \cosh\left(\frac{tr\ (\tau - \alpha)}{2\ \sigma_M}\right) - (1 + f^2) = 0 \tag{21}$$

$$D_p = <\lambda>\ \frac{\partial \Phi}{\partial \tau}\ ,\ \overset{\circ}{\alpha} = <\lambda>\ R\ (\tau - \alpha) \tag{22}$$

$$\dot{\sigma}_M = b\,h\,\frac{(\tau - \alpha):D_p}{\sigma_M\,(1-f)} \quad , \qquad \dot{f} = (1-f)\,\mathrm{tr}\,D_p \tag{23}$$

$$\dot{\tau} = (L - \frac{(L:\frac{\partial\Phi}{\partial\tau})\otimes(L:\frac{\partial\Phi}{\partial\tau})}{H + \frac{\partial\Phi}{\partial\tau}:L:\frac{\partial\Phi}{\partial\tau}}):D \tag{24}$$

$$R = (1-b)\,\frac{H + \frac{\partial\Phi}{\partial\tau}\,(1-f)\,\mathrm{tr}\,\frac{\partial\Phi}{\partial\tau}\,\frac{\sigma_y}{\sigma_M}}{\frac{\partial\Phi}{\partial\tau}:(\tau - \alpha)} \tag{25}$$

$$H = \frac{h}{(1-f)\,\sigma_M^2}\,(\frac{\partial\Phi}{\partial\tau}:(\tau - \alpha)^2 - (1-f)\,\frac{\sigma_e}{\sigma_M}\,\frac{\partial\Phi}{\partial f}\,\mathrm{tr}\,\frac{\partial\Phi}{\partial\tau}) \tag{26}$$

σ_y — initial yield stress of the matrix material
σ_M — yield stress of the matrix material
σ_e^M — measure of the matrix flow stress for isotropic hardening
b — portion of hardening to be isotropic
h — plastic hardening modulus
$h_\alpha = (1-b)h$ — kinematic hardening modulus.
The equations (21-26) are based on the extension of the Gurson model
for the kinematic hardening predented in [2-4]. The main difference lies
in the equation (22)$_2$ that corresponds to the projection of the increment of kinematic hardening parameter calculated according to the Prager
law on the direction $(\tau - \alpha)$, pertinent to the Ziegler law. Such an
approach appears more suitable for the application of the return mapping
algorithm.

5. Discussion of the numerical approach

The detail presentation of the numerical approach and computational implementation for finite plastic deformations of porous materials with
combined isotropic-kinematic hardening is given in [10] . The integration of the constitutive equations (21-26) was based on the operator
splitting methodology and in this context the return mapping algorithm
was applied. The algorithm was formulated soleley by means of the yield
function, the normal to the yield surface, the direction of plastic flow
and the tangent elastic moduli without involving their gradients. The
starting point was the principle of virtual work in the Lagrangean formulation with respect to the initial configuration:

$$G = \int_B (F\pi_k):\mathrm{Grad}\delta u\ dV - \int_{\partial B_t} t\,\delta u\ dA = 0 \tag{27}$$

where δu denotes virtual displacement and t corresponds to the external
traction. The solution of the linearized form of (27):

$$\Delta \, G(\phi) \, u = - \, G(\phi) \, , \quad \phi \, - \, \text{configuration} \, , \qquad (28)$$

within the context of the finite element method is accomplished by an iterative scheme based on the Newton's method. Accordingly, one solves a sequence of lineari ed problems given by (28) until the residual $G(\phi)$ vanishes (to within a prescribed tolerance).

As an example some results obtained from the numerical calculations of the deformation of the cylindrical bar under tension with rigid grips are displayed in Fig.1. This problem was studied in [10] for porous material with combined isotropic-kinematic hardening. The realistic description of the stress-strain curve was based on the experimental data for mild steel presented in [30]. In Fig.1 the consequences of the application of two different objective rates for the numerical computations of inhomogeneous plastic deformations are depicted. The difference in the prediction of material behaviour increases for larger strains pertinent to the developing necking. This difference should be considered qualitatively, for the influence of finite element mesh appears quite strong. Nevertheless, this effect is remarkably stronger in comparison with the onset of localization in the homogeneous conditions where the both rates provide similar results (cf. [2]).

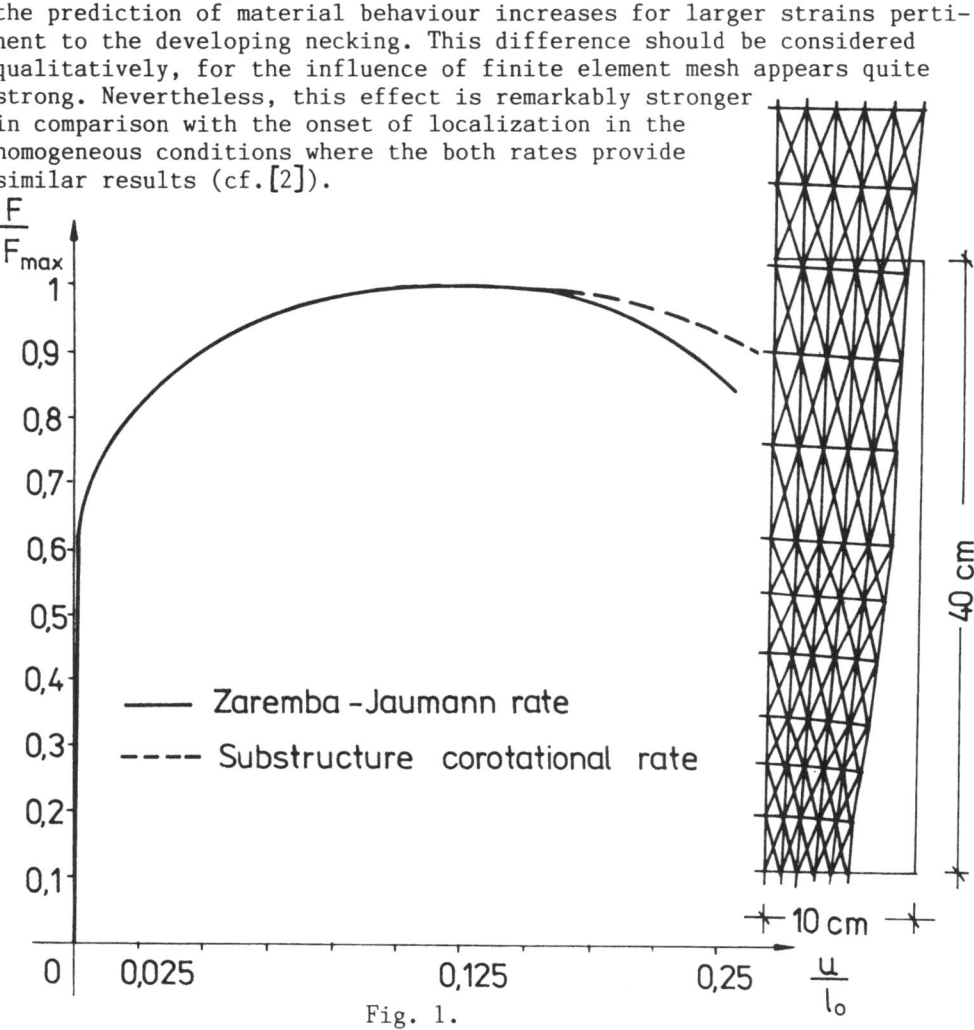

Fig. 1.

18

6. References

1. T. J. R. HUGHES, Proc. Workshop on the Theoretical Foundation for Large-Scale Computations of Nonlinear Material Behaviour, Evanston, Illinois, October 24, 25, and 26, 1983, S. Nemat-Nasser et. al. (eds.), Martinus Nijhoff Publ., Dodrecht-Boston-Lancaster, p.29,1984.
2. M. E. MEAR and J. W. HUTCHINSON, Mech. Mater., 4, p. 395, 1985.
3. V. TVERGAARD, J. Mech. Phys. Solids, 35, p. 43, 1987.
4. R. BECKER and A. NEEDLEMAN, J. Appl. Mech., 53, p. 491, 1986.
5. M. DUSZEK and P. PERZYNA, Z.A.M.P., 1988, to appear.
6. J. E. PAULUN and R. B. PĘCHERSKI, Arch. Mech.,37, p. 661, 1985.
7. J. E. PAULUN and R. B. PĘCHERSKI, Int. J. Plasticity, 3, p.303,1987.
8. R. B. PĘCHERSKI, Arch. Mech., 40, 1988 - in print.
9. R. LAMMERING, Beitrage zur Theorie und Numerik grosser plastischer und kleiner elastischer Deformationen mit Schadigungseinfluss, Ph. D. Thesis, Hanover University, 1987.
10. R. LAMMERING, R. B. PĘCHERSKI and E. STEIN, The 4th Bilateral Symposium PRL-BRD on Mechanics of Solids and Structures, 13-19 September, 1987, Krakow-Mogilany - to appear in Arch. Mech.
11. A. NEEDLEMAN and V. TVERGAARD, Finite Elements. Special Problems in Solid Mechanics, vol. 5, J. T. Oden and G. F. Carey, (eds.), Prentice-Hall, New Jersey, p. 94, 1984.
12. M. ORTIZ, P. M. PINSKY and R. L. TAYLOR, Comp. Meth. Appl. Mech., Eng., 39, p. 137, 1983.
13. M. L. WILKINS, Meth. Comp. Phys., vol.3, B. Adler et. al. (eds.), Academic Press, 1964.
14. M ORTIZ and J. C. SIMO, Int. J. Num. Meth. Eng., 23, p. 353, 1986.
15. A. KORBEL and P. MARTIN, Acta Metall., 34, p. 1905, 1986.
16. A. KORBEL, J. D. EMBURY, M. HATHERLEY, P. L. MARTIN and W. ERBSLOH, Acta Metall., 34, p. 1999, 1986.
17. M. SZCZERBA and A. KORBEL, Acta Metall., 35, p. 1129, 1987.
18. A. KORBEL and P. MARTIN, Acta Metall. - to appear.
19. V. V. RYBIN, Phys. Met. Metall., 44, p. 139, 1978.
20. R. B. PĘCHERSKI, Eng. Fracture Mech., 21, p. 767, 1985.
21. J. MANDEL, Int. J. Solids Structure, 9, p. 725, 1973.
22. R. W. LARDNER, Mathematical Theory of Dislocations and Fracture, University of Toronto Press, 1974.
23. R. B. PĘCHERSKI, Proc. 2nd. Int. Conf. on Numerical Methods in Industrial Forming Processes-NUMIFORM'88, 25-29 August, 1986, Gothenburg, K. Mattiasson et al., (eds.), A. A. BALKEMA, Rotterdam-Boston, p. 145, 1986.
24. R. ASARO, Advances of Appl. Mech., vol. 23, p.1, 1983.
25. B. LORET, Mech. Mater., 2, p. 287, 1983.
26. Y. F. DAFALIAS, J. Appl. Mech., 52, p. 865, 1985.
27. Y. F. DAFALIAS, Acta Mech., 69, p. 119, 1987.
28. B. RANIECKI and H. V. NGUYEN, Arch. Mech., 36, p. 687, 1984.
29. A. AGAH-TEHRANI, E. H. LEE, R. L. MALLET and E. T. ONAT, J. Mech. Phys. Solids, 35, p. 519, 1987.
30. H. W. SWIFT, Engineering, 163, p. 253, 1947.
31. J. E. PAULUN and R. B. PĘCHERSKI, The 4th Bilateral Symposium PRL-BRD, on Mechanics of Solids and Structures, 13-19 Sept., 1987, Krakow-Mogilany - to appear in Arch. Mech.

SIMULATION OF IMPACT TENSION DEFORMATION OF METALS BY FEM

Günter Dackweiler and Essam El-Magd
Lehr- und Forschungsgebiet Werkstoffkunde
RWTH Aachen, West-Germany

1. Introduction

Under dynamic loading, stress waves propagate in metallic bodies. These waves can be steepened and focussed by changes in cross-section and by reflections. Plastic deformation can occur and even internal cracks can be initiated. For the calculation of elastic-plastic wave propagation and the resulting stress and strain distributions, the material behaviour has to be described by suitable constitutive equations with parameters being determined experimentally.

In this paper, the material behaviour of X2 CrNi 18 9 is determined by a combination of experiments and computations, which analyse the data obtained from high speed tests.

2. Description of Material Behaviour

An accurate description of the mechanical behaviour of metals under dynamic loading depends not only on the applied numerical computation procedure but also on the accuracy of the material laws used in this computations. Assuming viscoplastic behaviour, the Perzyna-equation /1/ can be used:

$$\dot{\varepsilon}_{ij} = \frac{\overset{\bullet}{s}_{ij}}{2G} + \frac{1-2\nu}{3E} \dot{\sigma}_{kk}\delta_{ij} \ + 2\gamma \ \langle\Phi(F)\rangle \ \frac{\partial f}{\partial\sigma_{ij}} \tag{1}$$

where f is square root of the second invariant of the stress deviator and $F = (f/\varkappa - 1)$ the relative difference between f and the shear flow stress $\varkappa = \sigma_{\pmb{r}}/\sqrt{3}$. In this equation $\langle\Phi(F)\rangle$ is zero for $F \leq 0$ and $F > 0$ the value of the function $\Phi(F)$ is obtained.

In contrary to quasi-static loading, metals show a high strain rate sensitivity under impact conditions. The process is approximately adiabatic and the influence of temperature changes should be taken into consideration.

Even in the simple case of monotonic proportional loading, a ma-

19

J.L. Chenot and E. Oñate (eds.), Modelling of Metal Forming Processes, 19–26.
© 1988 by Kluwer Academic Publishers.

terial law used for the impact range should represent a relation between stress σ, strain ε, strain rate $\dot{\varepsilon}$ and temperature T /2/.

A simple description is based on the assumption that $\Phi(F) = F$ and leads to the linear relation:

$$\sigma = \sigma_f(\varepsilon,T) + \eta(\varepsilon,T)\dot{\varepsilon} \tag{2}$$

which is used for high strain rates. For the flow stress σ_r, the linear statement

$$\sigma_f = \sigma_y + H\varepsilon \tag{3}$$

or the exponential statement /3/

$$\sigma_f = A(B + \varepsilon)^n \tag{4}$$

can be used for the calculation.

Other relations based on structure-mechanical models for thermal activated processes were introduced allowing a combined description of the influences of temperature and strain rate.

Macherauch and Vöhringer /4/ deduced the relation, which is based on a Seeger-formulation:

$$\sigma = \sigma_G(\varepsilon) + \sigma_0^*(\varepsilon) \left[1 - \left(\frac{kT}{\Delta G_0} \ln \frac{\dot{\varepsilon}^*}{\dot{\varepsilon}} \right)^q \right]^p \tag{5}$$

in which σ_G is the athermal part of the flow stress and k is the Boltzman's constant. The expression $kT \cdot \ln(\dot{\varepsilon}^*/\dot{\varepsilon})$ represents the activation energy. These and other relations are discussed in /5/.

3. Experimental Tests

The experimental tests were made by using a modified Split-Hopkinson-tension-test. Figure 1 shows the set-up of the tension apparatus. An impact produced by a hammer at the end of the sleeve creates an elastic wave, which propagates through the sleeve with the velocity of sound. Reaching the end of the yoke the wave will be reflected and a part of it propagates through the input bar and induces a plastic deformation in the specimen. The plastic wave leaves the specimen in form of an elastic wave through the output bar.

The measurement carried out during the test regards the elastic strain in input (1) and output (2) bar in front of and behind specimen as functions of time from the beginning of specimen deformation till the unloading wave reflected from the free end of the output bar reaches the strain gauge (2) fixed on it. Figure 2 shows the relative voltage change for the input and output signal over time.

From these signals the particle velocities V_1 (Figure 3) and V_2 in front of and behind the specimen can be determined as time function

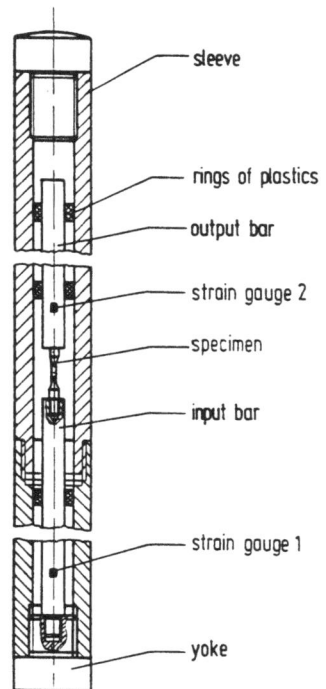

Figure 1: Set-up of tension impact apparatus

- sleeve
- rings of plastics
- output bar
- strain gauge 2
- specimen
- input bar
- strain gauge 1
- yoke

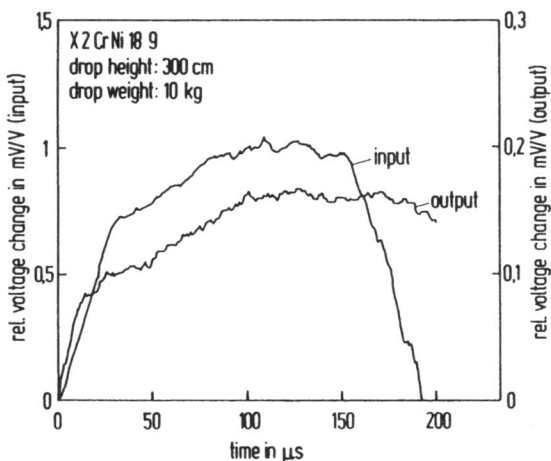

Figure 2: Relative voltage change of gauge bridge circuits over time in impact test

$$V_1 = c_1\varepsilon_1(t) \quad ; \quad V_2 = c_2\,\varepsilon_2(t) \tag{6}$$

with c_1 and c_2 the elastic wave velocities, $c = \sqrt{E/\rho}$ for tension or compression with E the modulus of elasticity and the density ρ

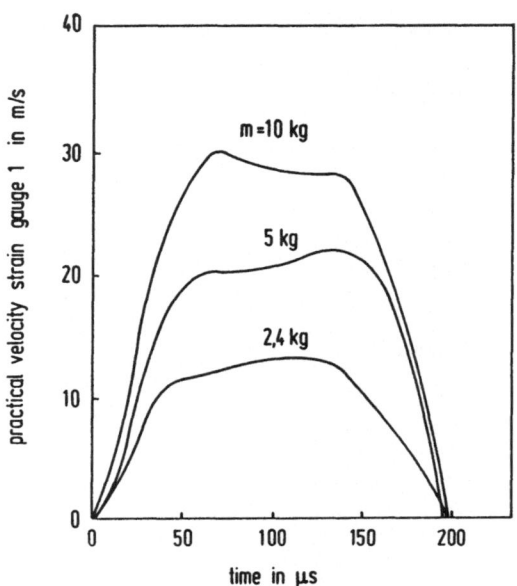

Figure 3: Particle velocity in the input bar

The current value of the flow stress in the output bar, the strain rate and strain are determined from the particle velocities:

$$\sigma(t) = K\,E_2\varepsilon_2(t) \tag{7}$$

$$\dot{\varepsilon}(t) = \frac{1}{L}\,(K_1 c_1\varepsilon_1 - K_2 c_2\varepsilon_2) \tag{8}$$

$$\varepsilon(t) = \int \dot{\varepsilon}(t)\,dt \tag{9}$$

where K, K_1 and K_2 are factors introduced to consider the wave reflections and geometrical changes at the ends of the specimen. They are functions of the diameters, moduli and density of specimen and bars. Figure 4 shows experimentally determined stress-strain curves for the material X2 CrNi 18 9 by changing the drop height and the specimen lengths.

23

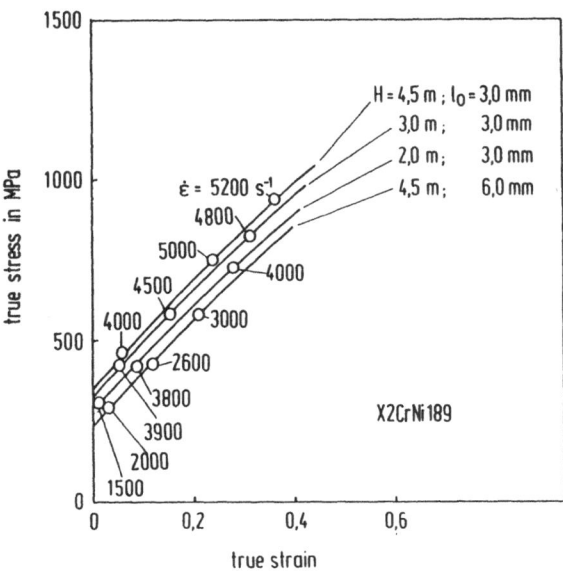

Figure 4: Stress-strain curve of austenitic steel X2 CrNi 18 9

4. Finite Element Computation

By the evaluation of the indirect impact tension test it was assumed a
uniform distribution of the plastic deformation and an uniaxial stress
rate neglecting shape effects on the inertia forces due to radial
acceleration during plastic lateral contraction. The two dimensional
Finite Element computation was carried out to improve these assumptions and predictions of the material constants. Further the calculation was used to optimize the test apparatus and the specimen geometry.

The FE-program for transient dynamic problems /6/ used here, was
applied to   axisymmetrical   bodies loaded in axial direction using
the principle of virtual work with the following equation:

$$\int_{\Omega} [\delta u]^T [b - \rho \ddot{u} - c \dot{u}] \, d\Omega + \int_{\Omega} [\delta \varepsilon]^T \sigma d\Omega = 0 \qquad (10)$$

The FE-program used an explicit time step procedure shown in Figure 5 and isoparametric 9-node ring elements.

The single scalar equations are decoupled by the diagonalizing
of mass matrix M, so that no big global system of equations needs to
be solved. The damping matrix C, the influence of which is very small in
the case of metals, is neglected. The critical time step for this condi-

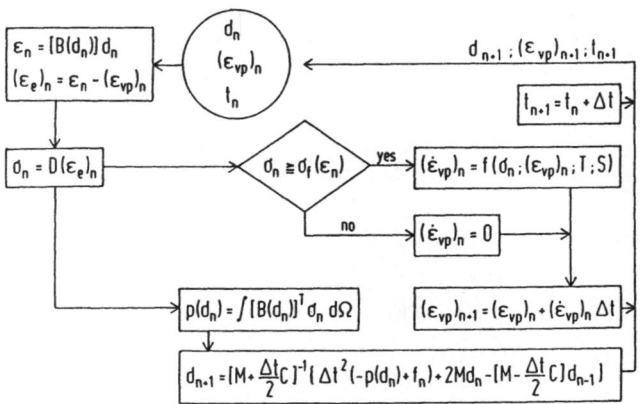

Figure 5: Calculation procedure during a time step

tionally stable explicit procedure must be worked out for each material and each constitutive equation separately

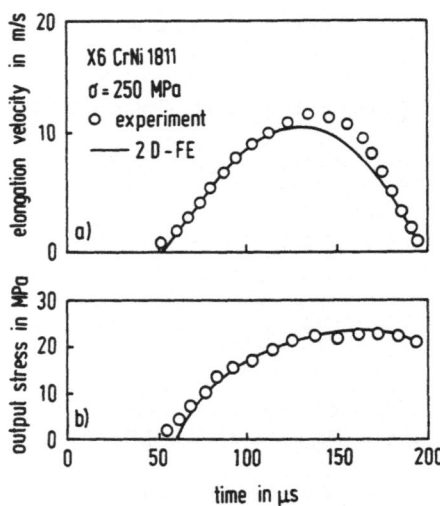

Figure 6: Comparison between experimental and calculation results
a) Elongation velocity over time
b) Stress in output bar over time

The deformation process is computed using only the time function $\varepsilon_1(t)$ of the elastic strain in the input bar. The stress rate and the deformation distribution are determined. A comparison will show the differences of the computed time function $\varepsilon_2(t)$ of the elastic strain in the output bar and of the elongation velocity with the measured.

In the computation for the material X2 CrNi 18 9 a description like equations (2) and (4) were used. Figure 6 shows a comparison for calculated and measured time curves of the stress in the output bar and the elongation velocity of the specimen.

The distribution of the stress and strain in the specimen for a special time are shown in figure 7.

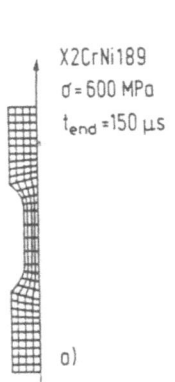

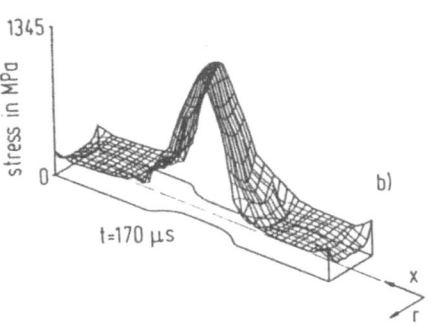

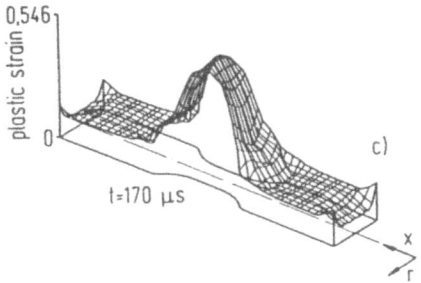

Figure 7: a) FE-idealization
b) calculated stress distribution in the axial section of specimen
c) calculated strain distribution in the axial section of specimen

5. References

/1/ Perzyna, P.: Quart. App. Math. 20(1963), p. 231/232

/2/ El-Magd, E.; Abdel-Ghany, W. and Homayun, M.: "Material constants and mechanical behaviour during plastic wave propagation", Proc. of the Numeta '85 Conference/Swansea, 1985, p. 437/445

/3/ Swift, M.W.: "Plastic Instability under Plane Stress", J. Mech. Phys. Solids 1(1952), p. 1/8

/4/ Macherauch, E. and Vöhringer, O.: Z. Werkstofftechnik 9(1978) No. 11, p. 370/391

/5/ El-Magd, E.; Abdel-Ghany, W. and Homayun, M.: "Application of the FEM to elasticplastic wave propagation in metals", Eng. Comput. 2(1985), p. 114/120

/6/ Owen, D.R.J. and Hinton, E.: "Finite Elements in Plasticity", Pineridge Press Limited, 1980

FINITE ELEMENTS ANALYSIS OF LARGE PLASTIC DEFORMATION IN METALS

G.J. CREUS*, A.G. GROEHS
Curso de Pós-Graduação em Engenharia Civil, UFRGS
Porto Alegre, R.S. Brasil

ABSTRACT: The paper describes a code used for the analysis of finite plastic deformations. A consistent large deformation theory is developed in rate form, which is then specialized for the case of small elastic strains and von Mises material. The finite element method, associated with an updated Lagrangian procedure is used for the numerical solution. Two examples of application of the code are shown.

1. INTRODUCTION

This paper describes a finite element code ("ESFINGE"), developed for the analysis of finite plastic deformations, with application to metal forming processes.

In the Section 2, a constitutive theory [1], [2], [3], that includes the consideration of finite elastic as well as plastic deformations of initially isotropic materials that remain elastically isotropic after plastic deformation, is reviewed. This formulation is coordinate free and based on quantities (such as the rate of plastic deformation) and properties (such as symmetries of the yield surface) that are, in principle, measurable. The state and orientation of stress free elements is described by means of irreducible even rank tensors. Then, the representation is specialized for the case of small elastic (deviatoric) strains, which is usually adequate to model metal deformations. In this way, the formulation becomes simpler for its implementation into a finite elements program. Also at this stage, we introduce the von Mises yield condition and hardening rules of the isotropic-kinematic type.

In Section 3, the equilibrium relations are introduced through Hill's functional for an actualized Lagrangian incremental process. Using the stiffness formulation for finite elements, a non linear system of equations is established, that the code can solve with incremental, iterative or combined techniques. Loading

---

* Presently visiting professor, ETS de Ingenieros de Caminos, Canales y Puertos, UPC, Barcelona

J.L. Chenot and E. Oñate (eds.), Modelling of Metal Forming Processes, 27–36.
© *1988 by Kluwer Academic Publishers.*

and unloading are possible with any kind of load as well as with prescribed displacements.

The code is written in FORTRAN. The comunication with the user is through a problem oriented language [4], which incorporates facilities as automatic generation of coordinates, connectivities, loads and properties. Most control parameters used in the nonlinear analysis have "default" values to help the less experienced users, but may be overruled when convenient. In Section 4, two examples are shown to validate the code, corresponding to finite plastic deformations in axisymmetric situations. Good correlation is obtained with other numerical as well as theoretical and experimental results.

2. CONSTITUTIVE RELATIONS

2.1 Kinematics

We use the well-known multiplicative decomposition [5], [6].

$$F = F_e \, F_p \quad ; \quad F_e = F_e^T \tag{2.1}$$

where F is the deformation gradient, subindices e and p indicate elastic and plastic parts and superindex T indicates transposition. We define the velocity gradient

$$L = \dot{F} \, F^{-1} = D + \Omega \tag{2.2}$$

where D and Ω are the rates of strain and rotation respectively and the elastic strain is (note that the Cauchy-Green tensors associated with F_e are equal to each other)

$$B_e = C_e = F_e^2 \tag{2.3}$$

From (2.1) we obtain the decomposition

$$D_e + \Omega_e = \dot{F}_e \, F_e^{-1} = D + \Omega - F_e(D_p + \Omega_p) \, F_e^{-1} \tag{2.4}$$

while from (2.3) and (2.4) we obtain

$$\dot{B}_e = (D + \Omega)B_e + B_e(D - \Omega) - 2F_e \, D_p \, F_e \tag{2.5}$$

Defining the alternative measure of plastic strain rate $\hat{D}_p$ by means of

$$\hat{D}_p \, B_e + B_e \, \hat{D}_p = 2F_e \, D_p \, F_e; \qquad \hat{D}_p = \hat{D}_p^T \tag{2.6}$$

we may write

$$\dot{B}_e = (D - \hat{D}_p + \Omega)B_e + B_e(D - \hat{D}_p - \Omega) \tag{2.7}$$

Moreover, from (2.4) we obtain

$$(\Omega - \Omega_p)F_e + F_e(\Omega - \Omega_p) + (D + D_p)F_e - F_e(D + D_p) = 0 \tag{2.8}$$

which shows that it is

$$\Omega - \Omega_p = \mathbf{W}\,(D + D_p); \qquad \mathbf{W} = \mathbf{W}\,(F_e) \tag{2.9}$$

Eq. (2.7) is a decomposition that may be interpreted as the finite strain equivalent to the small strain additive decomposition $\varepsilon_e = \varepsilon - \varepsilon_p$. As for (9) it means that $\Omega - \Omega_p$ may be obtained from $D + D_p, \mathbf{W}$ being a forth order linear operator which depends solely on F_e. Thus, once $F_e, D + D_p$ and Ω are known, we may obtain Ω_p as a consequence.

2.2. State and orientation of stress free previously deformed elements

The elastic behaviour of a stress free element, and particularly its elastic range will depend on its internal state and orientation, which we assume can be represented through n parameters s_n in $S \in \Sigma \subset \mathbf{R}^n$.

The state and orientation of a stress free element rigidly rotated by Q is indicated by $P_Q\,S$. It is easy to see that the composition law

$$P_{QR}\,S = P_Q\,P_R\,S \tag{2.10}$$

for all rigid body rotations must be obeyed. This determines that S must be composed of irreducible even rank tensors $S = (s_1, \ldots s_n)$ with $P_Q\,S = (P_Q\,s_1, \ldots P_Q\,s_n)$, where $P_Q\,s_i$ are the linear tensor transformations appropriate to the rank of s_i. For example, for a second rank tensor a

$$P_Q\,a = Q\,a\,Q^\mathsf{T} \tag{2.11}$$

The derivative of $P_Q S$ under a time dependent rigid body rotation $Q(t)$ is

$$\frac{d}{dt}\,P_{Q(t)}\,S; \qquad Q(t) \in 0^+(3); \qquad S = \text{const} \tag{2.12}$$

If $Q(t) = I$ and $\dot{Q}(t) = \Omega = -\Omega^\mathsf{T}$, we write

$$\frac{d}{dt}\,P_Q\,S = T_\Omega\,S \tag{2.13}$$

For a second rank tensor

$$T_\omega S = \Omega S - S\Omega \tag{2.14}$$

2.3. Elastic response, elastic range and yield surface

According to the initial assumptions, all stress free elements will exhibit the same isotropic elastic behaviour. If Ψ denotes the strain energy stored per unit mass of a generic stress free element, it may be written as a positive valued smooth function of the invariants I_i of the left Cauchy-Green tensor B_e

$$\Psi = \Psi(I_1, I_2, I_3) \quad ; I_i = tr(B_e^i) \quad i = 1, 2, 3 \tag{2.15}$$

The Cauchy stress σ caused by the spinless deformation F_e is given by

$$\sigma = 2\rho \left(\frac{\partial \Psi}{\partial I_1} B_e + 2\frac{\partial \Psi}{\partial I_2} B_e^2 + 3\frac{\partial \Psi}{\partial I_3} B_e^3\right) \tag{2.16}$$

where ρ is a current density of the material. Note that because mass conservation

$$\rho = \rho_0/\det B_e \tag{2.17}$$

where, as we assume isochoric plastic deformations, the density of the stress free elements is the same density as the virgin material.
From (2.15) and (2.16) we find

$$I_i = 2i\ B_e^i(D - \hat{D}_\rho) = 2i\ tr\ B_e^i(D - \hat{D}_p) \tag{2.18}$$

and, from

$$\dot{\sigma} = -\sigma\ tr\ D + (D - \hat{D}_p + \Omega)\sigma + \sigma(D - \hat{D}_p - \Omega) + K(D - \hat{D}_p) \tag{2.19}$$

where the elastic modulus K has the components

$$
\begin{aligned}
K_{ijkl} = 2\rho[&2\Psi,_2\left(B_{ik}\ B_{lj} + B_{il}\ B_{kj}\right) \\
+ &3\Psi,_3\left(B_{ik}^2\ B_{lj} + B_{ik}\ Blj^2 + B_{il}^2\ B_{kj} + B_{il}\ B_{kj}^2\right) \\
+ &2\Psi,_{11}\ B_{ij}\ B_{kl} + 4\Psi,_{12}\left(B_{ij}\ B_{kl}^2 + B_{ij}^2\ B_{kl}\right) \\
+ &8\ \Psi,_{22}\left(B_{ij}^2\ B_{kl}^2\right) + 12\ \Psi,_{23}\left(B_{ij}^2\ B_{kl}^3 + B_{ij}^3\ B_{kl}^2\right) + 18\ \Psi,_{33}\ B_{ij}^3\ B_{kl}^3]
\end{aligned}
\tag{2.20}
$$

where, for simplicity of writing, the subscript e of B_e is omitted.
As we are interested in volume preserving plastic deformations, with $tr\ \hat{D}_p = 0$ we may write (2.19) also as

$$\hat{\sigma} = \Omega\sigma - \sigma\Omega + K_e(D - \hat{D}_\rho) \tag{2.21}$$

The elastic range E_s is defined in the stress space by means of a real valued yield function $Y(S,\sigma)$ as

$$E_s = \{\sigma : Y(S,\sigma) \leq 0\} \tag{2.22}$$

For a fixed S, the boundary of E_s is composed of stress states satisfying $Y(S,\sigma) = 0$; when $Y(S,\sigma) < 0$, σ is outside E_s. A rigid body rotation Q of the element together with its surface tractions cannot change the yield status of the element. Thus, we must have

$$Y(P_qS, Q\sigma Q^T) = Y(S,\sigma) \tag{2.23}$$

and then

$$\frac{d}{dt} Y(P_{Q(t)}S, Q(t) \sigma Q^T(t)) = 0 \tag{2.24}$$

For the case $Q(t) = I, \dot{Q}(t) = \omega = -\omega^T$ this gives rise to the useful identity

$$\frac{\partial Y}{\partial S} \cdot T_\omega S + \frac{\partial Y}{\partial \dot{\sigma}}(\omega\sigma - \sigma\omega) = 0 \tag{2.25}$$

On the other hand, for a continuous plastic process, we must have the consistency condition

$$\dot{Y} = \frac{\partial Y}{\partial S} \cdot \dot{S} + \frac{\partial Y}{\partial \sigma} \cdot \dot{\sigma} \tag{2.26}$$

2.4. Elastic-plastic behaviour

The state and orientation of a deforming material element at time t is defined by the pair (S, σ) where S is the state and orientation of the associated stress free element. The material will behave elastically for

$$Y(S, \sigma) < 0 \quad \text{or} \quad Y(S, \sigma) = 0, \quad \text{but} \quad \dot{Y} = \frac{\partial Y}{\partial S} \cdot \dot{S} + \frac{\partial Y}{\partial \sigma} \cdot \dot{\sigma} \leq 0 \tag{2.27}$$

In this case

$$D = \hat{D}_p = 0$$

$$\Omega_p = \Omega - \mathbf{W} D$$

$$S(t) = P_{Q(t)} S(t) \quad \text{with} \quad Q(t) = I, \frac{dQ}{dt} = \Omega_p \tag{2.28}$$

$$\frac{dS(t)}{dt} = T_{\Omega_p}S = -T_{\mathbf{W}D} S(t) + T_\Omega S(t)$$

that shows that $dS(t)/dt$ may be nonzero even when $\Omega = 0$. Now we can calculate, substituting the rates $\dot{S}, \dot{\sigma}$ from (2.28), (2.21)

$$\dot{Y} = \frac{\partial Y}{\partial S} \cdot \dot{S} + \frac{\partial Y}{\partial \sigma} \cdot \dot{\sigma}$$

$$= \frac{\partial Y}{\partial S}(-T_{\mathbf{w}D}S + T_\Omega S) + \frac{\partial Y}{\partial \sigma}(K_e D + \Omega\sigma - \sigma\Omega) \tag{2.29}$$

$$= \frac{\partial Y}{\partial S}[K_e D + (\mathbf{W}D)\sigma - \sigma(\mathbf{W}D)]$$

by application of (2.25). The right hand side of the last (2.29) defines a function $\mathbf{Z}$ such that

$$\dot{Y} = \mathbf{Z}(S, \sigma) \cdot D \tag{2.30}$$

during episodes of elastic deformation.

The material will behave plastically if at the instant of interest we have

$$Y(\sigma, S) = 0 \quad ; \quad \mathbf{Z}(S, \sigma) \cdot D > 0 \tag{2.31}$$

where (S, σ) is calculated by the laws of evolution for elastic behaviour. For plastic processes, we assume as in classical plasticity

$$\hat{D}_p = \mu(S, \sigma) \frac{\partial Y}{\partial \sigma} (\mathbf{Z} \cdot D) \tag{2.32}$$

where μ is a positive valued function of its arguments.

2.5. Formulation for small elastic strains

In ductile metals yield is reached with small elastic shear strains, as relation between elastic modulus and yield stress is of the order of a thousand. In such a case, the formulation above simplifies through the use of the following approximations.

We assume that $F_e \sim I + \varepsilon_e$ and thus from (2.4), (2.6) and (2.9)

$$D = D_e + D_p$$

$$D_p = \hat{D}_p \tag{2.33}$$

$$\Omega = \Omega_p + \omega$$

From (2.19), as $(D - \hat{D}_p)$ is small compared with Ω

$$\dot{\sigma} = -\sigma tr D + \Omega \sigma - \sigma \Omega + \mathbf{E}(D - D_p) \tag{2.34}$$

where $\mathbf{E}$ has constant coefficients and may be taken as in the small strain case

$$\mathbf{E}_{ijkl} = 2\mu \delta_{ik} \delta_{lj} + \lambda \delta_{ij} \delta_{kl} \tag{2.35}$$

The von Mises yield function with isotropic-kinematic hardening is adopted in the form

$$Y = \sqrt{\frac{3}{2}(s_{ij} - \alpha_{ij})(s_{ij} - \alpha_{ij})} - k = 0 \tag{2.36}$$

with

$$s_{ij} = \sigma_{ij} - \frac{1}{3} \delta_{ij} \sigma_{kk} \tag{2.37}$$

The growth laws for α and k are

$$\dot{\alpha} = \Omega_p\, \alpha - \alpha\Omega_p + h(\alpha, D_p)$$

$$= T_{\dot{R}(t)} + (\omega\alpha - \alpha\omega) + \Theta^\alpha\, D_p \tag{2.38}$$

$$\dot{k} = \Theta^k\, D_p$$

where functions Θ^α and Θ^p are to be determined in each case for the specific material.
From (2.26), (2.32), (2.33), (2.34), (2.35), (2.38) we obtain

$$\sigma_{ij}^* = \left\{ 2\mu\, \delta_{ik}\, \delta_{lj} + \lambda\delta_{ij}\, \delta_{kl} - \right.$$

$$\left. - \frac{\frac{\partial Y}{\partial \sigma_{rt}}(2\mu\delta_{rk}\delta_{lt} + \lambda\delta_{rt}\delta_{kl})(2\mu\delta_{ip}\delta_{qj} + \lambda\delta_{ip}\delta_{pq})\frac{\partial Y}{\partial \sigma_{pq}}}{\frac{\partial Y}{\partial \sigma_{rs}}(2\mu\delta_{rp}\delta_{qs} + \lambda\delta_{rs}\delta_{pq})\frac{\partial Y}{\partial \alpha_{pq}} - \Theta^\alpha\frac{\partial Y}{\partial \alpha_{rs}}\frac{\partial Y}{\partial \sigma_{rs}} - \Theta^k\frac{\partial Y}{\partial k}} \right\} D_{kl} \tag{2.39}$$

with

$$\sigma^* = \dot{\sigma} + \sigma(trD) + \sigma\Omega - \Omega\sigma \tag{2.40}$$

3. FINITE ELEMENT FORMULATION

The finite elements formulation of elastoplastic problems is well known. In this work we use the variational relation [7]

$$\int_v [\sigma_{ij}^*\, \delta D_{ij} - \frac{1}{2}\, \delta(2D_{ik}\, D_{kj} - L_{ki}\, L_{kj})]dv = \int_s t_i\delta v_i dS + \int_v b_i\delta v_i dv \tag{3.1}$$

where the contact force t and the mass force b are measured with relation to the reference (current) areas and integration is performed on the current configuration. The interpolation functions are collected into a matrix N, so that, at element level,

$$\dot{u}_i = N_{ij}\, \dot{U}_j \tag{3.2}$$

where $\dot{u}_i$ and $\dot{u}_j$ are the displacement rates inside the element and at the nodes, respectively. From this we obtain

$$D_i = B_{ij}\, \dot{U}_j \tag{3.3.}$$

$$L_{ki} = N_{kj,i}\, \dot{U}_j \tag{3.4}$$

Substitution of (3.3) (3.4) into (3.1) leads to the determination of the element stiffness matrices

$$K = \int_v B^T \mathbf{E} \; B \; dv \tag{3.5.}$$

$$\bar{K} = \int_v (N_{k,i}^T \; \sigma_{ij} \; N_{k,j} - 2 \; B_{ki} \; \sigma_{ij} \; B_{rj})dv \tag{3.6}$$

where $\mathbf{E}$ is the constitutive matrix that relates D or $(D - D_p)$ and σ^* [8]. More details may be found in [9].

4. EXAMPLES

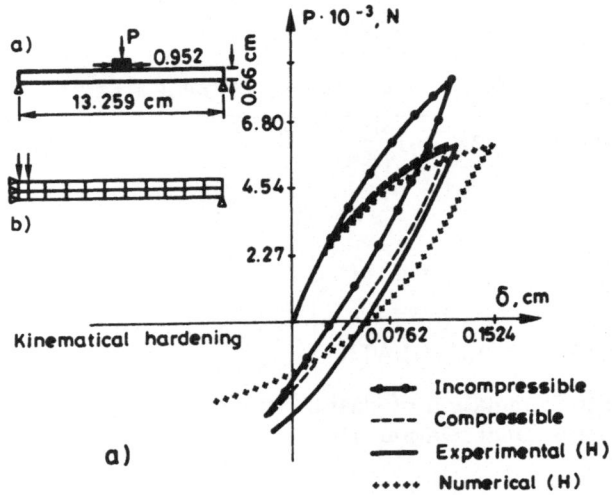

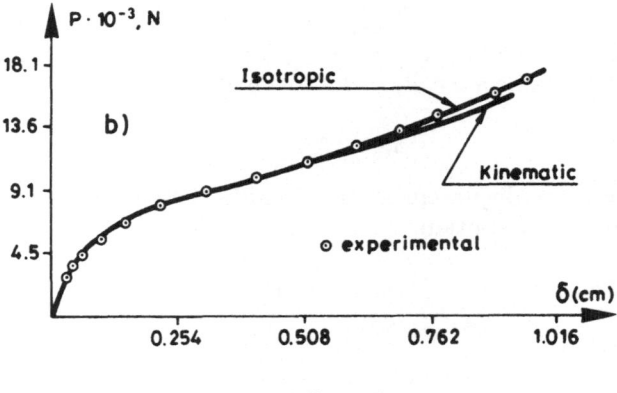

Figure 1.

Fig. 1(a) indicates some results corresponding to the analysis of a simply supported circular plate loaded at the center; the example is taken from reference [10], where the complete data and numerical and experimental results are   given . The curve labeled "incompressible" corresponds to the normal procedure and shows a gross error due to incompressibility "locking". The curve labeled "compressible" corresponds to the analysis using the technique proposed in reference [11] and isoparametric four nodes elements. In Fig. 1(b), the results are plotted for larger loads, with the "compressible" formulation. We notice a good approximation of the experimental results.

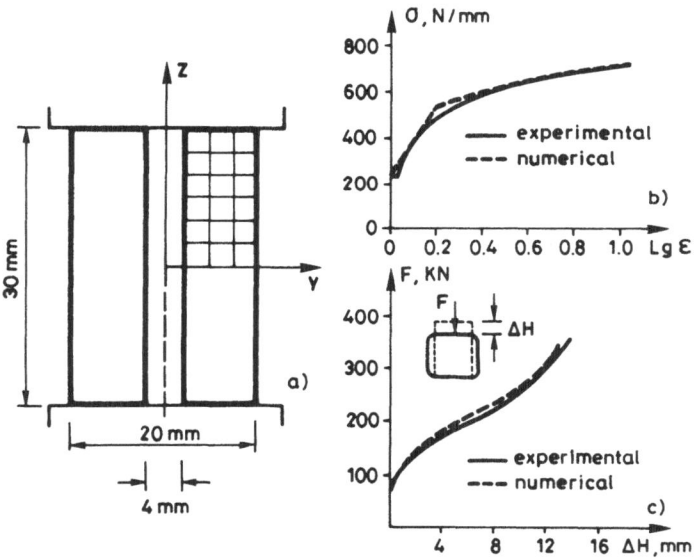

Figure 2

In Fig. 2, results are given for an "upsetting" test. Experimental data were taken from reference [12]. Again, four nodes isoparametric axisymmetric elements with the compressible formulation were used.

ACKNOWLEDGMENT

This work is part of a research program on nonlinear analysis, financially supported by FINEP, CNPq and CAPES (Brasil), and developed with the collaboration of E.T. Onat, from Yale University. This paper was written in the congenial environment of the University of Barcelona.

REFERENCES

[1]   G.J.Creus, A.G. Groehs, E.T. Onat, "Constitutive equations for finite deformation of elastic-plastic solids", *Yale Technical Report* , (1984).

36

Also presented to the *Symposium on Plasticity: Foundations and Future Directions*, University of Florida, Gainsville, (1987).

[2] E.T. Onat, G.J. Creus, A.G. Groehs, "Representation of elastic- plastic behaviour in the presence of finite deformations and anisotropy", in *Proc. Conf. on Structural Analysis and Design of Nuclear Power Plants*, Porto Alegre, Brasil, (1984).

[3] A. Agah-Tehrani, E.H. Lee, R.L. Mallet, E.T. Onat, "The theory of elastic-plastic deformation at finite strain with induced anisotropy modeled as combined isotropic-kinematic hardening", *J. Mech. Phys. Solids*, Vol. **35**, 950, pp. 519–539, (1987).

[4] A.J. Ferrante et alii, "LORANE language, for the linear analysis of structures" (in portuguese), *CPGEC*, Porto Alegre, Brasil, (1977).

[5] E.H. Lee, "Elastic-plastic deformations at finite strains", *Journal of Applied Mechanics*, New York, **36**, pp. 1–6, (1969).

[6] F. Fardshisheh, E.T. Onat, "Representation of elastoplastic behaviour by means of state variables", *Problems of Plasticity, Noordhoff, Leyden*, pp. 89–115, (1974).

[7] R. Hill, "Some basic principles in the mechanics of solids without a natural time", *J. Mech. Phys. Solids*, **7**, pp. 209, (1959).

[8] R.M. Mc Meeking, J.R. Rice, "Finite-element formulations for problems of large elastic-plastic deformation", *Int. J. Solids Structures*, **11**, pp. 601–616, (1975).

[9] A.G.Groehs, "ESFINGE, a computer system for the analysis of finite elastic-plastic deformations" (in portuguese), Ph D Theses, COPPE, UFRJ, Rio de Janeiro, Brasil, (1983).

[10] B. Hunsaker, W.E. Haisler, J.A. Stricklin, "On the use of two hardening rules of plasticity in incremental and pseudo force analyses", *Constitutive Equations in Viscoplasticity: Computational and Engineering Aspects*, New York, ASME, pp. 139–170, (1979).

[11] J.C. Nagtegaal, D.M. Parks, J.R. Rice, "On numerically accurate finite elements solutions in the fully plastic range", *Computer Meth. in Appl. Mech. and Engng.*, *4*, **2**, pp. 153, (1974).

[12] R. Herbertz, " Zur Braucharbeit eines starr-viscoplastischen Materialgesetzes der Lösung umformtechnischer mit Hilfe der Finite Elemente Methode, *Ph D Theses, Technischen Höchschule Aachen*, (1982).

MODELLING OF STRUCTURE FORMATION AND RELATION TO MECHANICAL PROPERTIES OF NODULAR CAST IRON

Erik Lundbäck, Ingvar L Svensson and Per–Eric Persson

Dept. Casting of Metals, The Royal Institute of Technology,
Fack, S–100 44 Stockholm, Sweden

ABSTRACT

The computer simulation of shaped castings is now in progress and introduced to foundry industry. The simulation of solidification and cooling processes in a casting is a useful tool in the process of designing a casting optimally with respect to several phenomena like microstructure development and mechanical properties. Today the heat transfer calculations can be done in a FEM or FDM environment including solidification and solid state relations in shaped castings. The macroscopic heat transfer have been coupled to kinetic growth laws to describe the transformation rate. The macro and micro models of solidification and solid state transformation have been used to calculate the fraction and coarseness in nodular cast iron. The coarseness of the solidified structure is an input to the austenite transformation to ferrite and pearlite. The amount of structure has been related to mechanical properties. the mechanical properties of nodular cast iron depend on the cooling rate and other metallurgical factors. It is of great interest to calculate the variation of mechanical properties in a casting. Iso hardness graphs have been plotted for experimental castings. Good agreement has been achieved between experiments and calculations.

1. INTRODUCTION

Just now a new technology is under development, CAD/CAM. May be the CAD/CAM will be most used in the foundry and casting production in combination with "solidification and casting process simulation" due to the complex shapes which are difficult to manufacture by other methods. At casting production, the designer makes a drawing with help of a graphical terminal (CAD). To optimize the construction a simulation of the casting process can be done by a computer of some important phenomena e.g. mold filling, feeding, solidification structure formation, and mechanical properties as a function of structure. If defects are detected by the casting simulation, the geometry, casting process and casting material can be changed and a new simulation be done by the computer until a satisfactory result has been obtained. When the casting is optimized, the pattern can be manufactured by NC–tooling.

J. L. Chenot and E. Oñate (eds.), Modelling of Metal Forming Processes, 37–46.
© *1988 by Kluwer Academic Publishers.*

2. SOLIDIFICATION MODELS

To calculate the solidification of cast–iron is difficult due to the complex growth. At the same time it is important to do this, to control and get the right solidification and transformation structure in casting, as it is the most used casting material for shaped castings. The solidification of cast–iron can be summarized as follows

Stable eutectic solidification: The cast–iron solidifies by forming an austenite-graphite eutectic, grey structure. The solidification is controlled by the eutectic growth. The graphite can grow in different modifications; flake, compacted and spheres, depending on the metallurgical treatment.

The graphite–nodule growth is controlled by carbon–diffusion through the austenitic shell.

The inoculation and cooling rate are very important for the solidification behavior of cast iron. The inoculation controls the number of growing eutectic cells and the temperature growth occur. If the cooling rate is too high, ledeburite can be formed, which is growing much faster than the graphite–eutectic

2.1 Kinetic solidification models

For growth of nodular cast–iron the model for binary Fe–C system derived by Wetterfall, Fredriksson and Hillert in ref. 1, has been modified to take into account the effect of silicon. The model is based on the carbon diffusion through an austenitic shell and the carbon gradient is estimated from the phase diagrams in ref. 2. No solid diffusion of silicon is assumed.

The growth of the graphite nodule is calculated by (ref.1);

$$\frac{dR_g}{dt} = D_c^\gamma \; \frac{V_m^{gr}}{V_m^\gamma} \; \frac{1}{R_g(1-R_g/R)} \; \frac{X_c^{\gamma/L}-X_c^{\gamma/gr}}{X_c^{gr}-X_c^{\gamma/gr}}$$

(1)

This equation can be simplified to:

$$\frac{dR_g}{dt} = \frac{C}{R} \; \frac{\Delta T}{} = \frac{C}{R_g} \; (T_{eut} - T^*) \tag{2}$$

The eutectic temperature depends on the silicon content and is calculated by equ (3). This has been investigated by several and in this paper we will use (ref.1);

$$T_{eut} = 1153 + 5.75 \; C_{Si}^\ell \tag{3}$$

During solidification the silicon will segregate and the silicon content will decrease in the melt, this is calculated by a modified Scheil equation.(ref.3)

$$dC_{Si}^{\ell} = K \quad C_{Si}^{\ell}(1/f_s)^{-1} \quad df^{\gamma} \qquad 0 < f_s < 0.85 \tag{4}$$

The partition coefficient is given by Kagawa et al (ref.4).

$$K = 1.7 - 0.3 \ C_{Si}^{\ell} + 0.05 \ (C_{Si}^{\ell})^2 \tag{5}$$

By using Johnson–Mehls equation the amount of solidified metals can be calculated, the number of growing cells are calculated by a simple nucleation law evaluated from experiments.

$$N = 1.4 \ 10^{11} \ \Delta T_m \quad R = 2.3 \ R_g \quad fs = 1 - \exp(-\tfrac{4}{3} \ \pi \ NR^3) \tag{6abc}$$

The derivative, gives the solidification rate,

$$dfs/dt = (1-fs) \ 4\pi \ NR^2 \ dR/dt \tag{7}$$

The liberation of solidification heat,

$$dQ/dt = L \ df/dt \tag{8}$$

This set of equations are used in a FEM–program to calculate heat transfer and solidification.

Figure 1 shows the segregation of silicon during solidification.. The silicon content was measured by dot analysis by WDS (Wave length Dispersive Spectrometer) in a SEM. (Scanning Electron Microscope).

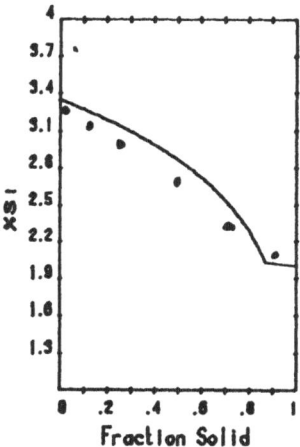

Fig.1 Segregation according equation (3). Dots measured by WDS.

3. MODELS FOR THE SOLID STATE TRANSFORMATION

During   cooling, temperature will reach the stable eutectoid temperature and austenite will start to   decompose into ferrite and graphite. The growth rate depends on carbon diffusion through a ferritic shell. If the growth rate is too low, or the growth distance is too long the temperature will reach the metastable temperature before all austenite is transformed, and pearlite will start growth in competition with the ferritic transformation.

For all calculations it's assumed that the earlier result from solidification and cooling could be taken as initial values for the solid state transformation e.g. silicon gradient.

The influence of silicon for solid state transformation have been discussed by Ågren (ref.7), Stefanescu et al. (ref.8) and Kanetkar (ref.14). The computer simulation of solidification and solid state transformation for lamellar grey cast iron has been treated by Lundbäck and Svensson in ref.(11)

3.1 Austenite transformation to ferrite

To calculate the transformation of austenite into ferrite, the following expressions have been used:

$$\frac{d\ell}{dt} = \frac{\Delta X_c^\alpha \, D_c^\alpha}{X_c^\gamma \, \ell^\alpha} = f(\ell^\alpha, Si) \qquad (9)$$

The stable austenite transformation temperature is calculated by (ref.8), and taking into account the influence of Mn. (eq 10)

$$T_{eut}^{\alpha/gr} = 738 + 32.7 \, \%Si - 33\%Mn \qquad (10)$$

For an alloy with 2.7% Si is the silicon gradient approximated by a straight line

$$. \quad \%Si = 3.4 - 1.3 \, \frac{r - r_g}{r_{max} - r_g} \qquad r_g < r < r_m \qquad (11)$$

The amount of transformed ferrite is calculated by:

$$f^\alpha = \frac{(r_g + \ell^\alpha)^3 - r_g^3}{r_{max}^3 - r_g^3} \qquad \ell^\alpha < r_{max} - r_g \qquad f^\alpha < 1 \qquad (12)$$

3.2 Austenite transformation to pearlite.

The metastable transformation of eutectoid steels have been studied for a long time. In the area of solid state transformation and modelling of cast iron, Heine (ref.6,7) Stefanescu (ref.8) and Kanetkar (ref.14) have made applications to cast irons. In this paper following growth laws (ref.6,8) have been used.

$$dR_p/dt = 2.86 \ 10^{-2} \ \Delta F \ \Delta T \ exp \ (-Q_d/RT^*) \qquad (13)$$

$$\Delta F = -1.46 \ (T^* - T_E) = 1.46 \ \Delta T \qquad (14)$$

$$T_E^p = 723 + 22.7 \ \%Si - 33 \ \%Mn \qquad (15)$$

By using Johnson–Mehls equation and a constant number of pearlitic cells, the amount of transformed pearlite can be calculated.

$$f^p = (1 - exp(-4\pi N^p.R_p^3/3)) \ (1 - f^\alpha) \qquad (16)$$

The liberation of transformation heat

$$dQ/dt = L^\alpha \ df^\alpha/dt + L^p \ df^p/dt \qquad (17)$$

4. NUMERICAL TREATMENT

The FEM program solves the partial differential equations of heat conduction with phase change in two and three dimensions by the Finite Element method. The principal features are outlined below, taking the model

$$\rho Cp \ dT/dt = div \ (kgrad \ T) + Q \qquad (18)$$

$$df/dt \quad = g \ (T,f) \qquad (19)$$

$$Q \quad = L \ df/dt \qquad (20)$$

The model is supplemented by suitable boundary conditions, expressing heat flow by Boltzmann radiation and heat exchange,

$$k \ dT/dn = \sigma\epsilon \ (T_e^4 - T^4) + h(T_e - T) \qquad (21)$$

The partial differential equation (eq.13) is discretized by conventional finite elements, three node triangles in two dimensional and rotationally symmetric models, and eight node trilinear "bricks" in solid problems. The ordinary differential equations (eq.14) and (eq.15) are not subjected to the Galerkin procedure, but the nodal values of f and T are made to satisfy the equations:

$$MD \ dT/dt = S(T) \ T+Q \tag{22}$$

$$dfi/dt \quad = g(Ti,fi) \tag{23}$$

$$Qi \quad = L \ g(Ti,fi), \ i=1,2 \ .. \ N \tag{24}$$

D is a diagonal matrix containing nodal values of $dH(T)/dT$. The time stepping procedure uses implicit (backward differentiation formulae) methods of order one and two, and step size and order control based on estimated error incurred over the latest step. A more detailed description is given in ref. 5, 12 and 13.

5. DIFFUSION OF CARBON IN SOLID STATE

During cooling the solubility of carbon in austenite decreases and carbone has to diffuse into the graphite.

For the binary system it is rather simple to solve numerically the partial differential equation for diffusion of carbon through an austenitic shell.

$$D_c^\gamma(t) \ (\frac{\partial^2 c}{\partial r^2} + \frac{2}{r} \frac{\partial c}{\partial r}) = \frac{\partial c}{\partial t} \qquad r_g < r < r_{max} \tag{25}$$

with boundary conditions $C(r_g,t) = C(T)$ $C(T) =$ given by the phase diagram and $(\partial c/\partial r)_{r=R_{max}} = 0$

Equation (25) is solved numerically by explicit Euler forward integration. The figure below shows such a calculation for the binary case with input values according to the experiments.

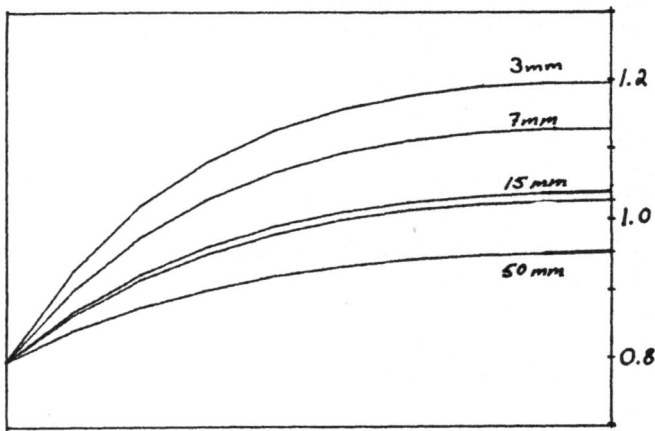

Figure 2   Shows the carbongradient for the binary case, with input values according experiment.

6. THE RELATION BETWEEN MICROSTRUCTURE AND MECHANICAL PROPERTIES.

Nodular cast irons can be distinguished according to the type of matrix, e.g. ferrite, ferrite-pearlite and pearlite type. By alloying and heat treatment the properties can be varied in a wide range.

The mechanical properties are depending on the amount and type of microstructure. To calculate the mechanical properties the following relations have been used (ref.16) H_B^I from experiment.

$$H_B^I = 153 + 1.0 \ (\%\text{Pearlite}) \tag{26a}$$

$$H_B^{II} = 329 \ /(2.06 - f^P) \tag{26b}$$

$$\sigma_B = 3.39 \ H_B^{II} - 73 \tag{27}$$

$$\sigma_S = 1.95 \ H_B^{II} + 6 \tag{28}$$

$$\delta_f = f(H_B) = 775 \ /(H_B^{II} - 115) \tag{29}$$

7. EXPERIMENTAL AND CALCULATED RESULTS

Plates with different dimensions and with an eutectic composition and silicon content about 2.7% have been cast in sodium silicate bounded sand moulds. Several parameters have been measured, such as cooling curves, number of growing cells, amount of pearlite and ferrite as cast and hardness. Some of the measured parameters are given in table (1) and figure 3.

Thickness (mm)	$N_A(1/mm^3)$	$N_V(1/m^3)$	%Pearlite	H_B^I	r_g μm
3	601	$8.7 \ 10^{12}$	80%	231	15
7	393	$3.0 \ 10^{12}$	40%	201	21
15	225	$1.8 \ 10^{12}$	20%	176	27
30	173	$1.0 \ 10^{12}$	15%	168	36
50	126	$6.5 \ 10^{11}$	7.5%	161	36

Table 1 Measured nodule number, pearlite amount ,hardness and maximum radius of graphite nodule in the plates, at thermocouple position.

44

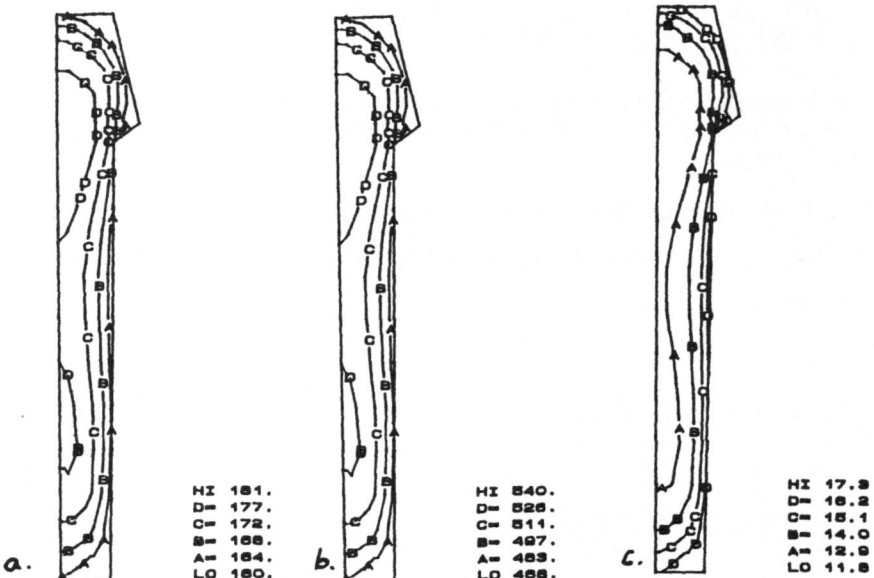

Figure 3   Shows calculated Iso– a) –hardness, b) –tensile strenght,
         c) –tensile strain graphs t=30 mm

The figures below shows calculated cooling curves compared with measured.

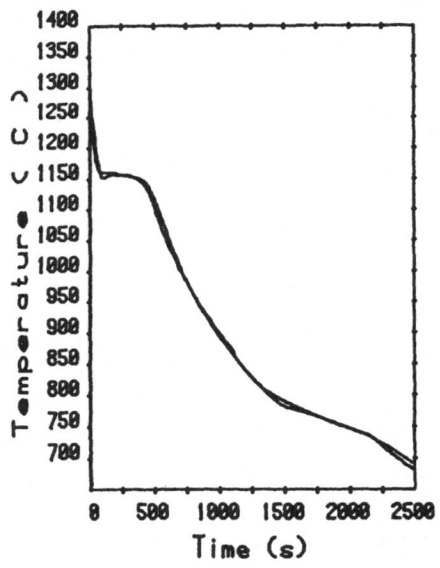

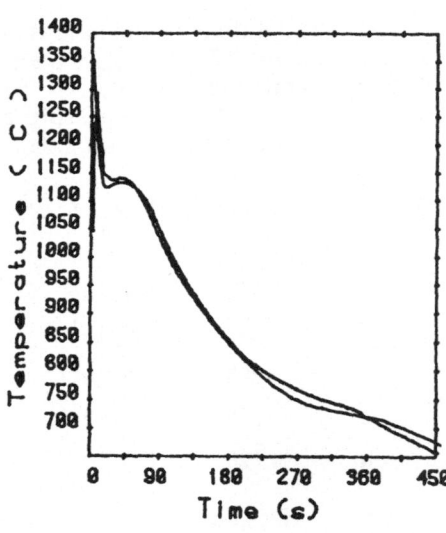

Figure 4   Calculated and measured
cooling curves for t=30 mm

Figure 5 Calculated and measured
cooling curves for t=7mm

8. DISCUSSION AND CONCLUSIONS

A comparison between calculated and measured results shows that it's possible to simulate how the microstructure develops during solidification and solid state transformation. However, there is some deviation between calculations and measurements mainly due to inaccuracies on physical properties of the metal and mould material, and secondarily to the solid state model. This is a first step towards integration of solidification and solid state transformation into a FEM calculation, and combination of calculated structure with a prediction of the mechanical properties of a casting.

9. ACKNOWLEDGEMENTS

The authors want to thank the Swedish Board of Technical Development and Nordic Industrial Fund for financial support. The project belongs to the cooperative nordic HUBERT–project. The authors also wants to thank Nils Lange for the help with SEM–measurements of the silicon and carbon.

10. REFERENCES

1. WETTERFALL S.E., FREDRIKSSON H., HILLERT M.: J. Iron and Steel Institute (1972)323
2. LACAZE J.: Internal communication 1987.
2a. LACAZE J., FREDRIKSSON H. AND LESOULT G.: Eutectic solidification of S.G. Fe–C–Si Alloys. Presented at Sheffield conferences ,sept. 1987
2b. LACAZE J., SUNDMAN B.: An assessment of the Fe–Si system with ordering of the bcc phase (under preparation)
2c. Lacaze J.: Study of the Fe–C–Si system. (under preparation).
3. SCHEIL: Metallforschung. 2(1947), p.69
4. KAGAWA A., OKAMOTO T.: Metal Science (1980)519
5. H. FREDRIKSSON, I.L. SVENSSON: "Computer simulation of the structure formed during solidification of cast iron." The physical Metallurgy of Cast Iron. Material Research Society 34 (1984)273.
6. R.W. HEINE: "The Carbon Equivalent Fe–C–Si Diagram and its Application to Cast Irons." AFS Cast Metals Research Journal, June 1971, p. 49.
7. R.W. HEINE: "The Fe–C Transformation Diagram Related to the Solidification of Cast Irons." AFS Transactions, 1970, vol 78, p.187.
8. J.ÅGREN: "A Thermodynamical Analysis of the Fe–Si–C System." between 973 and 1373 k. TRITAMAC – 0125. October 1977.
9. D.M. STEFANESCU: "Computer modeling of the solidification of eutectic alloys". The case of cast iron. Proceeding Society for Metals Materials Science Division Computer Simulation. The Metallurgical Society in Toronto, Canada, October 13–17, 1985.
10. J. Ågren: "Computer simulation of the Austenite/Ferrite diffusional transformation in low alloyed steels". TRITAMAC, April 1981.

11.   E. LUNDBÄCK, I.L. SVENSSON, J. THORGRIMSSON: "On calculation of structure and mechanical properties in grey cast–iron". Proceeding of "Solidification Processing 1987", Sheffield, England 21–24 September 1987.

12.   I.L. SVENSSON, E. LUNDBÄCK: " Computer simulation of castings" MRS–symposium: State of the art of computer simulation of casting and solidification processes. Strasbourg, France 17–19 June 1986.

13.   I.L. SVENSSON, E. LUNDBÄCK: "Simulation of grey castiron in a shaped casting". Proceeding of the NUMIFORM'86 Conference, Göteborg, 25–29 August 1986.

14.   C.S. KANETKAR: "Computer simulation of microstructural evolution during solidification of cast iron and aluminium–silicon alloys." Thesis 1988, University of Alabama, Tuscaloosa, Alabama, USA.

15.   E.O. LISELL: Segjärn. Några synpunkter på egenskaper och användningsområden. Teknisk information, Verkstadsteknik nr 1–1955,p.9.

16.   L.E. BJÖRKEGREN: Intern communication 1980. Swedish Foundry Association, Jönköping.

17.   CRANK J: The Mathematics of diffusion, second ed. 1975.

11. NOMENCLATURE

R =	Radius of austenite (m)	
R_g =	Radius of graphite (m)	
t =	Time (s)	
D_c^γ =	Diffusion coefficient (m^2/s) $1.5 \ 10^{-10}$m^2/s	
V_m^{gr} =	Molar volume of graphite (m^3/mol) $7 \ 10^6$	
V_m^γ =	Molar volume of austenite (m^3/mol) $5.5 \ 10^6$	
$X_c^{\gamma/\ell}$ =	Carbon concentration in austenite in contact with melt (mol/Σmol)	
$X_c^{\gamma/gr}$=	Carbon concentration in austenite in contact with graphite.	
X_c^{gr} =	Concentration of carbon in carbon	
T_{eut} =	Stable eutectic temp (C)	
T^* =	Local temperature [$^\circ$C or K]	
C_{Si}^ℓ =	Weight %Silicon i liquid	
ΔT =	Supercooling ($^\circ$C)	
ΔT_m =	Max supercooling ($^\circ$C)	
N =	Number of growing cells in each Mode (1/m^3)	
f_s =	Fraction solid	
f^γ =	Fraction austenite	
K =	Partion coefficient $(K = \dfrac{X^s}{X^\ell})$	
H_B^I =	Hardness Brinell	
H_B^{II} =	Hardness Brinell according strandard	
σ_B =	Tensile Strenght(MPa)	

ρ_f =	Tensile strain (%)
r =	Radius (m)
r_{max} =	Outer radius austenite (m) = 2.3 r_o
f^α =	Fraction ferrite
ℓ =	Length of the ferrite bord.
R_p =	Radius of pearlite nodule [m]
ΔF =	Free Energy
T_e^p =	Eutectoid temperature for pearlite
f^p =	Fraction pearlite
N_p =	Number of pearlite nodules (1/m^3)
L^α =	Latent heat for formation of ferrite
L^p =	" " " " of pearlite
Q_d^γ =	Activation energy for carbon diffusion in austenite
ρ =	Density of the material [kg/m^3]
Cp =	Specific heat [J/kg$^\circ$C]
K =	Heat conductivity [$^W/_m$$^\circ$C]
Q =	Latent heat release [W]
σ =	Bolzmann constant
ϵ =	Emissivity
T_e =	Ambient temperature [$^\circ$C]
h =	Heat exchange [$^W/_m$2°C]
Q_i =	Nodal heat source
M =	Mass matrice
S =	Stiffness matrice
$\frac{dT}{dn}$ =	Outward normal derivativ.

PART 2

NUMERICAL TECHNIQUES

H AND P MESH REFINEMENT IN THE METAL-FORMING F.E.M. ANALYSIS

L. Cannizzaro, E. Lo Valvo, F. Micari[*],
R. Riccobono
Dipartimento di Tecnologia e Prod. Meccanica
Università degli Studi di Palermo, Italy
* C.N.R. App. Researcher

Abstract

In this paper a comparison between H and P refinement techniques in the metal-forming F.E.M. analysis is carried out in order to evaluate their computational efficiency. The results are compared using a particular error estimator which locally allows determining the workpiece zones where the refinement is necessary.

Numerical examples are reported both in axisymmetrical and threedimensional analysis.

Introduction

The finite element method has been applied in the past years to a large number of problems of different type. In particular several applications have been performed in metal forming processes [1][2][3].

In fact the old methods based on the continuum analysis do not always allow determining all the process parameters because they give informations only on the loads and on the pressure distributions, while they cannot analyze the deformation mechanic; furthermore to apply them it is necessary to assume simplified hypothesis which give rise to crude approximations.

On the contrary the finite element method is able to describe the deformation process evaluating the stresses and strains states in the body under deformation. So this method gives to the manufacturing process designer all the informations that he needs to determine the process parameters. In fact only from the knowledge of the velocity field, the stress and strain field, the pressure distribution and the loads it is possible to state the type and the power of the machines, the shape and the stiffness of the tools and the geometry of the predeformed workpiece.

However in metal forming F.E.M. analysis the determination of the error which arises in the numerical solution of the problem has not been enough investigated.

On the contrary an accurate evaluation of this parameter is very

49

© 1988 by Kluwer Academic Publishers.

important in the non linear plastic problems due to the heavier calculation with respect to the linear elastic analysis; in fact the estimation of
the error can lead to the more appropriate choice of the mesh employed
obtaining a compromise solution between CPU time and accuracy level
of the results.

Moreover in the elastic analysis the evaluation of the error in
some cases can be performed comparing the numerical results with the
exact values obtained by the continuum analysis and so it is possible
to transfer the mesh employed with good results to similar problems of
which there is not the exact solution. On the contrary, in the plastic
analysis the exact solutions are limited to very few cases, so it is more
important to individuate an estimator of the error able to give global
and local informations and then representing an important tool for the
analyst to decide if and where a refinement of the mesh is necessary.

In this paper the Authors, employing the error estimator defined
in two previous works [4][5],compare two different strategies of mesh
refinement: H and P refinement.

The first one consists in a several step analysis increasing at
each step the number of the elements but taking constant the order of
the shape functions; in the second one, on the contrary, the same
mesh is employed but the order of the interpolating polinomial functions is progressively increased.

The error estimator

The numerical metal forming model employed by the Authors is a
variational approach based on the upper-bound theorem and the discretization of the workpiece into finite elements. This approach leads to
the minimization of a discrete functional which depends on the constitutive law of the material

$$\sum_m \phi_i = \sum_m \left[\int_{V_i} \bar{\sigma} \sqrt{\frac{2}{3} \left(\{u\}^T [K] \{u\} \right)^{1/2}} \, dV_i - \{u\}^T \int_{S_{ti}} [G]^T \{T\} \, dS_{ti} \right]$$

where $[K] = [B]^T [B]$ is the stiffness matrix, $[G]$ is the interpolating
function relative to the surface S_{ti} of the i-element, $\{T\}$ is the vector
of the known external unit forces and m is the number of the elements.

The minimization, respecting the incompressibility and boundary
conditions, is performed by Lagrange multipliers technique leading to a
nonlinear equation system [6][7].

Thus from the derived displacement field the deviatoric stress
field is obtained with the compatibility and constitutive equations:

$$\{\dot{\varepsilon}\} = [B]\{u\} \quad \text{and} \quad \{\hat{\sigma}'\} = \frac{2\bar{\sigma}}{3\bar{\varepsilon}} \{\dot{\varepsilon}\}$$

The deviatoric stress field so obtained is a discontinuous one due
to the use of interpolation functions with C_0 continuity. Then an error
estimator can be evaluated comparing this field with a continuous deviatoric stress one $\{\sigma^{*}\}$ obtained applying the same interpolation functions
used for the velocity field on the vector of the nodal equivalent

deviatoric stresses.

$$\{\sigma^{*\prime}\} = [N]\{\bar{\sigma}^{*\prime}\}$$

where the vector $\{\bar{\sigma}^{*\prime}\}$ is given by the equations

$$\int_V [N]^T (\{\bar{\sigma}^{*\prime}\} - \{\hat{\sigma}^{\prime}\}) \, dV = 0$$

which leads to

$$\{\bar{\sigma}^{*\prime}\} = \left[\int_V [N]^T [N] \, dV\right]^{-1} \int_V [N]^T \{\hat{\sigma}^{\prime}\} \, dV$$

Then the local error is defined by

$$\{e_{\sigma^{\prime}}\} = \{\sigma^{*\prime}\} - \{\hat{\sigma}^{\prime}\}$$

This measure of the error is defined in each integration point inside the elements giving an important help to the analyst highlighting the zones of the workpiece where the refinement is more necessary.

Furthermore the global error estimate must be defined to evaluate the level of precision reached in the whole workpiece, weighting the local value of the error above calculated with respect to the element size.

A norm of the error is then defined as

$$\left\|e_{\sigma^{\prime}}\right\|^2 = \int_V \{e_{\sigma^{\prime}}\}^T \{e_{\sigma^{\prime}}\} \, dV$$

or better in percentage of the deviatoric stresses norm

$$\eta = \frac{\|e_{\sigma^{\prime}}\|}{\|\sigma^{*\prime}\|} \times 100$$

H and P refinement

H and P are two different techniques used to refine the mesh employed in order to improve the precision of the results.

In this paper the error estimator above defined is used to evaluate the performance reached with these two different strategies.

H refinement is based on the progressive increasing of the number of the elements choosing the zones where the local error estimator has given the higher values; of course, the mean dimension (h) of the single element becomes smaller.

The order of the shape function remains unchanged and the heavier calculations is due only to the increased number of the degrees of freedom.

P refinement, on the contrary, is a technique based on the progressive increasing of the order of the interpolating shape function, while the mesh employed is taken unchanged.

Following this technique the number of the elements can be taken relatively low with a limited value of the error estimated, but the calculation is heavier due to the greater number of the operations made inside each element.

This is due both to the more complex shape functions and to the higher necessary number of points in the Gauss numerical integration process.

Applications

The comparison between the two refinement methods has been carried out both in the axisymmetrical condition and in the threedimensional analysis.

For the H refinement the quadratic uncomplete isoparametric element has been employed with a shape function of the form [8]

$$N_i = \frac{1}{4}(1+\xi\xi_i)(1+\eta\eta_i)$$

Following the P technique, in the second refinement step the above element have been substituted by a cubic uncomplete isoparametrical one with a shape function of the form [8]

$$N_i = \frac{1}{4}(1+\xi\xi_i)(1+\eta\eta_i)(\xi\xi_i+\eta\eta_i-1) \quad \text{for Corner nodes}$$

$$\xi_i = 0, \qquad N_i = \frac{1}{2}(1-\xi^2)(1+\eta\eta_i)$$
$$\text{for Mid-side nodes}$$
$$\eta_i = 0, \qquad N_i = \frac{1}{2}(1+\xi\xi_i)(1-\eta^2)$$

Thus, with the P-technique, the refinement involves an 8-node element in axisymmetrical condition or a 20-node element in threedimensional one, while the elements employed in H-technique are 4-node or 8-node respectively.

These two different techniques have been applied to the analysis of a case of axisymmetrical extrusion-forging and to a simple threedimensional upsetting with a yield stess equal to 6.3 kg/mm^2.
For the first example, employing the H-refinement technique, various meshes have been carried out.

In the figs. 1-4 the deformed meshs, with 56,72,100,204 elements at the first step of the punch displacement are reported.

The H-refinement has been performed particularly in the zones where the largest values of the local error have been estimated. In the fig.5   the trend of the global error versus the number of the elements is reported.

The P-technique has been carried out refining the mesh of fig.1 with 56 elements by means of the same discretization   , but employing 28 cubic elements (Fig.6).The global error falls from η =21.0 to η =18.5.

The same technique, for the 72 element mesh refined to 36 cubic elements (Fig.7) has shown a decay of the error from η =20.6 to η =18.2.

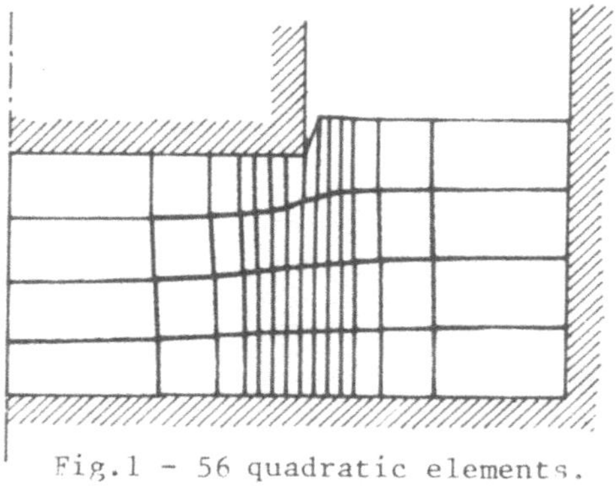

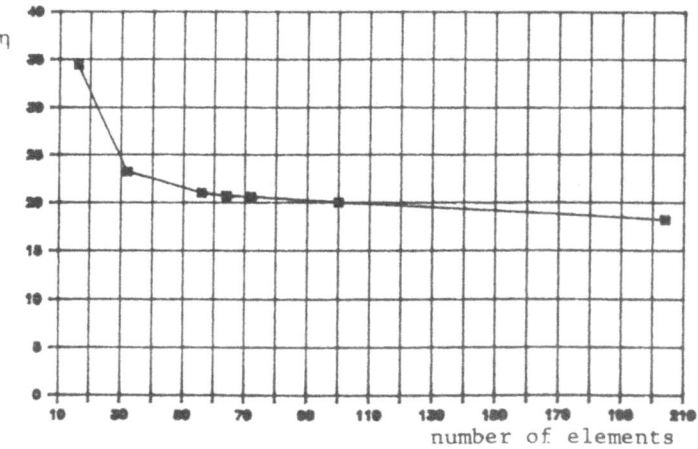

Fig.1 - 56 quadratic elements. 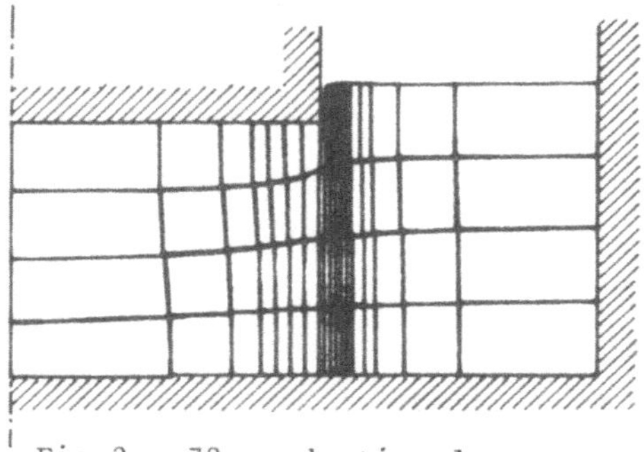 Fig.2 - 72 quadratic elements.

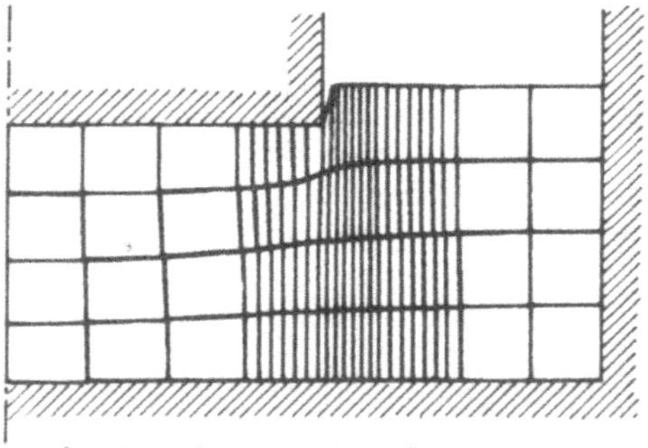

Fig.3 - 100 quadratic elements. 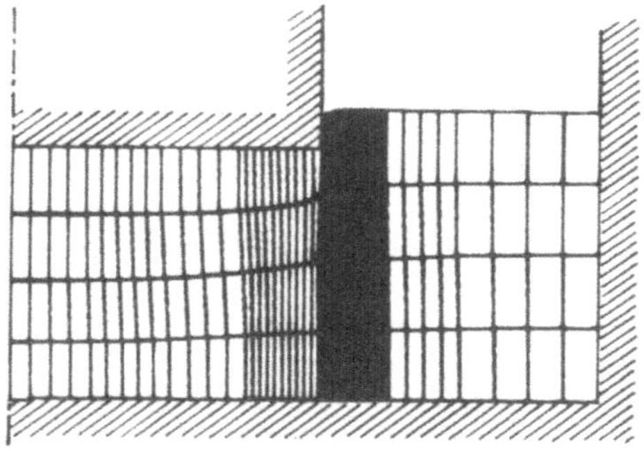 Fig.4 - 204 quadratic elements.

number of elements

Fig.5 - Error trend versus the number of elements.

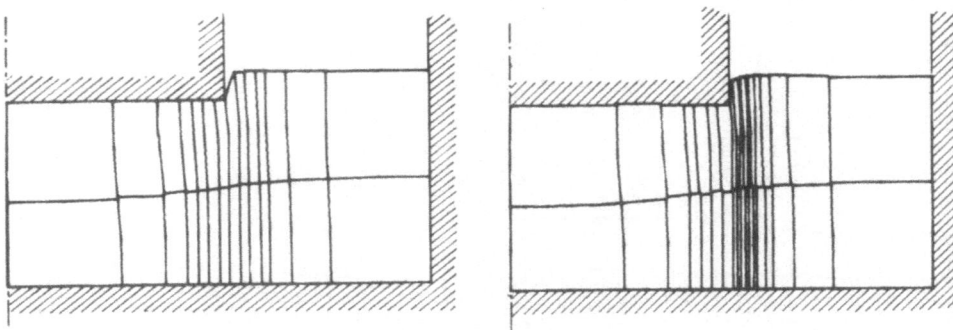

Fig.6 – 28 cubic elements.   Fig.7 – 36 cubic elements.

It is very important to notice that, in order to obtain the same value of the global error employing the H-refinement technique, it was necessary to draw a mesh with 204 quadratic elements.

In the fig.8 the p trend versus the radius of the workpiece is reported, employing the 36 cubic element mesh, in order to show that in the free surface the smoothed calculated value is strictly close to the exact value easily predictable, in this case, by the boundary conditions.

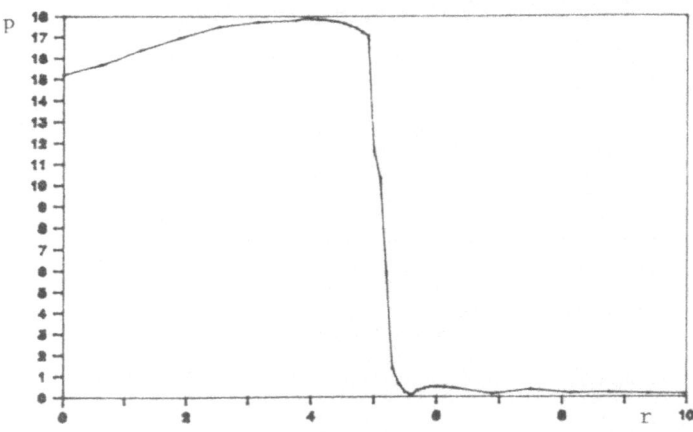

Fig.8 – p trend versus the radius of the workpiece.

The same techniques have been applied to a simple case of thr-eedimensional upsetting, employing 8-nodes or 20-nodes element mesh.

The values of the global error were η =4.80 with 32 trilinear elements and η =4.13 with 8 cubic elements. In the figs. 9-10 the discontinuous and the continuous pressure fields are reported.

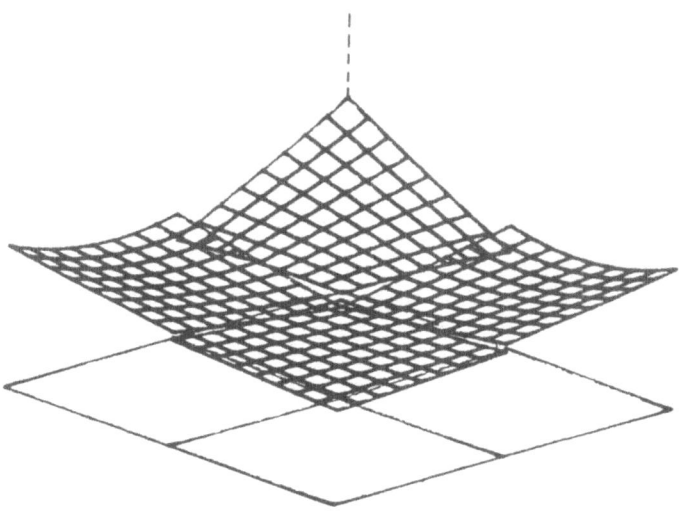

Fig.9 - Discontinous pressure distribution.

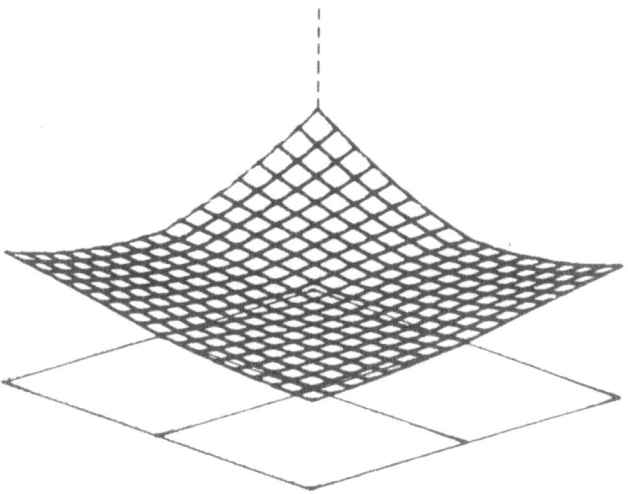

Fig.10 - Continous pressure distribution.

Conclusions

The H and P techniques, applied to various examples represent two efficient methods to reduce the global error in metal forming analysis.

Moreover the choice between H and P technique is strictly due to the computational tools available. H-refinement requires an automatic mesh generator in order to perform the progressive steps of the discretization, while the more limited number of the elements required by the P-technique in order to obtain a suitable global error, doesn't impose the use of an advanced preprocessor.

Despite this consideration P technique requires a larger amount of calculations, involving consequently a higher CPU time; moreover, since the RAM memory required is larger than the H technique one, the above considerations show that the P refinement technique cannot be employed easily on Personal Computers.

References

[1] C.H. Lee, S. Kobayashi (1973) "New solutions to rigid plastic deformation problem using a matrix method", Jnl. of Eng. for Ind., Aug.1973, pp.865-870
[2] S. Kobayashi (1981) "Metalworking process modelling and the finite element method", 9th NAMRC, pp.16-21
[3] N. Alberti, L. Cannizzaro, R. Riccobono (1987) "A new numerical method for axisymmetrical forming processes", Annals of CIRP, 36/1, pp.131-133
[4] L. Cannizzaro, F. Micari, R. Riccobono (1987) "Impiego di uno stimatore dell'errore nell'analisi ad elementi finiti di problemi di formatura dei metalli", 2nd Convegno di Meccanica Computazionale, pp.101-105
[5] L. Cannizzaro, E. Lo Valvo, F. Micari, R.Riccobono (1988) "Error estimates and automatic adaptive mesh refinement for the metal forming F.E.M. analysis", 27th MATADOR Conference
[6] L. Cannizzaro, M. Enea (1978) "Applicazione del metodo di soluzione frontale allo studio delle lavorazioni plastiche dei metalli", 4th AIMETA Conference, vol.3, pp.1-12
[7] N. Alberti, L. Cannizzaro, F. Micari, R. Riccobono (1986) "Analisi tridimensionale di problemi di formatura", 8th AIMETA Conference, pp. 45-48
[8] O. C. Zienkiewicz (1977) "The finite element method" - McGraw-Hill

This work has been supported by M.P.I. (Ministero della Pubblica Istruzione).

A MIXED EULERIAN-LAGRANGIAN FINITE ELEMENT METHOD FOR SIMULATION OF THERMO-MECHANICAL FORMING PROCESSES

J. Huétink, J. van der Lugt and P.T. Vreede
University Twente, Dept. of Mechanical Engineering
P.O. Box 217, 7500 AE Enschede, The Netherlands

Summary

A mixed Eulerian-Lagrangian finite element method has been developed to adapt nodal point locations independently from the actual material displacements.

Temperature and elastic-plastic material behaviour are included. Hardening and other deformation path dependent properties are determined by incremental treatment of convective terms.

A special contact element is developed to describe the thermal and mechanical boundary and interface behaviour.

Applications are shown by simulations of an upsetting problem, a cold rolling process with time dependent material properties and a deep drawing process.

1. Introduction

Simulation of metal forming processes by the finite element method using the updated Lagrange method is limited with respect to the deformation range because the element mesh may be completely distorted after a number of steps. Besides boundary conditions in general have to be adapted after each step or after a number of steps due to the changing contact between tool and material. Therefore a procedure has been developed in which nodal point locations can be (incrementally) adapted independent of the material displacement increments. Conditions for free or forced surface movements can be satisfied. An updated Lagrange approach as well as an Eulerian approach can be regarded as special cases of the procedure. Therefore it is called the mixed Eulerian-Lagrangian formulation [1,2].

Note: vectors are denoted by subscript wiggles and tensors by subscript bars.

J. L. Chenot and E. Oñate (eds.), Modelling of Metal Forming Processes, 57–64.
© *1988 by Kluwer Academic Publishers.*

2. Mathematical Modelling

The mathematical model is based on the general principles of continuum mechanics, viz: equilibrium, compatibility and constitutive relations. In this paper we will present a brief description of a, two-dimensional, FEM model suited for thermo-mechanical analysis and time dependent material properties.

2.1. CONSTITUTIVE EQUATIONS

The constitutive equation for elastic-plastic material can be written as [2,3,4,5,6,7,8]

$$\overset{\triangledown}{\underline{\sigma}} = \frac{\dot{e}}{\rho}\,\underline{\sigma} + \underline{D}:\underline{d} + \underline{K}\dot{T} - \underline{\Phi}^t \tag{1}$$

Here $\overset{\triangledown}{\underline{\sigma}}$ is the Jaumann rate of the Cauchy stress tensor, $\underline{D}$ is the elastic plastic "tangent modulus" tensor depending on the material properties, the stress and the deformation history, the tensor $\underline{\Phi}^t$ represents the dissipation related to the time dependent material behaviour, T is the absolute temperature, the tensor $\underline{K}$ denotes the temperature dependence of the material properties including the thermal expansion and ρ is the mass density. The tensor $\underline{d}$ is the rate of deformation tensor.

The equilibrium conditions yield

$$\underline{\sigma}.\underline{\nabla} + \rho\underline{f} = \underline{\varnothing} \tag{2}$$

and

$$\rho c\dot{T} - \overset{\rightarrow}{\underline{\nabla}}\cdot(\underline{\lambda}\cdot\overset{\rightarrow}{\nabla T}) - q = \varnothing \tag{3}$$

where f represents an external body force per unit of mass, $\underline{\lambda}$ the thermal conductivity tensor and q the mechanical energy rate.

2.2. CONTACT AND FRICTION

A finite element formulation to model the contact conditions, especially in the multi-body case has been presented by the authors [9,10]. The contact description is inspired on the assumption of a friction layer, between the contacting bodies, with a small but finite thickness. Stick-slip constitutive behaviour is introduced as non-associated elasto plasticity.

The instantaneous situation in the contact region is shown in Fig. 1.
    The relative surface velocity, corrected for rotation of the reference surface (S_c) is defined by

$$\underline{d} = \Delta\underline{\dot{u}} - \underline{\omega}\cdot\Delta\underline{x} \tag{4}$$

where $\underline{\omega}$ denotes rate of rotation of the local basis ($\underline{e}_i$), which in this paper is defined to equal the rate of rotation of reference surface S_c.

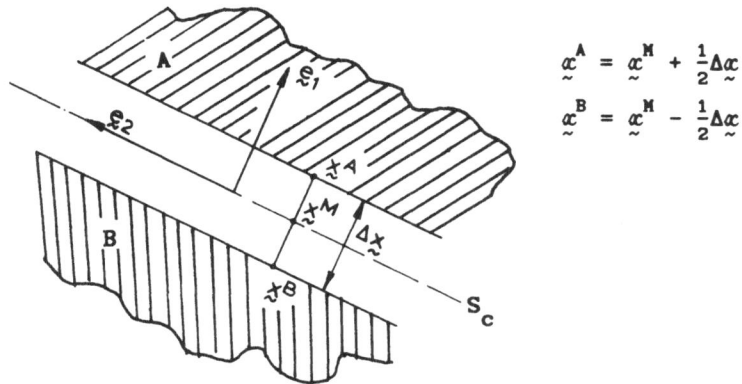

$$\underset{\sim}{x}^A = \underset{\sim}{x}^M + \tfrac{1}{2}\Delta\underset{\sim}{x}$$

$$\underset{\sim}{x}^B = \underset{\sim}{x}^M - \tfrac{1}{2}\Delta\underset{\sim}{x}$$

Fig. 1   *Position of related points.*

The relative surface velocity is assumed to consist of a reversible (elastic) part and a non- reversible (plastic, c.q. slip) part [11]. The relative displacement in normal direction is strictly elastic. The non-elastic rate of deformation has the direction of the tangential stress. The slip criterion used is Coulomb friction and μ denotes the friction coefficient. The constitutive contact behaviour results into

$$\overset{\triangledown}{\underset{\sim}{\tau}} = \underline{E}\cdot\underset{\sim}{d} - \underline{Y}\cdot\underset{\sim}{d} \text{ , where } \underline{E} = \begin{bmatrix} E_{11} & \varnothing \\ \varnothing & E_{22} \end{bmatrix}, \underline{Y} = \begin{bmatrix} \varnothing & \varnothing \\ \mu sgn(\tau_2)E_{11} & E_{22} \end{bmatrix} \qquad (5)$$

This relation is valid in case of slip (plastic) deformation. For the elastic situation: $\underline{Y} = \varnothing$, for the open situation: $\underline{Y} = \underline{E} = \varnothing$.

3. Some Remarks on the Finite Element Formulation

For the two bodies A, B (Fig. 1) the 'weak form' of the mechanical equilibrium yields

$$\int_{V_A + V_B}\delta\underline{d}:\underline{\sigma}dV - \int_V\delta\underset{\sim}{v}\cdot\rho\underset{\sim}{f}dV + \int_{S_C}\delta(\Delta\underset{\sim}{v})\cdot\underset{\sim}{\tau}dS + \int_S\delta\underset{\sim}{v}\cdot\underset{\sim}{F}dS = \varnothing \qquad \forall\delta\underset{\sim}{v} \qquad (6)$$

where the vector $\underset{\sim}{F}$ represents the surface traction per unit surface.

The 'weak form' of the thermal equilibrium results in

$$\int_V\left[\delta\dot{T}\rho c\dot{T} + \overset{\rightarrow}{\nabla\delta\dot{T}}\cdot\underline{\lambda}\cdot\overset{\rightarrow}{\nabla T} - \delta\dot{T}\underline{\sigma}:\underline{d}\right]dV + \int_S\delta\dot{T}\underset{\sim}{\psi}\cdot\underset{\sim}{n}dS = \varnothing \qquad \forall\delta\dot{T} \qquad (7)$$

In thermo-mechanical problems the 'weak form' of the mechanical equilibrium as well as the 'weak form' of the thermal equilibrium have to be satisfied. The finite element formulation is derived in the usual way [5,6,12].

4. Incremental Stress Formulation

In the mixed Eulerian Lagrangian formulation material displacement and nodal (grid) point displacement are uncoupled. Uncoupling of material and grid point displacement implies that in addition to the incremental calculation as in the Updated method, convection must be taken into account in order to be able to update the state at the grid points. The stress increment is given by [1,2]

$$\Delta\underset{\sim}{\sigma} = \overset{\wedge}{\underset{\sim}{\dot\sigma}}\Delta t + (\Delta\underset{\sim}{x} - \Delta\underset{\sim}{u})\cdot\nabla\underset{\sim}{\sigma} \qquad (8)$$

where Δu is the material displacement increment derived from the solved nodal point velocities and Δx the nodal displacement increment. The first term on the right hand side of (8) equals the stress increment as in the Updated method, the second term represents the convective stress increment.

The convective stress increment is calculated from the differences between the values in adjacent elements of each material-associated quantity respectively. Use is made of a local least square smoothing [13] and a weighted global smoothing. The weight factor depends on the nodal point displacement increment $\Delta u - \Delta x$ and the element size [2]. Nodal equilibrium can be achieved using an iteration method related to the Hu-Washizu principle [12].

5. Surface Movement

Movement of (contact) surface points can be taken into account by adapting the locations of nodal surface points in such a way that they remain on the moving surface. The procedure is illustrated in Fig. 2. It is observed that the new position of nodal surface points is not exactly on the surface found by the element boundaries if the material displacement increments are followed. However, if the new position of the nodal points were chosen on these element boundaries, material is lost at every increment. When using a spline, the amount of lost material is more or less in equilibrium with the amount of added material.

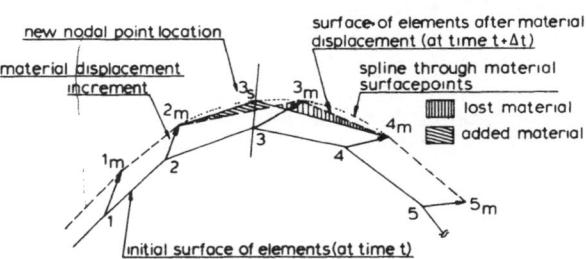

Fig. 2   *Adaptation of nodal surface point location.*

6. Simulations of Forming Processes

6.1 SIMULATION OF AN UPSETTING PROCESS

The aim of this simulation was to compare the updated Lagrange method with the mixed Eulerian-Lagrangian method.

A steel billet is deformed by a punch, the initial shape is the same as used by Doltsinis. [14]. The material is assumed to be isotropic and hardening is taken into account. The element mesh shown in Fig. 3, consists of 50 4-noded isoparametric and 14 contact elements. The Coulomb friction coefficient is 0.2

Fig. 4 shows the distorted grid using the updated Lagrange method after 47% deformation. The following difficulties occur:

-Apparent punch size increase occurs. Change of geometry requires a special treatment of boundary conditions.

-The element mesh distorts due to large deformations. Some elements are turned inside out. A continuation of the simulation is senseless.

In contrast with the updated Lagrange method the calculation based on the mixed Eulerian-Lagrangian method does not show these problems and can easily be continued (Fig. 5 and 6).

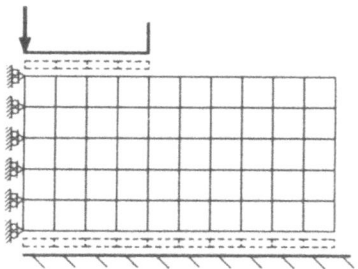

Fig. 3   Element mesh of the
upsetting simulation.

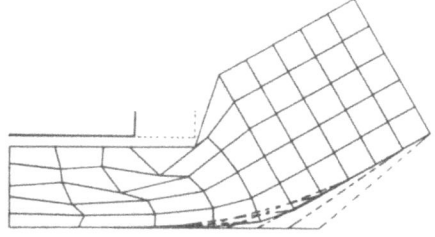

Fig. 4   Distorted grid using
updated Lagrange method.

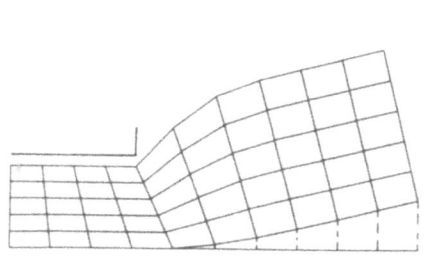

Fig. 5   Deformed mesh using
mixed method.

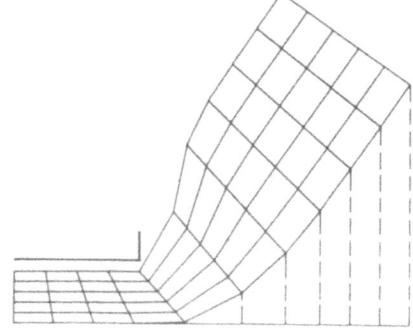

Fig. 6   Continuation of 'mixed'
simulation.

6.2 THE SIMULATION OF A COLD ROLLING PROCESS

A thermo-mechanical simulation, including time dependent material behaviour was carried out for stand 3 in a 5-stands cold rolling mill at HOOGOVENS IJmuiden BV., The Netherlands.

The rolled material is low carbon steel. The coefficient of friction is 0.035. The thickness of the rolled material is reduced from 0.7 mm. to 0.4 mm. The initial temperature is 62 $^\circ$C for the roll and 100 $^\circ$C for the rolled material. The roll has a diameter of 500 mm.

Isoparametric 4-noded plane strain elements were used. In order to predict roll deformation the roll is taken into account as an elastic structure. The contact elements are located between roll and rolled material. The finite element mesh is shown in Fig. 7. Because of symmetry only half of the problem is modelled.

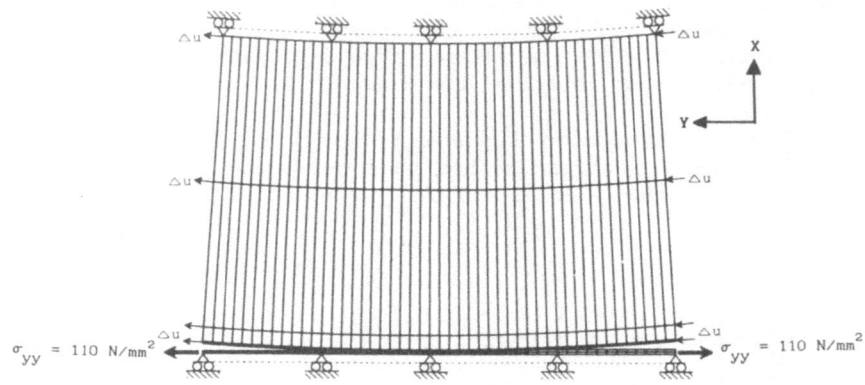

Fig. 7   *Element mesh of the cold rolling simulation.*

The simulation has been performed in 1500 steps with a displacement increment of 0.0001·R. An automatic adjustment procedure was applied to control the exit thickness. The results are shown in Figs. 8-11.

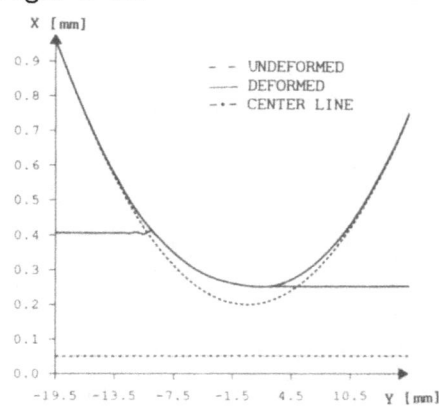

Fig. 8   *Undeformed and deformed roll shape.*

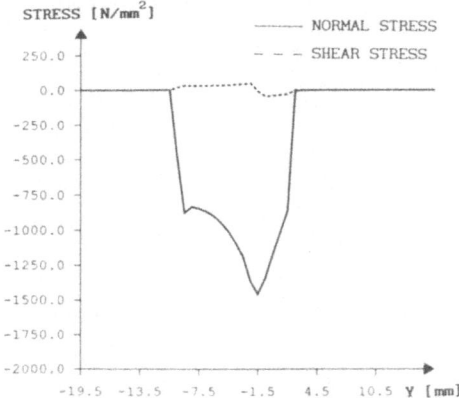

Fig. 9   *Stress distribution on contact interface.*

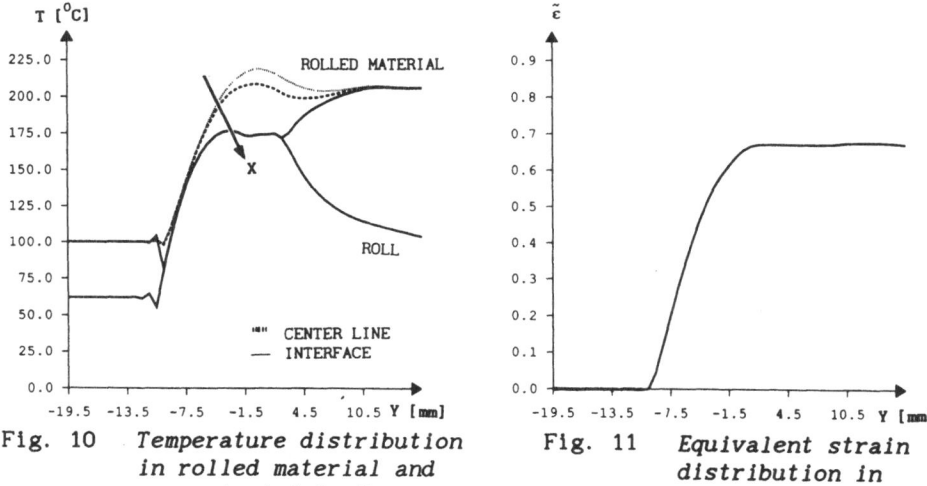

Fig. 10 *Temperature distribution
in rolled material and
on contact interface.*

Fig. 11 *Equivalent strain
distribution in
rolled material.*

6.3 THE SIMULATION OF A DEEPDRAWING PROCESS

In cooperation with HOOGOVENS IJmuiden BV. The Netherlands recently a
start is made to simulate an axisymmetric deepdrawing process.

Because of symmetry half of the problem is modelled. The element
mesh shown in Fig. 12, consists of 50 4-noded isoparametric and 73
contact elements. The displacement of the bulk elements is calculated
according to the updated Lagrange method. On one side the contact
elements are connected with the bulk elements and on the other side
they move in such a way along punch (die or blankholder) that they
remain rectangular. Fig. 13 shows some steps of the simulation.

For future simulations special shell elements are under
development.

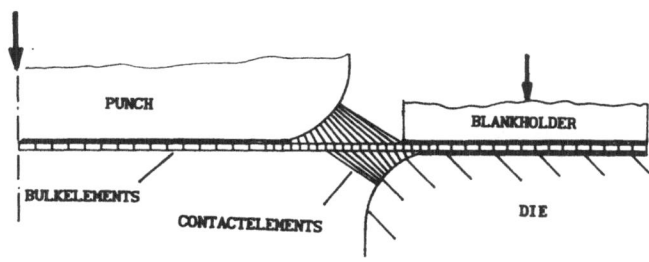

Fig. 12   *Element mesh of deepdrawing simulation.*

64

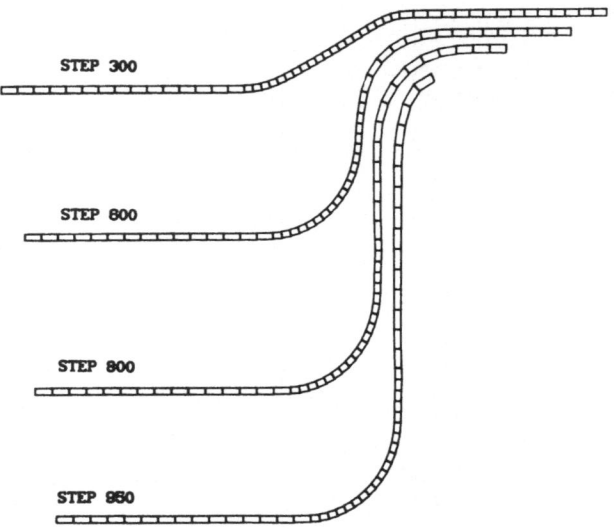

STEP 300

STEP 600

STEP 800

STEP 950

Fig. 13   *Result of some steps of the deepdrawing simulation.*

References

1.   Huétink, J., 'Analysis of metal forming processes based on a combined Eulerian-Lagrangian finite element formulation', *Proc. Int. Conf. Num. Meth. Industr. Forming Processes*, 501-509, 1982.
2.   Huétink, J., 'On the simulation of thermo-mechanical forming processes',Dissertation, University of Twente, The Netherlands, 1986.
3.   Lugt, J. van der and Huétink, J.,'Thermal-mechanically coupled finite element analysis in metal forming processes', *Comp. Meth. Appl. Mech. Eng.*, 54 (1986), 145-160.
4.   Prager, W., **Introduction to Mechanics of Continua**, Ginn, New York, 1961.
5.   Nagtegaal, J.C., Parks, D.M. and Rice, J.C., 'On numerically accurate finite element solutions in the fully plastic range', *Comp. Meth. Appl. Mech. Eng.*, 4 (1974), 153-177.
6.   Nagtegaal, J.C. and Jong, J.E. de, 'Some computational aspects of elastic plastic large strain analysis', *Int. J. Num. Meth. Eng.*, 17, (1981), 15-41.
7.   Besseling, J.F., 'A thermodynamic approach to rheology', *Proc. IUTAM symp.* on irreversible aspects of continuum mechanics, Springer-Verlag, Wien (1968) 16-53.
8.   Besseling J.F., Models of metal plasticity; Theory and experiment, **Plasticity today**, edited by A. Sawczuk and G. Bianchi, Elsevier (1985), 97-113.
9.   Huétink, J., Lugt, J. van der, Miedema, J.R., 'A mixed Eulerian- Lagrangian contact element to describe boundary and interface behaviour in forming processes', *Proc. Int. Conf. Num. Meth. Eng.* (NUMETA), vol 1 paper D17 (1987), Martinus Nijhof publ., Dordrecht, The Netherlands.
10. Lugt, J. van der, 'A finite element method for simulation of thermo-mechanical contact problems in forming processes', Dissertation, University Twente, The Netherlands, to be published Nov. 1988.
11. Fredriksson, B., 'Finite element solution of surface nonlinearities in structural mechanics with special emphasis to contact and fracture mechanics problems', *Comp. & Struct.* 6 (1976), 281-290.
12. Zienkiewicz, O.C., **The Finite Element Method**, McGraw-Hill, New York, 1983.
13. Hinton, E., and Campbell, J.S., Local and global smoothing of discontinuous finite element functions using least square method, *Int. J. Num. Meth. Eng.*, 8 (1974), 461-480.
14. Doltsinis, J. St. and Luginsland, J., 'Unformspezifische Rechenverfahren Kontakt und Reibung, variabele Diskretisierung', presented at Workshop: *Numerische Methoden der Plastomechanik*, Hannover (W-Germany), 6-8 Nov. 1986.

A METHOD TO REDUCE COST OF MESH DEFORMATION IN EULERIAN–LAGRANGIAN FORMULATION

J.P. PONTHOT

IRSIA Research Engineer

L.T.A.S. THERMOMECANIQUE

UNIVERSITE DE LIEGE, LIEGE, BELGIUM

Summary

In the past few years, the Eulerian-Lagrangian formulation in simulating forming processes has been develloped by several authors in order to overcome problems met by using purely Eulerian or purely Lagrangian formulation. The principal drawback of this new method is the computational time required in handling the moving finite element mesh. In the present paper, we present an explicit, low computer cost procedure for mesh adaptation based on the transfinite mapping method.

Introduction

While modelling Metal Forming Processes, an Updated Lagrangian displacement formulation is often used. In such a description, a given reference volume is associated with the same set of material particles at all stages of the deformation. So, in the Lagrangian formulation, the reference system is attached to the body. As Schreurs [1], we shall name it the Material Reference System (MRS). In a finite element model, this means that there is no material motion relative to the convected mesh and consequently the discretization of the structure must be chosen ab initio. This is a large drawback for the method since the element mesh can be completely entangled after a number of steps according to the deformation pattern. Besides, if material-associated boundary conditions are quite easy to take into account, non-material-associated boundary conditions must generally be adapted after each step or after every few steps because of the changing contact between tool and nodal material points of the body to be formed.

Despite the popularity of the Updated Lagrangian description, some solid mechanics problems are described most naturally by an Eulerian description (i.e. problems with

J. L. Chenot and E. Oñate (eds.), Modelling of Metal Forming Processes, 65–74.
© *1988 by Kluwer Academic Publishers.*

fixed boundaries). In this formulation, the reference system, called the Spatial Reference System (SRS), is fixed in space and the particles associated with a given reference volume change throughout the deformation. In a finite element model, this means that material particles have motion relative to the mesh and can move across element boundaries. Eulerian descriptions afford very large material distorsions but it is a rather tedious work to take into account material- associated boundary conditions.

In order to overcome the difficulties mentioned above, some authors [1,2,3] have developped a combined Eulerian-Lagrangian finite element formulation resulting into finite elements with nodal points displacements that are uncoupled from the material displacements so that matter can flow through the elements (as in a Eulerian formulation) whereas the shape of the moving boundaries can be controlled very easily by the material-associated boundary conditions as in a Lagrangian formulation.

Basically, the reference system used in the Eulerian-Lagrangian formulation is not a priori fixed in space or attached to the body. It is called the Computational Reference System (CRS). This provides much freedom in formulating the mathematical model. It is possible to fix the CRS in space which leads to an Eulerian formulation, or to attach it to the body , thus resulting in a Lagrangian formulation. It is also possible to allow the CRS to move independently of the material. In the latter case, the tangent stiffness matrix obtained from the linearization of the equilibrium equations is in general non-symmetric and rectangular. The rectangular nature of the matrix results from the fact that, for a three- dimensional problem, a reference location has six degrees of freedom (three Eulerian and three Lagrangian) but only three equations of equilibrium are provided by the principle of weighted residuals. A square matrix can be obtained by either eliminiting degrees of freedom or by adding supplementary constraint equations

Schreurs [1], for example, increases the number of equations by using a fictitious isotropic elastic body that deforms simultaneously with the real material. This fictitious material governs mesh deformation and tries to improve the shape of each element. The principal drawback of this method is that it leads to a system of equations about twice as large as in a standard Lagrangian formulation.

Cescutti and Chenot [3], for their part, introduce a functional quantifying the grid quality and the element volume adaptation with respect to some solution quantity (such as the strain energy). This functional has to be minimized through a Newton-Raphson method.

Our aim is different. We have tried to improve the computational efficiency of the Eulerian-Lagrangian formulation by decreasing the computational time required for handling the mesh. Therefore, we control mesh deformation by "Discrete Transfinite Mapping" which allows at the same time favourable element shapes and precise modelling of the boundaries.

Basic Equations

(i) Assuming that inertia effects are negligible and using the Weighted Residual Principle, the equilibrium equations in the current state are found to be:

$$\int_V \sigma_{ij} \frac{\partial \psi_k}{\partial x_j} \, dV = \int_V \psi_k \, q_i \, dV + \int_S \psi_k \, f_i \, dS \tag{1}$$

where the tensor components are referred to a fixed spatial coordinate system x_i and
σ_{ij} is the Cauchy stress tensor
V is the current volume of the material
S is the current surface of the material
q_i is the body force per unit volume
f_i is the surface force per unit area
ψ_k is the weight function.

Since the integrals over the unknown, state-dependent, volume V and boundary S are extremely difficult to evaluate, they are transformed into integrals over the known, invariable sets G and G^* via the change of variables $dV = J \ dG$ and $dS = J^* \ dG^*$

where
G is an invariable set containing coordinates of all CRS points
G^* is an invariable set containing coordinates of all CRS boundary points

Equations (1) become thus:

$$\int_G \sigma_{ij} \frac{\partial \psi_k}{\partial x_j} \, J \, dG = \int_G \psi_k \, q_i \, J \, dG + \int_{G^*} \psi_k \, f_i \, J^* \, dG^* \tag{2}$$

(ii) The other equations of interest are the constitutive equations. In finite strain plasticity calculations, the generally accepted constitutive rate equations are:

$$\sigma_{ij}^{\triangledown} = L_{ijkl} \cdot D_{kl} = \dot{\sigma}_{ij} + \Omega_{ik}^T \cdot \sigma_{kj} + \sigma_{ik} \cdot \Omega_{kj} \tag{3}$$

where a superimposed dot denotes time rate of change and:
$\sigma_{ij}^{\triangledown}$ is the Jaumann rate of the Cauchy stress tensor.
L_{ijkl} is a tensor depending on the material properties and the current state of stress.
D_{ij} is the deformation rate tensor.
Ω_{ij} is the spin tensor.

D_{ij} and Ω_{ij} are related to the jacobian deformation tensor F_{ij} by:

$$D_{ij} + \Omega_{ij} = \dot{F}_{ik} \cdot F_{kj}^{-1} \quad ; \quad D_{ij} = D_{ji} \quad ; \quad \Omega_{ij} = -\Omega_{ji} \tag{4}$$

Iterative solution process and the finite element method

Starting from a given state at t_0, we want to determine the state of the system at $t = t_0 + \Delta t$ such that (2), (3) and the boundary conditions be satisfied. Due to the non linearity of these equations, an iterative method must be used, yielding a sequence of approximate solutions for (2). The quantities σ_{ij}, $\frac{\partial}{\partial x_j}$, J and J^* depend on the unknown CRS and MRS displacements, while q_i and f_i are assumed to be prescribed.

In order to obtain a set of iterative linear vector equations, (2) is first linearized and then, the finite element method is employed. According to this method, the CRS —not the MRS!— is subdivised into regions of rather simple geometry, the elements. This leads to an algebraic system of equations, linear in $\mathbf{du}$, the structural material iterative displacement vector, and $\mathbf{dv}$, the structural nodal points iterative displacement vector (see figure 1):

$$\mathbf{K_1}\,\mathbf{du} + \mathbf{K_2}\,\mathbf{dv} = \mathbf{r} \tag{5}$$

Each line of which corresponding to a CRS nodal point, where

$\mathbf{K_1}$ is the tangent stiffness matrix with respect to material iterative displacement (MRS)
$\mathbf{K_2}$ is the tangent stiffness matrix with respect to mesh iterative displacement (CRS)
$\mathbf{r}$ is the residual vector

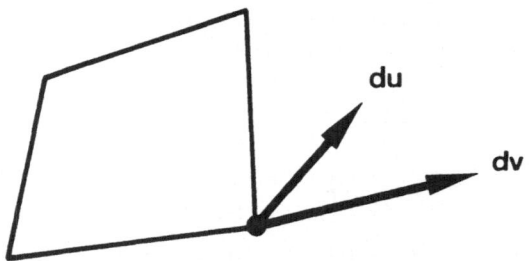

Fig 1.   Eulerian-Lagrangian formulation. Mesh nodal point x_j at time t, iteration i, with material iterative displacement $\mathbf{du}$ and mesh iterative displacement $\mathbf{dv}$

CRS Determination process

Each line of (5) for a nodal point contains the iterative displacement vectors $\mathbf{du}$ and $\mathbf{dv}$ which are both unknown. Thus, the number of unknowns in (5) will in general surpass the number of available equations. The set of unknowns is appropriately reduced to a solvable set of equations by taking into account the requirement that the boundaries of the CRS and the MRS must always coincide and specifying in addition the nodal point displacement $\mathbf{dv}$ through a CRS determination process.

Before deformation, a hierarchical partitioning of the initial mesh is imposed. This partitioning creates macroregions of simple form. Each of these regions is defined by its four (three) sides which are called "Master Lines" (ML). Discretization along these ML's will yield a second-level partition which is the finite element mesh topology (see fig. 2).

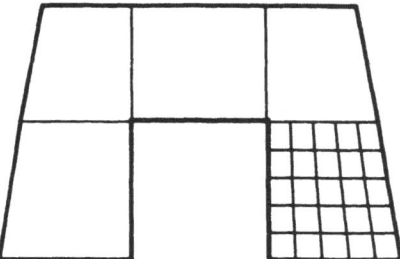

Fig 2. Hierarchical partitioning of structure

According to kinematical or material boundary conditions, each macroregion is assigned a attribute: Eulerian, Lagrangian or Eulerian-Lagrangian. The CRS determination process of a particular region will then depend on its attribute. This determination is trivial for Eulerian regions ($dv = 0$) or for Lagrangian ones ($dv = du$). In the case of an Eulerian-Lagrangian one , the CRS determination process is based on Transfinite Mapping Method as follows.

The transfinite mapping method

The transfinite mapping technique establishes a curvilinear coordinate system in arbitrary 2D domains. These mappings are described by adequate projectors. A projector is a linear operator which maps a true surface F onto a unit square. For example, the lofting projector $\wp$ performs a linear interpolation between two boundary curves, $\psi_1(\xi)$ and $\psi_2(\xi)$:

$$\wp[F] = (1 - \eta)\psi_1(\xi) + \eta\psi_2(\xi); \quad 0 \leq \xi \leq 1,\ 0 \leq \eta \leq 1 \qquad (6)$$

where ξ is a normalized parametric coordinate along ψ_1 and ψ_2 and, η is a normalized coordinate which has a value of zero on ψ_1 and unity on ψ_2, as shown on figure 3.

If more than 2 opposite sides of F are curvilinear, such a projector may be blended with another one of the same type in order to interpolate a region F bounded by four curves $\psi_1(\xi)$, $\psi_2(\xi)$, $\vartheta_1(\eta)$ and $\vartheta_2(\eta)$. This new projector matches exactly F on its entire boundary (see figure 4):

$$
\begin{aligned}
(\wp_1 \oplus \wp_2)[F] =& (1 - \eta)\psi_1(\xi) + \eta\psi_2(\xi) + (1 - \xi)\vartheta_1(\eta) + \xi\vartheta_2(\eta) - \xi\eta F(1,1) \\
& - (1 - \xi)(1 - \eta)F(0,0) - (1 - \xi)\eta F(0,1) - \xi(1 - \eta)F(1,0); \quad (7) \\
& 0 \leq \xi \leq 1,\ 0 \leq \eta \leq 1
\end{aligned}
$$

The latter may be called the transfinite bilinear Lagrange interpolant of F (for further details on this kind of projectors, see [5]).

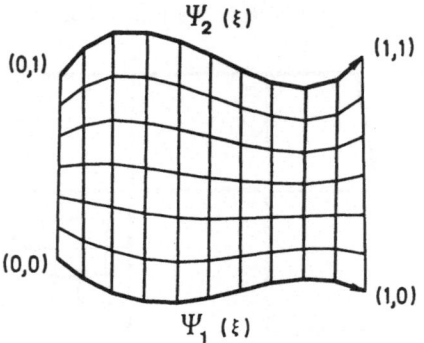

Fig 3. Lofting projector

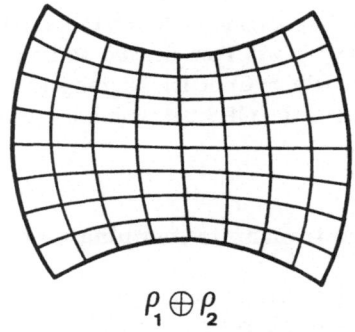

Fig 4. Bilinear projector: $\wp_1 \oplus \wp_2$

In the finite element framework, imposing discrete values to the reduced coordinates ξ and η generates very easily a mesh on surface F: e.g. equidistant reduced coordinates or values of ξ and η linked to a gradient of an unknown quantity (strain energy, local stress...).

Practical implementation of the iterative method

The flow chart for an iteration in the present method can be described as follows:

1st step - The tangent stiffness matrices K_1 and K_2 (see 5) are formed.

2nd step - The number of unknowns is reduced with the help of the transfinite mapping method (purely Eulerian and purely Lagrangian attributes are taken into account at elementary level by computing K_1 only or by adding K_1 and K_2) leaving the sole du vector as unknown.

3rd step - The iterative material displacements are computed by solving the resulting linear algebraic system.

4th step -The constitutive equations are integrated along material displacements. Results of these integrations are interpolated at the Gauss points of the CRS mesh.

5th step -Equilibrium is checked on the CRS domain.

Numerical results

The simulation of the coining process [1] clearly shows the merits of the Eulerian-Lagrangian formulation with respect to certain prescribed boundary conditions. Figure 5 shows the geometry and the initial mesh of the body to be deformed. The tool is rigid and its prescribed displacement will lead to a maximum height reduction of 70 %. All contact areas are assumed to be frictionless.

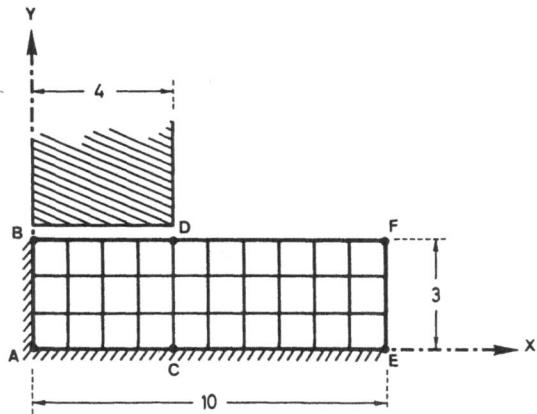

Figure 5. Geometry and initial mesh for the coining process

The material parameters are:

Young modulus	$E = 2.1 \cdot 10^5 \, N/mm^2$
Poisson ration	$\nu = 0.3$
Hardening parameter	$h = 10^4 \, N/mm^2$
Yield stress	$\sigma^0_{VM} = 250 N/mm^2$

Using the Lagrangian formulation, the simulation is rather tedious and necessitates a strongly refined mesh under the punch (especially near its edge). This is because the computational nodal points with prescribed tool displacement (y-direction) are also material points exhibiting a non-zero x-displacement. This leads to an changing number of nodal points involved with such boundary conditions.

On the contrary, the use of the Eulerian-Lagrangian formulation allows to prevent x-displacement of computational nodal points located under the tool, irrespective of the material flow.

The body to be formed is decomposed in two macroregions (fig. 5):

-region 1 defined by Master Lines : ACBD
-region 2 defined by Master Lines : CEFD

The following conditions are satisfied in region 1, which is of Eulerian type:

. x-displacement is equal to zero everywhere
. y-displacement is equal to the prescribed tool displacement on ML BD, equal to zero on ML AC and interpolated between those ML's using transfinite mapping for the remaining nodes.

In the first case (see figure 6), each nodal point of region 2 is imposed to be Lagrangian except those located on ML CD (for obvious reasons of compatibility). As a consequence, all elements of region 2 are purely Lagrangian ones except those contiguous to the Eulerian region which are Eulerian-Lagrangian. All the matter coming from region 1 flows through those elements. This leads to unrealistic shapes of the deformed body due to the large aspect ratio of the Eulerian-Lagrangian elements.

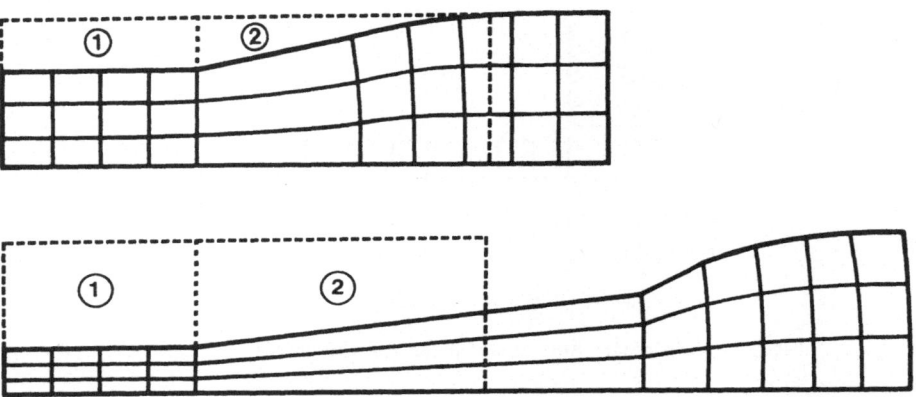

Fig 6.   (a) mesh after 35% height reduction (b) mesh after 70% height reduction
          dotted lines: initial configuration   solid lines: current configuration

In the second case, all nodal points of region 2 are Eulerian-Lagrangian and treated according to the transfinite mapping method. Figure 7 obviously shows mesh control improvement due to transfinite mapping which leads to a much more realistic deformation pattern, especially near the punch edge, even with a rather coarse mesh.

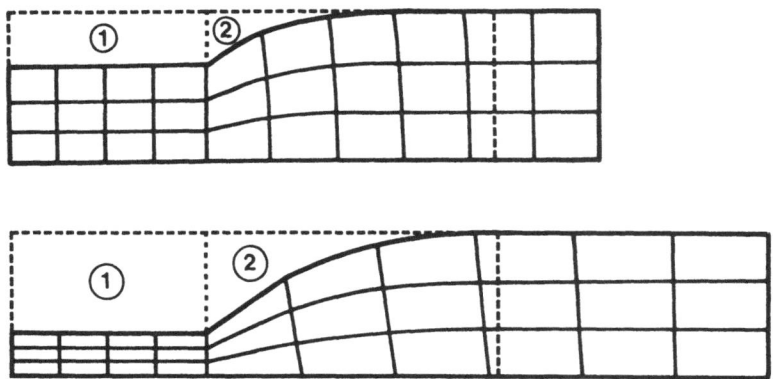

Fig 7.   *(a) mesh after 35% height reduction*   *(b) mesh after 70% height reduction*
         dotted lines: initial configuration   *solid lines: current configuration*

Conclusions

The Eulerian-Lagrangian formulation has been coupled with the transfinite mapping method in order to decrease computer cost required by mesh handling. This method also prevents mesh entangling due to the deformation pattern. The extra computer cost resulting from mesh handling and datas transfer from MRS to CRS mesh is only about 10% of the total cost.

Bibliography

[1] SCHREURS, P.G.J. (1983) "Numerical simulation of forming processes. The use of the Arbitrary-Eulerian-Lagrangian method." Thesis, Eindhoven University of technology, Eindhoven, The Netherlands.

[2] HUETINK, J. (1986) "On the simulation of thermo-mechanical forming processes. A mixed Eulerian-Lagrangian finite element method." Thesis, Twente University of Technology, Twente, The Netherlands.

[3] CESCUTTI, J.P. and CHENOT, J.L. (1987) "A geometrical continuous remeshing procedure for application to finite element calculation of non-steady state forming processes." S17, NUMETA 87, Swansea, U.K.

[4] NAGTEGAAL, J.C. and VELDPAUS F.E. (1984) "On the implementation of finite strain plasticity equations in a numerical model." Numerical Analysis of Forming Processes pp 351–371. Edited by J.F.T. PITTMAN, O.C. ZIENKIEWICZ, R.D. WOOD and J.M. ALEXANDER

[5] HABER, R., SHEPARD, M.S., ABEL, J.F., GALLAGHER, R.H. AND GREEN-BERG, D.P. (1981) "A general two-dimensional, graphical finite element preprocessor utilizing discrete transfinite mappings." Int. J. for Num. Methods in Engineering, vol 17 pp 1015–1044.

[6] HABER, R. (1984) "A mixed Eulerian-Lagrangian displacement model for large-deformation analysis in solid mechanics." Comp. Meth. in Applied Mech. and Engineering, vol 43 pp 277–292.

ERROR CONTROL, MESH UPDATING SCHEMES AND AUTOMATIC ADAPTIVE REMESHING FOR FINITE ELEMENT ANALYSIS OF UNSTEADY EXTRUSION PROCESSES

G.C. HUANG, Y.C. LIU and O.C. ZIENKIEWICZ
Institute for Numerical Methods in Engineering,
University College of Swansea, U.K.

Summary

Error estimation, adaptive remeshing as well as different special mesh updating schemes are applied in the finite element simulation of hot extrusions through shear die and tapered die. The material grid cooperated with the analytical mesh is introduced to simulate the real forming processes. The algorithms are presented.

1. Introduction

The flow formulation approach [1] and other large deformation finite element methods have made it possible to simulate the complicated forming processes. An intrinsic difficulty in the analysis is the constantly changing configuration of the deformed body. If simple updating is used, this results in a very much distorted mesh which either makes the further analysis impossible or introduces large approximation errors. These problems can be solved by remeshing or by using different mesh updating schemes, such as AEL (Arbitrary Eulerian Lagrangian) method [2,3], or by pseudo-concentration method [4]. The last two methods require a prior knowledge of the forming processes, therefore are not in general use. However, for some special cases they are simple and economical. Remeshing scheme generates a new mesh over the deformed domain with all the state variables transfered from the old mesh, therefore it has the merit in resuming the analysis and reducing the discretization error caused by the improper mesh. A simple error estimation and adaptive remeshing scheme was first suggested by Zienkiewicz and Zhu [6] for linear problems and later introduced in the forming analysis by Zienkiewicz, Liu and Huang [7]. The strategy of this method is to reduce the error to a specified level by using an adaptively generated mesh which is controlled by an error indicator. The degrees of freedom are kept as few as possible to achieve this aim. In the present paper the error control and different mesh updating schemes are described. The concept of material grid and analytical mesh is introduced, which can be explained schematically by a family tree in Fig. 1. The analytical mesh is for the use of the finite element analysis, while the material grid, as mentioned by Cheng [5], is only for output display purposes and for comparisons with the experiments.

75

J.L. Chenot and E. Oñate (eds.), Modelling of Metal Forming Processes, 75–83.
© *1988 by Kluwer Academic Publishers.*

Six-noded triangular element with a single pressure and the nodal
velocities is used for the analysis. The schemes are applied in some
unsteady extrusion problems.

2. Error Estimation and Adaptive Remeshing Scheme

Error (only discretisation error is concerned) is defined in terms of
the "energy" norm as

$$|| \underset{\sim}{e} ||_s^2 = \int_\Omega (\underset{\sim}{S} - \overset{\wedge}{\underset{\sim}{S}})^T (2\overset{\wedge}{\mu})^{-1}(\underset{\sim}{S} - \overset{\wedge}{\underset{\sim}{S}}) \, d\Omega \tag{1}$$

where $\underset{\sim}{S}$ stands for deviatoric stress vector. $\underset{\sim}{S}$ and $\hat{S}$ are the exact
solution and finite element solution respectively. $\overset{\wedge}{\mu}$ is the viscosity
obtained by finite element method. Since the exact solution of
deviatoric stress is not available, a smoothed one, $\underset{\sim}{S}$ (which is one
polynomial order higher than $\overset{\wedge}{\underset{\sim}{S}}$) is used instead of $\underset{\sim}{S}$. The least square
method [8] is adopted to obtain $\underset{\sim}{S}^*$ by the following equation

$$\left[\int_\Omega N^T N d\Omega \right] \overset{*}{\underset{\sim}{S}} = \int_\Omega N^T \overset{\wedge}{\underset{\sim}{S}} d\Omega \quad \text{and} \quad \underset{\sim}{S}^* = N\overline{\underset{\sim}{S}}^* \tag{2}$$

$\overline{S}^*$ being the smoothed nodal deviatoric stress components. Now the
predicted error norm can be written as,

$$|| \underset{\sim}{e} ||_{s^*}^2 = \int_\Omega (\underset{\sim}{S}^* - \overset{\wedge}{\underset{\sim}{S}})^T (2\overset{\wedge}{\mu})^{-1} (\underset{\sim}{S}^* - \overset{\wedge}{\underset{\sim}{S}}) \, d\Omega \tag{3}$$

The criterion for the remeshing is set up as

$$|| \underset{\sim}{e} ||_{s^*} / || \underset{\sim}{u} || = \eta^0 \le \overline{\eta} \quad \text{where} \quad || \underset{\sim}{u} ||^2 = \int_\Omega \overset{\wedge}{\underset{\sim}{S}}^T (2\overset{\wedge}{\mu})^{-1} \overset{\wedge}{\underset{\sim}{S}} d\Omega \tag{4,5}$$

$\overline{\eta}$ is the prescribed percentage error. Whenever Eq.(4) is unsatisfied
the remeshing is needed. The remeshing procedure is controlled by the
aiming percentage error $\overline{\eta}_{aim}$ which is the percentage error we desire to
reach after remeshing. The adaptive remeshing scheme is based on the
idea that after remeshing each element possesses the same error and the
total percentage error is equal to the aiming percentage error $\overline{\eta}_{aim}$,
i.e.

$$\left[|| \underset{\sim}{e} ||_{s^*}^2 \right]_{after} / || \underset{\sim}{u} ||^2 = \overline{\eta}_{aim}^2 \tag{6}$$

Since the square of the total error norm is the sum of the local element
contributions, i.e.

$$|| \underset{\sim}{e} ||_{s^*}^2 = \sum_{i=1}^{M} \left(|| \underset{\sim}{e} ||_{s^*}^2 \right)_i \tag{7}$$

If each element has the same error, then the aiming local error

$$\left[\left(|| \underset{\sim}{e} ||_{s^*} \right)_i \right]_{after} = || \underset{\sim}{u} || \, \overline{\eta}_{aim} / \sqrt{M} \tag{8}$$

Asymptotically the error is dependent on the size of the element, h [7],
i.e.

$$\left[|| \underset{\sim}{e} ||_{s^*} \right]_i \propto h_i^p \tag{9}$$

where p is the order of the polynomial shape function used in the analysis. The new size of the element can be predicted simply by

$$\left[\left[\left|\left|\underset{\sim}{e}\right|\right|_s^*\right]_i\right]_{\text{after}} \bigg/ \left[\left[\left|\left|\underset{\sim}{e}\right|\right|_s^*\right]_i\right]_{\text{before}} = h^p_{\text{after}} \bigg/ h^p_{\text{before}} = \xi_e$$

or $\quad h_{\text{after}} = h_{\text{before}} \; (\xi_e)^{1/p}$ \hfill (10)

It is obvious that if $\xi_e < 1$, the new mesh will be refined, otherwise, the new mesh will be "derefined". The algorithm of the mesh generation for this purpose is discussed in [12] by Peraire *et al.* h_{before} can be determined by different methods. In the present paper h_{before} is taken as the height on the longest side of a triangular element (see Fig.2). In order to avoid too large elements or too small elements, both upper and lower bounds of h-after are set up so that the present program can handle.

3. Mesh Updating Schemes

Instead of regenerating the whole mesh, we can manipulate the analytical mesh so that the mesh will not be too much distorted and the error can be limited. For some forming problems where the material flow is predictable, this method proves to be economical and efficient.

In the present work the updating is carried out explicitly by the following equations

$$\underset{\sim}{X}_{n+1} = \underset{\sim}{X}_n + \Delta t \; \underset{\sim}{u}_n \quad ; \quad \underset{\sim}{\varepsilon}_{n+1} = \underset{\sim}{\varepsilon}_n + \Delta t \; \underset{\sim}{\dot{\varepsilon}}_n \hfill (11)$$

3.1 MIXED MESH SCHEME

For some forming processes, such as high ratio extrusion, a large part of the billet does not deform very much, while a small part changes so fast that remeshing is inevitable every time step. To solve this problem, we divide the whole domain into three parts (see Fig.3). For the first part, the mesh is updated every time step according to Eq.(11). In the second part, the mesh is simply fixed, while in the third part a new mesh is generated to fit the extrudate. This forms a new mesh for the next analysis. The state variables $\underset{\sim}{u}$, T and $\underset{\sim}{\varepsilon}$ in part 2 and 3 are obtained by interpolations from the totally updated mesh.

3.2. PRESCRIBED MOVING MESH SCHEME

In this method, the analytical mesh is updated according to the prescribed velocities which are different from the real velocities obtained by finite element method, except the nodes on the free boundary. This method keeps the number of nodes, elements and the element connectivity unchanged. The aim is to avoid the distorted elements near the die corners. Again the interpolations are needed for the state variables.

3.3 SUCCESSIVE MESH UPDATING SCHEME

Dividing one time step into m subtime steps and assuming that within one time step, the velocity and the strain rate field are independent of time, i.e.

$$\underset{\sim}{u} = \underset{\sim}{u}(\underset{\sim}{x}, t) = \underset{\sim}{u}(\underset{\sim}{x}) \; ; \; \underset{\sim}{\dot{\varepsilon}} = \underset{\sim}{\dot{\varepsilon}}(\underset{\sim}{x}, t) = \underset{\sim}{\dot{\varepsilon}}(\underset{\sim}{x}) \tag{12}$$

we can write Eq.(11) as follows

$$\underset{\sim}{X}_{n+1} = \underset{\sim}{X}_n + \sum_{i=1}^{m} \Delta t_i \underset{\sim}{u}(\underset{\sim}{x}_i) \; ; \; \underset{\sim}{\varepsilon}_{n+1} = \underset{\sim}{\varepsilon}_n + \sum_{i=1}^{m} \Delta t_i \underset{\sim}{\dot{\varepsilon}}(\underset{\sim}{x}_i) \; ; \; \Delta t = \sum_{i=1}^{m} \Delta t_i \tag{13}$$

If subtime step Δt_i is equally distributed, then

$$\Delta t = m \Delta t_{n+1}; \; \underset{\sim}{X}_{n+1} = \underset{\sim}{X}_n + \frac{\Delta t}{m} \sum_{i=1}^{m} \underset{\sim}{u}(\underset{\sim}{x}_i); \; \underset{\sim}{\varepsilon}_{n+1} = \underset{\sim}{\varepsilon}_n + \frac{\Delta t}{m} \sum_{i=1}^{m} \underset{\sim}{\dot{\varepsilon}}(\underset{\sim}{x}_i) \tag{14}$$

where $\quad \underset{\sim}{x}_i = \underset{\sim}{X}_n + \frac{\Delta t}{m} \sum_{j=1}^{i} \underset{\sim}{u}(\underset{\sim}{x}_{j-1}) \; ; \; \underset{\sim}{x}_1 = \underset{\sim}{X}_n$

$\underset{\sim}{u}(\underset{\sim}{x}_i)$, $\underset{\sim}{\dot{\varepsilon}}(\underset{\sim}{x}_i)$ can be determined by interpolations, providing that $\underset{\sim}{\dot{\varepsilon}}$ is smoothed to the nodal position from the Gauss points by the least square method [8].

By using this scheme, the node at the die corner will be able to change its moving direction once it enters the exit area as shown in Fig.4.

3.4 MATERIAL MESH UPDATING

Material mesh is updated according to the calculated velocities for every time step so that the cumulative deformations can be visualised as observed in a real process. In the present example the mesh is updated by Eq.(14).

4. Mapping of State Variables

If a new mesh is formed in such a way that the nodes no longer represent the previous material particles, the mapping of state variables such as strains, temperatures and velocities is required for the further analysis. Variables, such as strains, which are obtained at Gauss points should be smoothed or extrapolated to the nodal positions in the old mesh which is obtained by Eq.(13) and is not very much distorted. If the old mesh is updated by the successive updating scheme, then all the state variables are at nodal positions, therefore the mapping is straightforward.

Mapping is carried out from the old mesh onto the new mesh by the following equation

$$\underset{\sim}{Y}_{new}(\underset{\sim}{X}) = \underset{\sim}{N}(\underset{\sim}{\xi}) \underset{\sim}{Y}_{old} \tag{15}$$

where $\underset{\sim}{Y}_{new}$ is a new nodal state variable at the global coordinates $\underset{\sim}{X}$, $\underset{\sim}{N}$

the shape function in the old mesh which is the function of the local coordinates $\underset{\sim}{\xi}$ and $\underset{\sim}{y}_{old}$ are the old nodal state variables.

To find out the corresponding local coordinates $\underset{\sim}{\xi}$, the following approximation is employed. Firstly the six-noded triangular elements are subdivided into three-noded linear triangular elements in the way shown in Fig.4. Linear interpolation is then used within a sub-triangle.

Extrapolation sometimes is needed for some boundary nodes and it is done by the closest element.

5. Numerical Examples

The presented algorithms are implemented in the following extrusion problems. The material is considered to be incompressible and obeys Von-Mises yielding criterion.

5.1 STRAIN HARDENING AXISYMMETRIC SHEAR DIE EXTRUSION

This example is shown in Fig.6 with the material equation

$$\bar{\sigma} = (1 + \bar{\varepsilon} / 0.05205)^{0.3}$$

piston speed 1.2 cm/sec, the extrusion ratio 400:1. Thermal effect is not considered. The sticking friction is assumed on both ends. The analytical mesh is obtained by a background mesh through the error estimation and the adaptive remeshing. The aiming percentage error is 20%. By using the mixed mesh scheme, this difficult problem can be simulated. The predicted error is around 20%. In Fig.6-5 the effective strain contour shows a severe strain gradient in the extrudate surface.

5.2 HOT AXISYMMETRIC TAPERED DIE EXTRUSION

The tapered die geometry is illustrated in Fig.7-1. The heat conditions and material constitutive equations are referred to [10]. Initial temperature for both the billet and the tool is 350°C (623°K), ram speed 1.2mm/sec. The sliding friction with friction factor 0.2 is assumed everywhere on the tool boundary. The successive updating scheme and the material grid are adopted. A regular triangular mesh is used with the prescribed moving mesh scheme to avoid element distortions (see Figs.7-2 and 7-3). The mesh movement near the exit corner is slowed down and the elements outside the exit are elongated, which does not affect the analysis very much. The procedure proves to be very economical. The predicted error is around 15%.

In the second test of the same example, the fully automatic error estimation and adaptive remeshing scheme is used (see Fig.8). Material mesh is merely updated and the analytical mesh is updated and regenerated through the whole procedure controlled by the prescribed percentage error and the aiming error. The chosen aiming error is 10%.

Whenever the predicted error exceeds 15% the remeshing is carried out. The evolution of the predicted error is shown in Fig.8-7. It is found that the remeshing scheme is very efficient. In spite of extra CPU time needed, the scheme is significant since more complicated forming processes can be dealt with. No severe effective strain gradient is observed on the extrudate surface (see Fig.8-6). In Figs.8-8 and 8-9 the comparison of the material grid with the experiment gives a reasonably good agreement.

6. Conclusions

The numerical results presented in this paper show that the application of error estimation, adaptive remeshing and special mesh updating schemes in the finite element analysis of complicated forming processes will not only make the further analysis possible but also improve the results. A code with error estimation and automatic adaptive mesh generation is now available for 2D forming problems from the authors. The introduction of the material mesh in the analysis makes the numerical simulation easier to compare with the real forming processes.

7. References

1. O.C. Zienkiewicz, Numerical analysis of forming processes, Ch.1, pp.1-44, ed. J.F.T. Pittman et al, J. Wiley and Sons (1984).
2. J. Huetink, Numerical methods in industrial forming processes, pp.501-506, ed. J.F.T. Pittman et al, Pineridge press (1982).
3. D.J.G. Schreurs, F.E. Veldpaus and W.A.M. Brekelmans, Numerical Methods in industrial forming processes, pp.491-500, ed. J.F.T. Pittman et al, Pineridge Press (1982).
4. E. Thompson, Proceedings of the NUMIFORM '86 Conference/Gothenburg, pp.65-69, ed. K. Mattiasson, A.A. Balkema (1986).
5. J. Cheng, Int. J. Num. Meth. Eng., **26**, pp.1-18, (1986).
6. O.C. Zienkiewicz and J.Z. Zhu, Int. J. Num. Meth. Eng., **24**, pp.337-357 (1987).
7. O.C. Zienkiewicz, Y.C. Liu and G.C. Huang, Int. J. Num. Meth. Eng., **25**, pp.23-42 (1988).
8. E. Hinton and J. Campbell, Int. J. Num. Meth. Eng., **8**, pp.461-480 (1974).
9. J. Peraire, M. Vahdati, K. Morgan and O.C. Zienkiewicz, Inst. Num. Meth. in Engng., CR/R/544/86.
10. P.L. Charpentier, B.C. Stone and J.F. Thomas, Metallurgical Transactions A, 17A, pp.2227-2237, Dec. 1986.

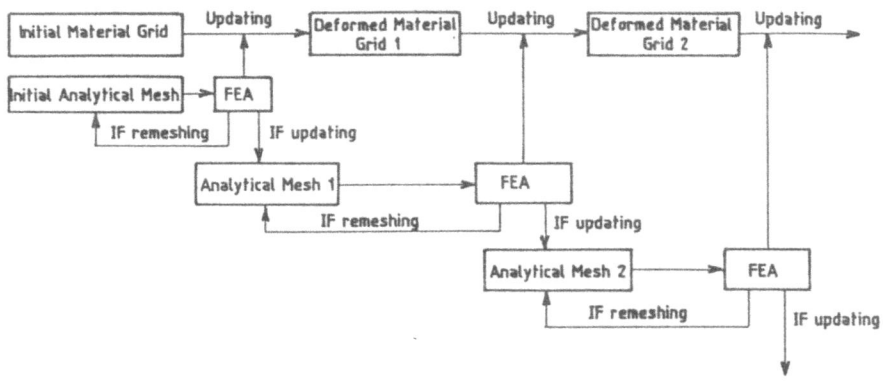

FIGURE 1   A FAMILY TREE FOR THE ANALYTICAL MESH AND
THE MATERIAL GRID

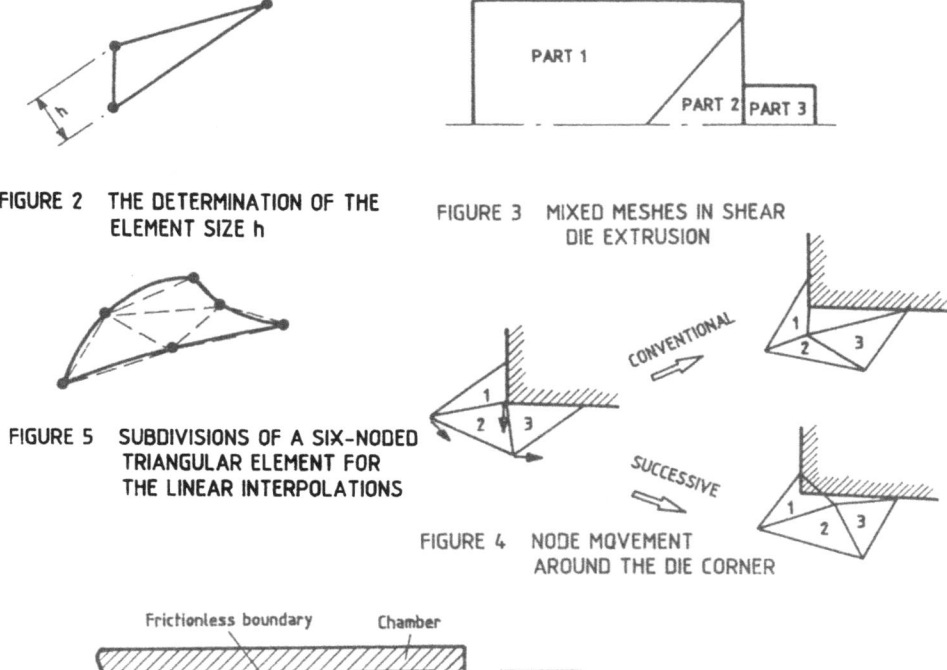

FIGURE 2   THE DETERMINATION OF THE
ELEMENT SIZE h

FIGURE 3   MIXED MESHES IN SHEAR
DIE EXTRUSION

FIGURE 5   SUBDIVISIONS OF A SIX-NODED
TRIANGULAR ELEMENT FOR
THE LINEAR INTERPOLATIONS

FIGURE 4   NODE MOVEMENT
AROUND THE DIE CORNER

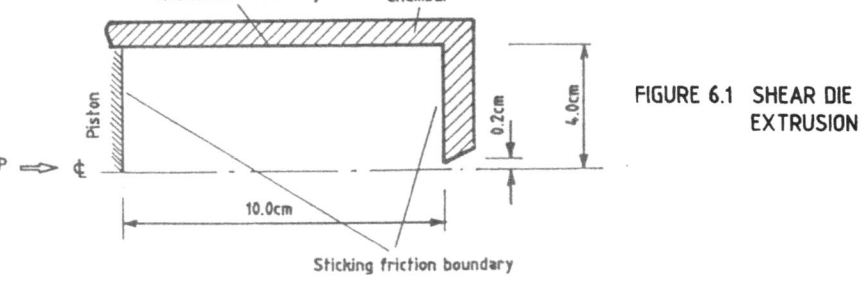

FIGURE 6.1   SHEAR DIE
EXTRUSION

82

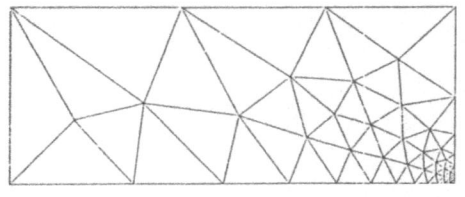

FIGURE 6.2   THE BACKGROUND MESH
η°= 27.8% , DOF = 338

FIGURE 6.3   THE REGENERATED MESH
η°= 19.2% , η_aim= 20.0% ,
DOF = 604)

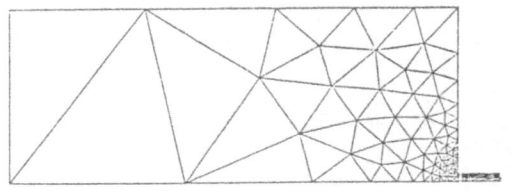

FIGURE 6.4   THE UPDATED MESH

FIGURE 6.5   THE EFFECTIVE STRAIN CONTOURS

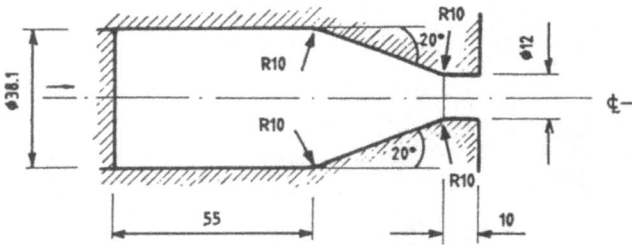

FIGURE 7.1   THE TAPERED DIE EXTRUSION

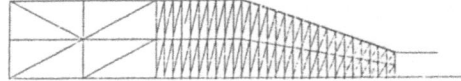

FIGURE 7.2   REGULAR TRIANGULAR-MESH AT INITIAL STAGE

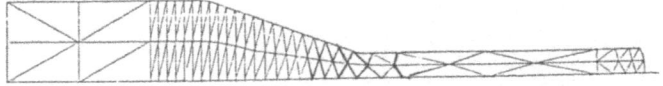

FIGURE 7.3   UPDATED MESH ACCORDING TO PRESCRIBED MESH VELOCITIES

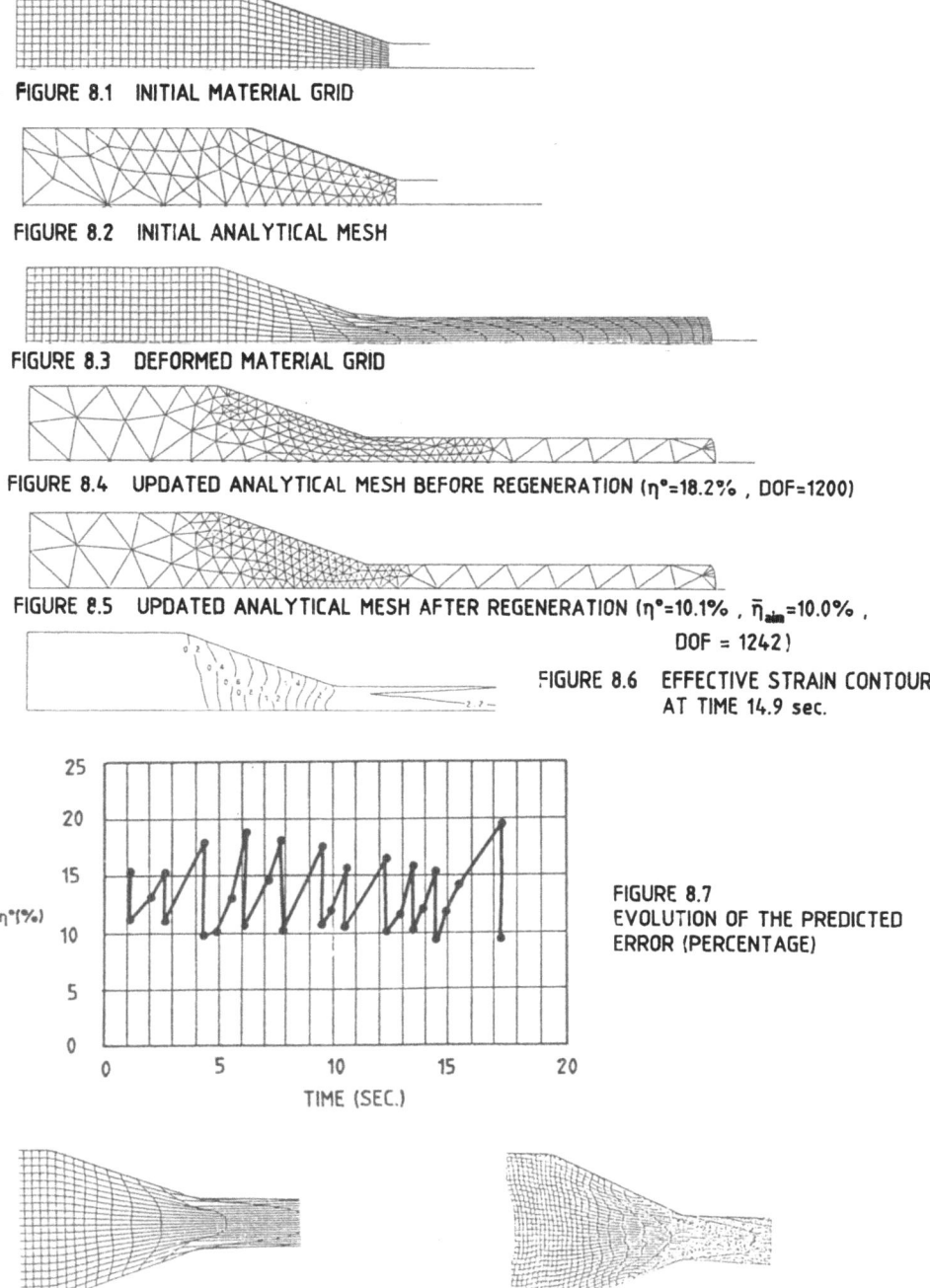

FIGURE 8.1   INITIAL MATERIAL GRID

FIGURE 8.2   INITIAL ANALYTICAL MESH

FIGURE 8.3   DEFORMED MATERIAL GRID

FIGURE 8.4   UPDATED ANALYTICAL MESH BEFORE REGENERATION ($\eta°$=18.2% , DOF=1200)

FIGURE 8.5   UPDATED ANALYTICAL MESH AFTER REGENERATION ($\eta°$=10.1% , $\bar{\eta}_{aim}$=10.0% ,
DOF = 1242)

FIGURE 8.6   EFFECTIVE STRAIN CONTOURS AT TIME 14.9 sec.

FIGURE 8.7
EVOLUTION OF THE PREDICTED ERROR (PERCENTAGE)

$\eta°$(%)

TIME (SEC.)

FIGURE 8.8   DEFORMED MATERIAL GRID

FIGURE 8.9   EXPERIMENTAL GRID

Modelling of Frictional Tool Surfaces in Finite–element Metalforming Analyses

I.Pillinger, P.Hartley and C.E.N.Sturgess
University of Birmingham, UK

1. Introduction

Metalforming can be viewed as a process of interaction between a yielding workpiece and essentially non–yielding tools. Clearly, any numerical simulation of a metalforming process must contain:

   i) a large–displacement flow formulation to predict how the material will flow in response to the changing boundary conditions

   ii) a method of determining these boundary conditions on an increment–by–increment basis.

Much work has been done in recent years in perfecting elastic–plastic displacement formulations and these are now very sophisticated. Less attention has been paid to the problem of determining boundary conditions, though their correct specification is vital to the success of the simulation. Some work has been published in this area, using a Coulomb friction model for example [1].

For a finite–element (FE) metalforming model to be really useful, particularly in an industrial context, it must satisfy two requirements:

   i) it should be fully–predictive. With the gross distortions usually found in metalforming, it is not known beforehand which parts of the workpiece will come into or out of contact with the tools during the deformation, nor the direction of the relative sliding of workpiece and tools in regions of contact. It would defeat the main purpose of using a numerical simulation if this information had to be provided by experimental trials.

   ii) it should be sufficiently general. The analysis of different metalforming operations should be possible simply by defining appropriate input data, not by re–writing the program. Ideally, the program should be able to simulate, for example, rolling and extrusion as well as forging.

As will be seen, the method of modelling tool surfaces described here satisfies both requirements.

J. L. Chenot and E. Oñate (eds.), Modelling of Metal Forming Processes, 85–92.
© 1988 by Kluwer Academic Publishers.

2. Displacement Formulation

A displacement formulation is used here in which the element stiffness equations relate component n of the change in position of node J of an element to component m of the change in the reaction at node I [2]:

$$\Delta f_{Im} = \left[K^{(\varepsilon)}_{ImJn} + K^{(\sigma)}_{ImJn} + K^{(\phi)}_{ImJn} \right] \Delta d_{Jn} \qquad (1)$$

The first term inside the parentheses has the same form as that obtained for the infinitesimal-deformation (small-strain) approach but uses a rotationally-invariant strain expression [3]. The second term ensures that the correct stress increment is used in the variational expression and the last term prevents the over-constraint of the mesh that arises from the condition of plastic incompressibility.

3. Modelling of Boundary Surfaces

Although the shapes of the tool surfaces used in metalforming operations may be very complex, it is generally possible to represent these shapes by a fairly small number of simple primitive surfaces. The advantage of this approach is that the determination of nodal contact and constraint is made very easy. Figure 1 shows the set of primitives that have been found to be adequate for a wide variety of metalforming simulations. Other possible simple geometric surfaces may easily be envisaged.

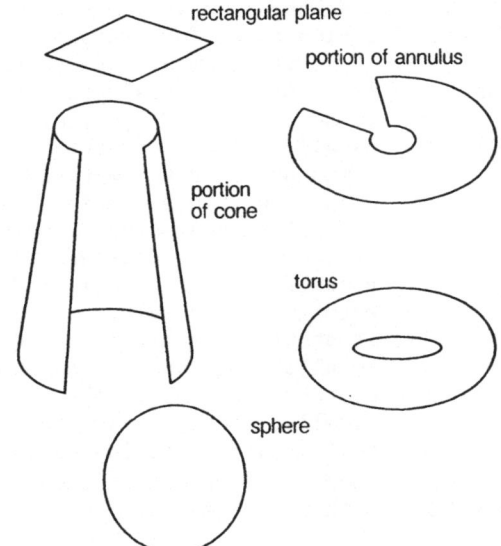

Figure 1. Types of primitive boundary surface.

Each primitive surface has associated with it an initial position and orientation, an incremental displacement and/or rotation, a friction factor, a temperature and a heat transfer coefficient.

Knowledge of the initial configuration and incremental movement of a primitive surface enables its position to be calculated at the start of each increment. It is therefore possible to check whether any of the external nodes of the FE mesh are in contact with this surface or have passed through it during the previous increment. If they have, they must be re-positioned at the nearest point on the surface and constrained to move tangentially to the surface for the next increment (figure 2). It should be noted that the size of the step that may be used in an FE metalforming analysis is not usually limited by the accuracy of the displacement formulation, but by the requirement that free-surface nodes do not pass too far through boundary surfaces during a given increment of the analysis.

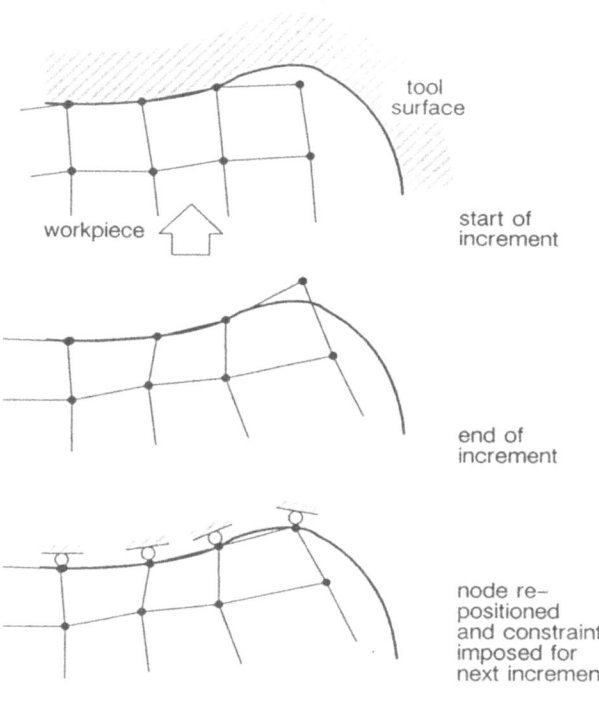

tool surface

workpiece

start of increment

end of increment

node re-positioned and constraints imposed for next increment

Figure 2. Contact between FE mesh and boundary surface during a deformation.

In the simplest case, when a tool surface is stationary, the component of incremental displacement of the node perpendicular to the surface at the point of contact is set to zero. The direction of this perpendicular is easily determined from the current configuration of the primitive surface. For the more general case of moving tool surfaces, the constraint imposed upon a node is less straightforward. Essentially, if P is the point of the boundary surface in contact with the node at the start of the increment then the node is constrained to move within a plane passing through the point P' that P will occupy at the end of the increment. The normal of this plane of constraint is the average of the normal to the original surface at P at the start of the increment and the normal to the surface at P' at the end of the increment.

This rather complicated method of constraint is necessary in order to be able to deal with rotation as well as displacement of boundary surfaces. For example, if a cylindrical surface is subject to a simple rotation about its axis, the plane of constraint is a chord to the surface passing through P and P'. However, if a surface does not rotate, the plane of constraint is simply a tangent to the displaced surface (figure 3).

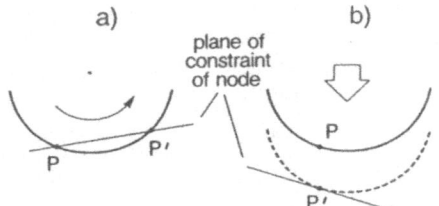

Figure 3. Constraint applied to nodes in contact with a boundary surface subject to a) pure rotation; b) pure displacement.

The procedure just described, in which the node is constrained to move within a plane, will apply whenever a free-surface node comes into contact with just one primitive boundary surface. If the node was originally subject to a constraining condition (such as node A in figure 4, which is constrained on a plane of symmetry) or if the node is found to be in contact with more than one boundary surface (such as node B in figure 4), then the constraint applied to the node will need to satisfy, if possible, all the conditions imposed upon it. This may result in the node being made to move along a line of intersection or, in the extreme case, being fixed in position upon the die surfaces.

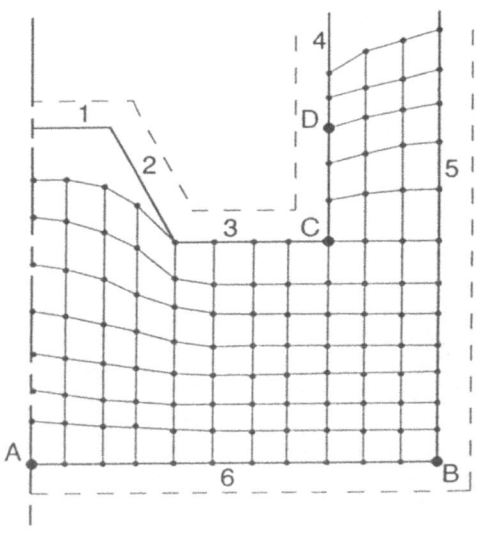

Figure 4. Boundary surfaces in a typical forging.

The situation at external corners is quite different. In the simulation illustrated in figure 4 for example, it is probably required that node C should begin to slide up the vertical wall of the punch (prior, presumably, to a re-meshing of the workpiece). This may be brought about by a judicious overlapping of the boundary surfaces at this corner so that the node is constrained on boundary surface 4 rather than boundary surface 3.

One further refinement to the contact algorithm is required. It may be that parts of an FE mesh are legitimately 'behind' a boundary surface in the geometric sense. For example, node D in figure 4 is behind surface 2. It is obviously important that such nodes are not re-positioned and constrained on the wrong tool surfaces. The problem can usually be avoided by specifying a distance to a secondary surface behind each boundary surface, beyond which nodes are not re-captured. The secondary surfaces are shown by the dashed lines in figure 4. The distance to a secondary surface must obviously be chosen with some care. If it is too small, nodes penetrating a boundary surface may pass all the way through in a single increment of the analysis. A distance equal to the average mesh spacing is normally about right.

The procedure outlined above is concerned with properly constraining nodes as they come into contact with tool surfaces. But the reverse process is equally important, and the FE model should allow nodes to move away from tool surfaces if the deformation requires it. An obvious example of this is the separation of nodes from a roll surface on exit from the roll gap. Therefore, at the start of each increment, the state of stress at each node in contact with a boundary surface is examined. If the stress acting normally to the surface of the mesh at a given node is tensile, the node is not constrained upon the die surface during the next increment.

4. Frictional Restraint

If the direction of sliding of the surface of the workpiece with respect to the die were known from the start, the imposition of a frictional traction force would be a simple matter. In a few simple geometries the direction is indeed known, and in others experiments may be carried out to determine this information, but this is not desirable if the FE program is to be generally applicable and fully predictive.

A method of applying a frictional restraint without prior knowledge of the pattern of flow is provided by the friction-layer technique [4]. This requires a layer of elements to be created at the interfaces between the workpiece and the dies (figure 5). The extra friction-layer nodes are fixed relative to the die surface. (If the dies are moving, of course, the nodes may have a superimposed motion.)

The stiffness matrix of each friction-layer element is then multiplied by a factor, the Stiffness-Matrix Multiplier (SMM), that is proportional to $m/(1-m)$, where m is the friction factor for the tool/workpiece interface. Since the friction-layer nodes cannot move parallel to the die surface, the effect is to apply a shear force to the interface nodes acting in the opposite direction to their movement. As m tends to zero, the SMM, and hence this shear force, also tends to zero. As m tends to one, the SMM tends to infinity and the friction-layer elements become very stiff, essentially preventing any tangential movement of the interface nodes and thus sticking them to the die.

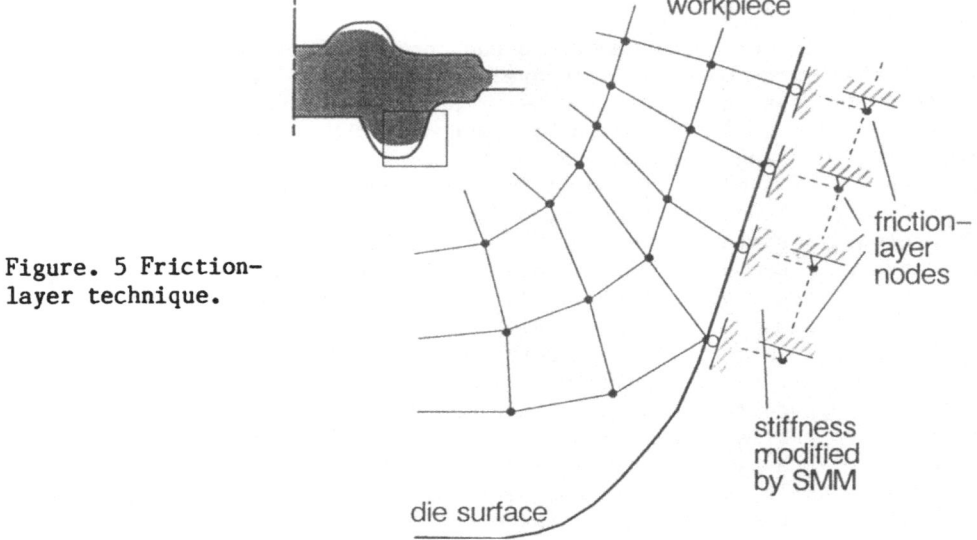

Figure. 5 Friction-layer technique.

The friction layer is not actually modelling a lubricant. It is merely a mechanism for modelling the effects of such a lubricant. Its existence in the computer model is also transitory. Each friction-layer element can be created when the stiffness matrix of the associated interface element is evaluated. The stiffness matrices of the interface element and the friction-layer element may be assembled together and a Gaussian reduction used to eliminate the friction-layer nodes from these equations (since the displacements of these nodes are known) before the equations of the interface element are assembled into the global matrix. Thus the friction-layer nodes need never form part of the main FE calculation.

5. Example

Figure 6 shows the initial FE mesh and initial positions of the primitive boundary surfaces used to model the forging of a commercially-pure aluminium section under plane-strain conditions. 18 boundary surfaces, rectangular planes and portions of cylinders, were required in this example. A friction factor of $m = 0.2$ was used for all the surfaces, simulating reasonably well-lubricated tools.

The material flow velocity vectors predicted by the FE analysis are also illustrated in figure 6. At the start of the analysis, surface 6 is not in contact with the mesh and the entire upper lobe of the section is pushed sideways by the extrusion of material from under the central punch (surface 1). At about 3% reduction in height of the central portion, the mesh contacts boundary surface 6 and this mode of deformation is replaced by one in which material from the centre flows into the upper and lower lobes of the section.

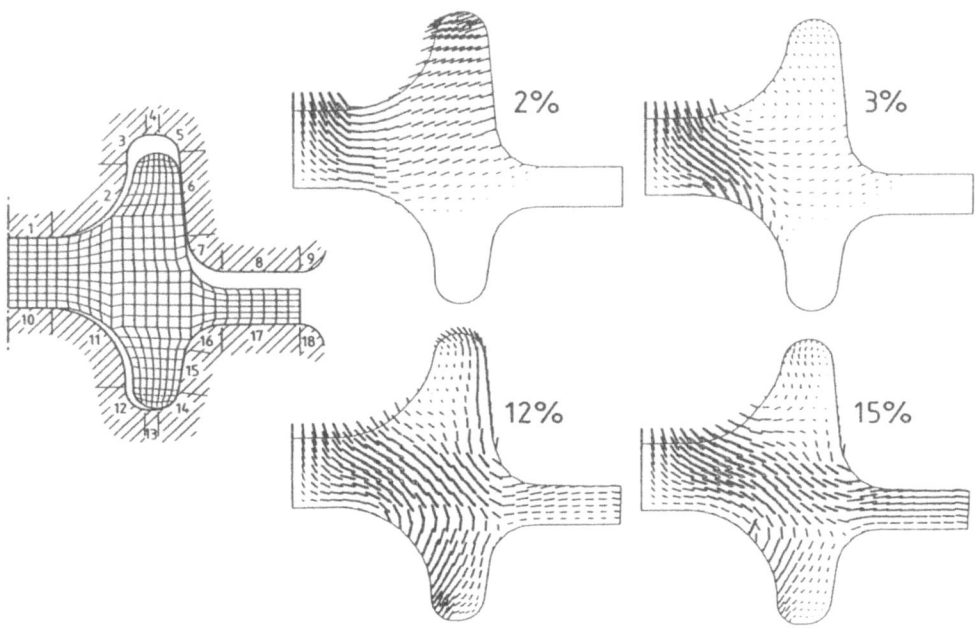

Figure 6. Boundary surfaces and velocity vectors in H-section forging.

By the time 12% reduction has been achieved, the flow of material into the lower lobe has intensified, while the flow into the upper lobe has been halted by the contact of the mesh with boundary surfaces 3, 4 and 5 and the influence of friction along boundary surface 6. This results in material in the outer part of the upper lobe moving downwards and towards the flash region.

At 15% reduction in height, the situation has changed yet again. Material in the outer part of the upper lobe is still being drawn downwards, but the additional contact of the mesh with boundary surface 2 has the effect of pushing more material away from the central region. Some of this is displaced upwards into the upper cavity, causing a very complex circulatory flow pattern, but most is pushed outwards into the flash.

As this example shows, fundamental changes in the pattern of material flow and deformation can take place during the course of forming operations. If the resulting component properties are to be predicted accurately, it is very important to model the boundary conditions correctly.

6. Future Developments

The major limitation of the techniques described in this paper is that the tool surfaces must be taken to be rigid. Quite often this is a reasonable assumption to make. In certain situations however, notably in the rolling of thin sheet, the elastic deformation of the tooling is very important in determining the resulting shape and properties of the workpiece.

One method of overcoming this limitation would be to use the values of stresses predicted by the FE metalforming program to estimate the elastic deformation of the tools. The metalforming analysis could then be performed again using the deformed shapes of the boundary surfaces. Clearly, this could not be expected to provide an acceptable solution the first time, and further assessments of tool deformation would probably be required.

A more satisfactory approach would be to model the tools by elastically-deforming finite elements or boundary elements. This would have the additional benefit that temperature gradients in the tools could also be modelled. The problem with this technique would be the difficulty in detecting contact between two FE meshes and the calculation of the resulting interaction. However, these difficulties are not insurmountable , and some work has already been carried out in connection with the friction-less contact of two deforming bodies [5].

Acknowledgements

The authors wish to thank the UK Science and Engineering Research Council who have funded the work described above and the Centre for Computing and Computer Science at the University of Birmingham for the use of computing facilities.

References

[1] B.Stok, A.Hudoklin and T.Rodic, 'Elasto-plastic solution of frictional contact problem by finite element method', Proc. Conf. Computational Plasticity, Barcelona, pp. 221-229 (1987).

[2] G.W.Rowe, C.E.N.Sturgess, P.Hartley and I.Pillinger, Finite-Element Plasticity and Metalforming Analysis, Cambridge University Press, 1988 (In Press).

[3] I.Pillinger, P.Hartley, C.E.N.Sturgess and G.W.Rowe, 'A new linearized expression for strain increment for the finite-element analysis of deformations involving finite rotation', Int. J. Mech. Sci. 28, pp. 253-262 (1986).

[4] P.Hartley, C.E.N.Sturgess and G.W.Rowe, 'Friction in finite-element analyses of metalforming processes', Int. J. Mech. Sci. 21, pp. 301-311 (1979).

[5] F.P.T.Baaijens, F.E.Veldpaus and W.A.M.Brekelmans, 'On the numerical simulation of contact problems in forming processes', Proc. 2nd Int. Conf. Num. Meth. Industrial Forming Proceses (NUMIFORM86), Gothenburg, pp. 85-90 (1986).

NUMERICAL MODELLING OF FRICTION FOR METAL FORMING PROCESSES

P. CHABRAND, Y. PINTO, M. RAOUS
Laboratoire de Mécanique et d'Acoustique
31, chemin Joseph Aiguier
13402 Marseille Cedex 9 - FRANCE

In metal forming processes, friction plays an important role. The quality of the final product depends on effective control of friction, and an accurate modelling of the friction effect is essential for a good description of the phenomenon. We focus, in this presentation, on the behaviour of sheet metal under a blank holder for metal forming problems encountered in the car industry.

In the case of dry contact, we have used a Coulomb friction law with unilateral conditions to take into account the evolution of the real contact area. These non-penetration conditions lead to a complementarity problem the variational form of which is a variational inequation. The introduction of friction leads to an implicit variational inequation problem including an undifferentiable term.

After presenting different numerical methods to solve this problem, we focus on the analysis of the blank holder function in metal forming processes.

1. The Mechanical Model

For the study of a blank holder without retaining line, the behaviour will be assumed to be elastic and the deformations to be small. Only the displacements will be large. Therefore, we set the problem as an elastic body lying in a domain Ω and submitted to external forces φ on a part Γ_2 on the boundary, to given displacements on a part Γ_1, and to contact and friction conditions on a part Γ_3.

On this part Γ_3, the displacements u and the contact forces F are split into tangential and normal components (n is the exterior local normal vector to the boundary Γ_3) :

$$u = u_N \cdot n + u_T \tag{1}$$

$$F = F_N \cdot n + F_T \tag{2}$$

The non-penetration conditions of the solid into the rigid obstacle are written :

93

J. L. Chenot and E. Oñate (eds.), Modelling of Metal Forming Processes, 93–99.
© 1988 by Kluwer Academic Publishers.

$$u_N \leq 0 \tag{3}$$

$$F_N \leq 0 \tag{4}$$

$$u_N \cdot F_N = 0 \tag{5}$$

These conditions caracterize the fact that a compression force F_N appears if and only if the contact occurs ($u_N=0$). Together with the equilibrium equations and the constitutive law, this leads to the well known Signorini problem ([1], [2]).

We now introduce a Coulomb friction law. It has been observed on different examples [3][4] that for dry metal contact the simple Coulomb friction law provides substantial agreement between theoretical and experimental results. It is written :

$$|F_T| \leq \mu \ |F_N| \tag{6}$$

with

$$|F_T| < \mu \ |F_N| \quad \Rightarrow \quad \dot{u}_T = 0 \tag{7}$$

$$|F_T| = \mu \ |F_N| \quad \Rightarrow \quad \dot{u}_T = -\lambda F_T \ , \ \lambda \geq 0 \tag{8}$$

An incremental formulation will be associated to the relations (6) (7) (8). It is clear that, for the general case, the solution is path dependant. A general variational formulation is given in (5). For the application concerned here, the loadings are monotone and a displacement formulation can be used in relations (7) and (8).

2. The Mathematical Problem

The variational form of the equilibrium equation, together with the constitutive equation and the contact and friction relations, leads to an implicit variational inequation [1][5]. We avoid the implicit character by using a fixed point method on the sliding limit. Because of the symmetry of the elasticity mapping, we write the problem as a minimization problem under constraints. It can be written :

Problem 1 : Find the sliding limit g such that :
$$g = \mu \ |F_N(u)| \tag{9}$$

where u is solution of Problem 2, depending on g

Problem 2 :
Find $u \in K = \left\{ v \in (H^1(\Omega))^3 ; \ v = 0 \text{ on } \Gamma_1 \text{ and } v \leq 0 \text{ on } \Gamma_3 \right\}$
such that : $J(u) < J(v) \quad \forall \ v \in K \tag{10}$
with : $J(v) = \dfrac{1}{2} \ a(v,v) - (f,v) + j(v) \tag{11}$

with :

$$a(v,v) = \int_\Omega \text{gradv.K.gradv } dx \tag{12}$$

K is the elasticity matrix

$$(f,v) = \int_{\Gamma_2} \varphi.v \; dl \tag{13}$$

$$j(v) = \int_{\Gamma_3} g.\,|v_T|\; dl \tag{14}$$

It is shown in [5] that the Coulomb problem (written on the velocities) can be solved by the use of a sequence of problems such as problem 1. This comes from the incremental formulation.

3. Numerical Methods

To solve this unilateral contact problem with friction, several different methods are used : either overrelaxation and conjugate gradient methods with projection techniques on the minimization problem with contraints, or mathematical programming methods (Lemke) on the initial complementarity problem [6]. We generally use condensation techniques for reducing the size of the finite element matrix concerned by the non linear solver. All these methods are presented in [5].

In general there is no one ideal method for solving unilateral contact problems with friction. In order to choose a good method among the ones that we have in our computer code PROTIS, we must consider the characteristics of the problem : the geometry (ratio of contact node number to the total node number ; band width ;...) the loading (static or evolutive process), the material behaviour (elasticity, viscoelasticity, viscoplasticity, etc...).

For the application presented in the next paragraph, the structure is very thin and we have a lot of contact nodes (one out of three) : we do not use a condensation process of course. In this case, the most efficient is a Cryer Christopherson method including the use of a sparse matrix storage technique. Cryer Christopherson method is an overrelaxation method with projection (see [5][7][8]). We have extended the algorithm to the friction case.

We use two different meshes : a 754 node one and a 1504 one. Because the respective solutions are in good accord with one another, most of the computations have been done on the 754 node mesh. For the first load steps the convergence is very fast and the solution is obtained in less than 3mn on a MicrovaxII. For the last ones, more iterations are necessary because a few unstabilities occur. Because of the shrinkage of the metal sheet, the contact between the sheet and the blank holder is barely lost and the numerical solution oscillates between a non contact state and a contact state with a very small normal contact force. This numerical oscillation does not affect the mechanical solution which remains the same in both cases.

4. Analysis of a Sheet Metal behaviour under Blank holders

In a metal forming process, a given load is applied by the press on the upper part of the blank holder. The deformations of this part remain small compared to those of the metal sheet. On another hand the contact can be lost between the sheet and the blank holder during the process. So, for the part of the sheet in contact with the blank holder, we use the following unilateral condition :

$$u_N \leq a \qquad\qquad (15)$$

where a is the given displacement of the blank holder. In this paper this displacement a is constant and corresponds to the total force in the initial situation of full contact. In the contract report [9] this displacement a is adjusted during the process according to the variations of the real contact surface so that the conservation of the total force is preserved.

The modelization with the condition (15) avoids the discretization of the blank holder and enables us to take into account the loss of contact which can occur in the zone. Experimental measurements on the sheet surface show that the repartition of the normal pressure is not uniform and confirm that a model using given loads on the sheet would be incorrect.

We consider the forming process of a long sheet for the manufacture of a main beam. This model is also helpful for the analysis of a more complex geometry in giving information about the behaviour of a straight part far from the angles of the structure. Only the part located under the blank holder is discretized. The purpose is to analyse the process in this part and to obtain a relationship between the tangential force and the tangential displacement at the exit of the blank holder. A plane strain hypothesis is used. The sheet is 0,7mm thick and 250mm long (under the blank holder). Because of the symmetry, the considered geometry is 0,35mm x 250mm. The following numerical results are related to a 754 nodes mesh with 251 contact nodes.

The given data is :
. Young modulus E - 21 000 daN/mm^2
. Poisson coefficient υ - 0.3
. normal displacement a - 5.10^{-5}mm
. friction coefficient μ - 0.15

The evolution of the solution is presented on figure 1 when a given tangential displacement Δ is increasing at the extremity.

The curves related to the normal and tangential forces are plotted. They are the lineic force densities by unit length (of the main beam). The quantity $\mu|F_N|$ is also plotted to show the accuracy obtained on the friction conditions (3) (4) (5).

At the beginning, only a sticking zone and a sliding zone appear (figures 1a, 1b). Then, a loss of contact occurs (figures 1c, 1d, 1e). These different states of the contact boundary are well characterized by the numerical results and can be observed on the figures. There is a stress singularity at the extremity of the stick zone which is progressing. Along the sliding zone, the tangential force modulus is equal to μ $|F_N|$. On the separated zone, both tangential and normal force components are zero. This loss of contact is very small and it has to be considered from a mechanical point of view as a contact without contact compression.

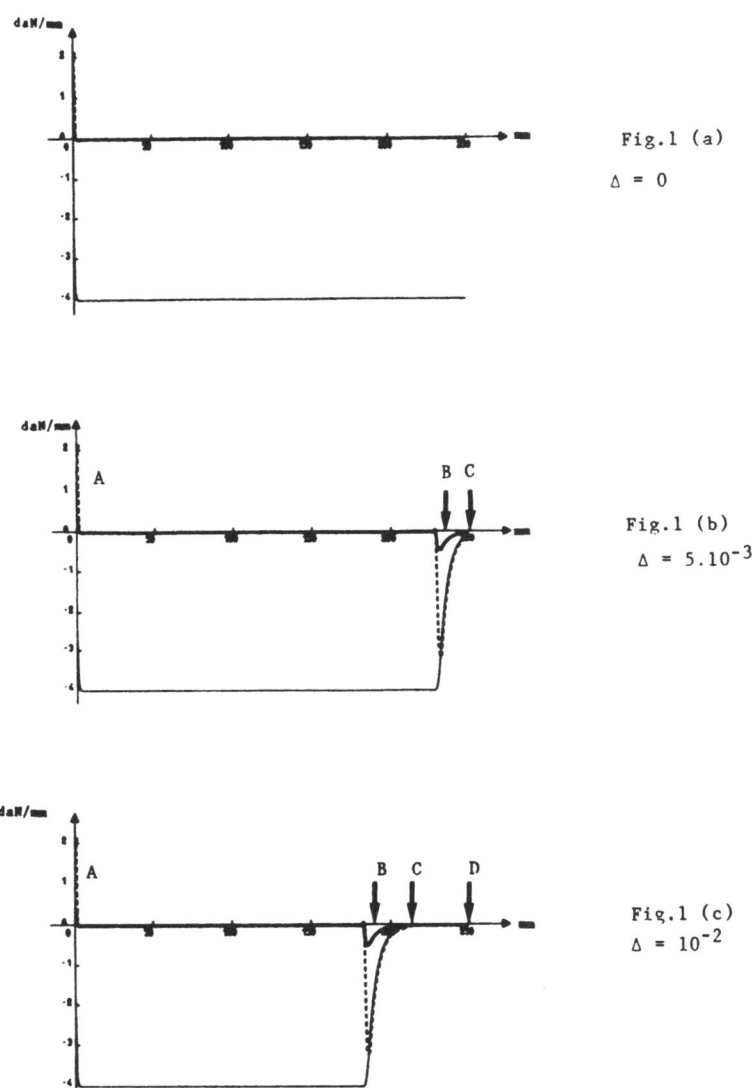

Fig.1 (a)

$\Delta = 0$

Fig.1 (b)

$\Delta = 5.10^{-3}$

Fig.1 (c)

$\Delta = 10^{-2}$

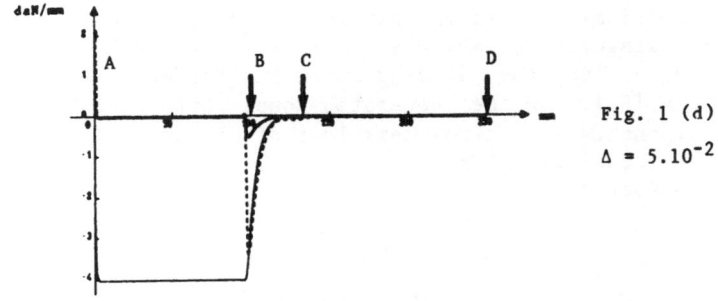

Fig. 1 (d)

$\Delta = 5.10^{-2}$

Fig.1 (e)

$\Delta = 9.10^{-2}$

Figure 1 : contact forces evolution
———————— normal force F_N
———————— tangential force F_T
– – – – – F_T/μ
AB stick zone
BC sliding zone
CD no contact zone.

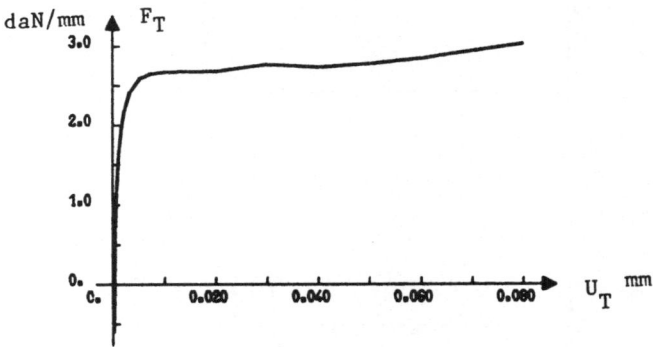

Figure 2 : relation ship between the tangential force
and the tangential displacement at the exit
of the blank holder

The global behaviour is given on figure 2.It represents the relationship between the tangential force and the tangential displacement at the exit of the blank holder.

5. Conclusion

An analysis of the friction phenomenon has been presented for a simple model connected with the blank holder function during a sheet metal forming process. Precise local information are given by the numerical method used to solve the involved variational inequation problem.

A more realistic technical analysis is done in [9] where the conservation of the total force is assured according to the variations of the real contact area.

The aim is to get a global behaviour law of this part of the sheet metal (as shown on figure 2) and to introduce it as a special boundary condition into a membrane or a shell model for the part located under the stamping tool.

A Blank holder with retaining lines is actually under study.

Acknowledgements : The present work has been partly supported by REGIENOV-RENAULT under contract H5-12-501.

REFERENCES

[1] DUVAUT G., LIONS J.L., Les inéquations en Mécanique et en Physique, Dunod, Paris, 1982.

[2] RAOUS M., Contacts unilatéraux avec frottement en viscoélasticité, in "Unilateral Problems in Structural Analysis", Ed. G. Del PIERO and F. MACERI, CISM Publisher, Springer-Verlag,Vienne, 1985.

[3] RAOUS M., TAALLAH F., VILLECHAISE B., Stress waves on the contact area of a solid sliding with friction on an obstacle, Proceedings Euromech 209, "Vibrations with unilateral constraints", 5-7 Juin 1986, Como, Italie.

[4] LEBON F., RAOUS M., Contact with friction modelling for the study of a bolted junction, Proceedings of 9th SMIRT, 17-21 August 1987, Lausanne, Suisse.

[5] RAOUS M., CHABRAND P., LEBON F., Numerical methods for frictional contact problems and applications, Journal de Mécanique Théorique et Appliquée, N° Spécial, 1988.

[6] RAOUS M., PINTO Y., Direct and Iterative methods for two body contact Problems, Proceedings COBEM 87, December, 1987, Florianopolis, Brésil.

[7] CRYER C.W., The solution of a quadratic programming problem using systematic overrelaxation, SIAM J.Control, 9, 1971, pp. 385-392.

[8] GLOWINSKI R., LIONS J.L., TREMOLIERES R., Analyse Numérique des Inéquations Variationnelles, Dunod, Paris, 1976.

[9] PINTO Y., RAOUS M., Modélisation de l'effet de serre flan avec ou sans jonc en emboutissage. REGIENOV/CNRS n°H5-12-501, Rapport intermédiaire, Juin 1988.

PART 3

SHEET METAL FORMING

MODELLING OF DEEP DRAWING PROCESSES BY MODEL TESTING AND DIMENSIONAL ANALYSIS

D. Bauer
Institute of Production Engineering
University Siegen
P.O. Box 101240, D - 5900 Siegen, FRG.

INTRODUCTION

The precise knowledge of force-stroke curves in deep drawing processes depending on plastic behaviour of sheet
material, friction, surface roughness, blank holder,
pressure and geometric tool design is of great importance
for the press shop to optimize sheet metal manufacturing
Using the well known theory of the yielding and plastic
flow of isotropic or anisotropic material, respectively,
the modelling of metal forming processes can principally
be described by a complete set of partial differential
equations. But in deep drawing processes it is not
possible to solve this set of equations without making
substantial simplifications. That is why a novel semi-
experimental or semi-theoretical modelling technique has
been developed for predicting force-stroke curves by means
of dimensional analysis and a special method of model
testing. In this work first results in case of an axis-
symmetric deep drawing process are presented, as shown in
Fig. 1.

BASIC THEORY

Starting point of the dimensional analysis is the so-
called π - theorem. It is based on the assumption that a
complete set of linear independent dimensionless products
π_i can be formed from the physical variables X_i that are
governing five points $Q_0...Q_4$ of the dimensionless force-
stroke curve, Fig. 2. It can now be shown that the π -
theorem of dimensional analysis is delivering the following complete set of linear independent dimensionless
products (ref. 1):

J. L. Chenot and E. Oñate (eds.), Modelling of Metal Forming Processes, 103–110.
© *1988 by Kluwer Academic Publishers.*

$$\pi_1 = \frac{100\, S_0}{d_1} \qquad \pi_2 = \frac{r_z}{s_0} \qquad \pi_3 = \frac{r_s}{S_0} \qquad \pi_4 = \frac{d_0}{d_1}$$

$$\pi_5 = \frac{U_z}{S_0} \qquad \pi_6 = \frac{C}{Y_{0,2}} \qquad \pi_7 = n \qquad \pi_8 = r$$

$$\pi_9 = \frac{\Delta T}{T_0} \qquad \pi_{10} = \frac{R_{ws}}{d_0} \qquad \pi_{11} = \frac{R_{ws}}{R_{wz}} \qquad \pi_{12} = \frac{\Delta H}{H_{wz}}$$

$$\pi_{13} = \frac{\Delta H}{P_n} \qquad \pi_{14} = \frac{\dot{\varepsilon}\eta}{P_n} \qquad \pi_{15} = \frac{Y_{0,2}}{P_n} \qquad \pi_{16} = \frac{h}{h_4}$$

$$\pi_{17} = \frac{\sigma_z}{Y_{0,2}}$$

In order to get this complete set of dimensionless products it is assumed that the physical variables regarded in Fig. 1 and the following ones, too are governing the deep drawing process totally:

$Y_{0,2}$ yield point of blank material   (ref. 2)

C   coefficient of blank material   (ref. 2)

n   coefficient of work-hardening

r   coefficient of plastic anisotropy

ΔT temperature increase by plastic deformation

T_0 temperature of blank material at beginning of deep drawing

R_{ws} surface roughness of blank

R_{wz} surface roughness of die

ΔH hardness difference between die and blank

H_{wz} hardness of die

$\dot{\varepsilon}$ effective strain rate

η dynamic coefficient of viscosity

105

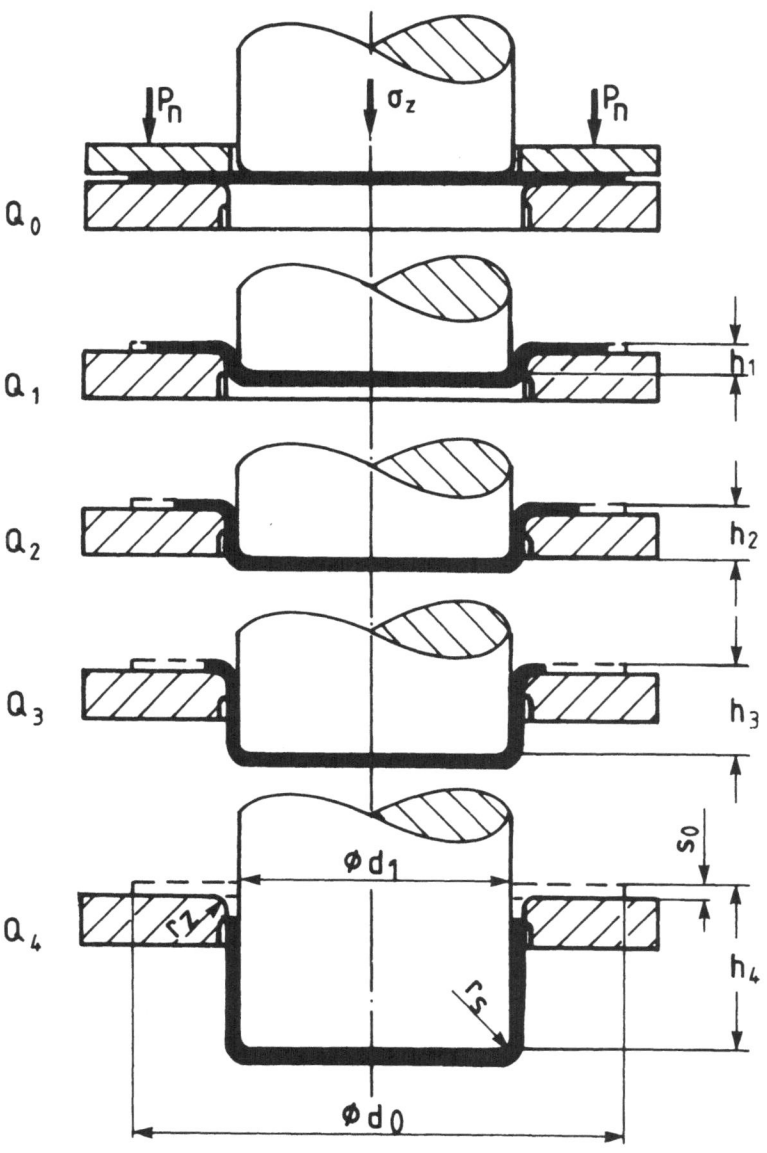

Fig. 1. Axis-symmetric deep drawing process.

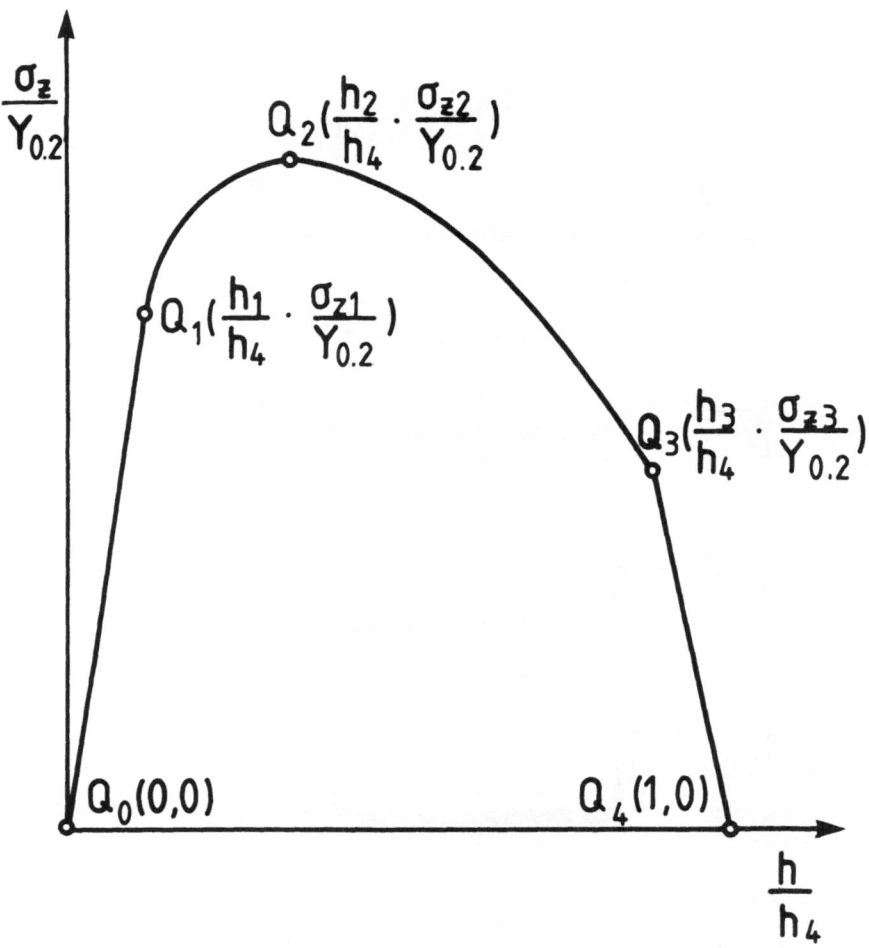

Fig. 2. Dimensionless force-stroke curve.

Using special conditions of experimental device (ref. 3) the dimensionless products π_4, π_5,...., π_{15} can be neglected. Hence, with respect to Fig. 2, each co-ordinate of point Q_i can now be described by a relationship of the 3 linear independent products π_1, π_2, π_3:

$$\sigma_{zi}/Y_{0,2} = \Phi_i\ (\pi_1,\ \pi_2,\ \pi_3)$$

$$h_i/h_4 \quad\quad = \Phi_i\ (\pi_1,\ \pi_2,\ \pi_3)$$

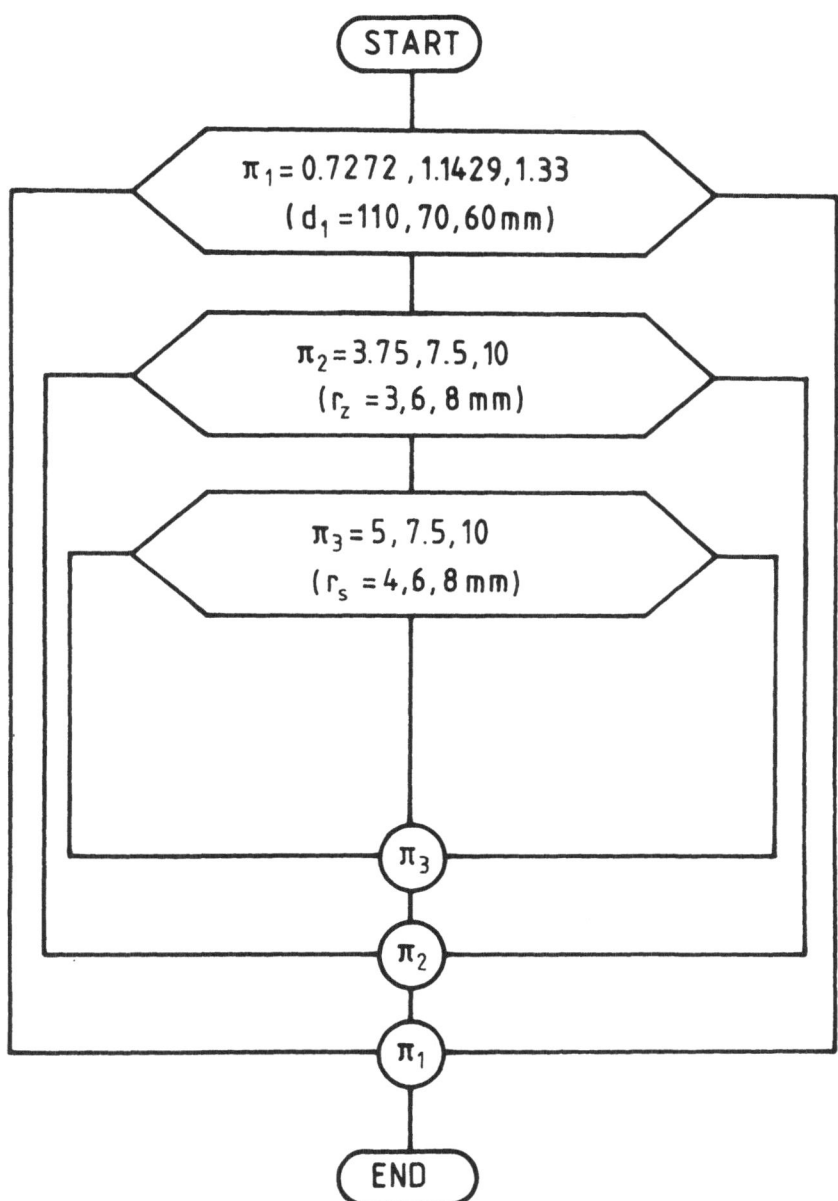

Fig. 3.   Model testing of autobody steel St 1405 m

$Y_{0,2} = 185$ N/mm^2 C = 542 N/mm^2 n = 0,22 r = 1,38

Regarding the well known laws of similarity (ref. 1) this relationships can be determined by model testing (ref. 1). Complete similarity exists if every product π has the same numerical value for the model as for the original. Though it is ovious, that complete similarity includes geometric similarity sheet material used in model testing should reasonably have the same thickness and mechanical proper- ties as the sheet material used in the press shop. Considering these assumptions complete similarity cannot be achieved in model testing. But this disadvantage can be changed into an advantage. For this purpose a special experimental technique (ref. 4) of model design and testing has been developed to produce a linear independent set of departures from complete similarity, Fig. 3. Evaluating these values measured by regression analysis, a special method of statistical mathematics (ref. 5), an approxi- mative equation can be formulated for each co-ordinate of the points Q considered in Fig. 2.

RESULTS

With the coefficients a_1, a_2, ..., a_{10} listed in Tab. 1+2, the new developed method submitted is delivering the following approximative equations, finally:

$$\sigma_{z0}/Y_{0,2} = 0 \quad , \quad h_0/h_4 = 0$$

$$\ln (\sigma_{z1}/Y_{0,2}) = A+B, \quad \ln (h_1/h_4) = C+D$$

$$\ln (\sigma_{z2}/Y_{0,2}) = A \quad , \quad \ln (h_2/h_4) = E+D$$

$$\ln (\sigma_{z3}/Y_{0,2}) = E+D, \quad \ln (h_3/h_4) = A$$

$$\sigma_{z5}/Y_{0,2} = 0 \quad , \quad h_4/h_4 = 1$$

$$A = \ln (a_1 (\pi_2+a_2)^{a_3}+a_4) + \ln (a_5 \pi_1^{a_6}+a_7 \pi_1+a_8)$$

$$B = \ln \pi_3 (a_9 \pi_1^{a_{10}})$$

$$C = \ln (a_1 \pi_1^{a_2} \pi_2^{a_3} \pi_1^{a_4} + a_5 \pi_1^{a_6})$$

$$D = \ln \pi_3 \, (a_7 \, \pi_1{}^{a_8})$$

$$E = \ln \, (a_1 \, \pi_1{}^{a_2} \pi_2{}^2 + a_3 \, \pi_1{}^{a_4} \pi_2 + a_5 \, \pi_1{}^{a_6})$$

These equations are valid for any arbitrary numerical value of π that can be calculated in the range investigated, Fig. 3.
Hence using these equations and providing for a suitable interpolation between the points Q_0, Q_2,..., Q_4 any arbitrary force-stroke curve can be predicted in the range regarded by model testing.

CONCLUSIONS

A novel modelling technique has been submitted for predicting force-stroke curves in axis-symmetric deep drawing processes by means of dimensional analysis, model testing and regression analysis. For this purpose a special experimental device of model design and testing has been developed to produce a linear independent set of departures from complete similarity. Evaluating these values measured by regression analysis and providing for a suitable method of interpolation an approximative equation can be formulated to predict any arbitrary force-stroke curve in the range investigated by model testing.

REFERENCES

1 H. Langhaar, Dimensional Analysis and Theory of Models, Chapman & Hall London 1964.

2 D. Bauer, Proc. Int. Conf. on Computational Methods for Predicting Material Processing Defects, Sept. 8-11, 1987, Cachan, France, E lsevier Amsterdam 1987, 9 - 18.

3 D. Bauer and H.-P. Landvogt, METALL 41 (1987), 624-627.

4 H.-P. Landvogt, Fortschritt-Berichte VDI, Reihe 2: Fertigungstechnik, Nr. 145, VDI-Verlag Düsseldorf 1987.

5 B. Flury and H. Riedwyl, Angewandte multivariate Statistik, Gustav Fischer Verlag Stuttgart 1983.

TABLE 1

Coefficients of approximative equations (Steel St 1405 m)

Coeff.	$\sigma_{z1}/Y_{0,2}$	$\sigma_{z2}/Y_{0,2}$	$\sigma_{z3}/Y_{0,2}$
a_1	+ 1,5871414	+ 2,3532610	+ 0,0030070
a_2	- 0,7280945	- 3,1621442	- 1,5395919
a_3	- 0,6263832	- 0,2118322	- 0,0617671
a_4	+ 1,6546157	+ 4,0743702	- 1,4904520
a_5	+ 0,2060578	+ 0,2152138	+ 1,0115244
a_6	+ 0,0064006	+ 0,5783150	- 0,3246084
a_7	+ 0,0244417	- 0,1561824	+ 0,1014320
a_8	+ 0,2504847	+ 0,1973827	+ 0,5513722
a_9	+ 0,0884436	0	0
a_{10}	- 0,5357391	0	0

TABLE 2

Coefficients of approximative equations (Steel St 1405 m)

Coeff.	h_1/h_4	h_2/h_4	h_3/h_4
a_1	+ 0,0259307	+ 0,0001903	+ 0,4849273
a_2	- 0,2437929	+ 3,0744203	- 3,6063517
a_3	+ 0,2911063	- 0,0069931	- 0,1279427
a_4	+ 0,8724075	+ 0,1742376	+ 1,0617999
a_5	+ 0,0365154	+ 0,3614157	+ 0,1597179
a_6	+ 2,4238987	- 0,0687747	- 1,1056502
a_7	+ 0,4599660	+ 0,1010354	+ 0,0757499
a_8	- 0,4601899	+ 0,5370978	+ 0,2890298

A FINITE DIFFERENCE MODEL AS A BASIS FOR DEVELOPING
NEW CONSTITUTIVE EQUATIONS FOR THE SHEET FORMING PROCESS

H. VEGTER - HOOGOVENS GROEP B.V., IJMUIDEN (THE NETHERLANDS)

For the simulation of the sheet forming process Finite Element Models are developed. In most cases the description of the material behaviour and friction is limited:
-no effect of strain rate
-isotropic material behaviour
-constant coefficient of friction
   The influence of taking into account these effects is hardly known until now. As a first step in the research program a finite difference model is developed in which constitutive equations can easily be changed. The results of model calculations show a great influence of the different descriptions of material behaviour and friction.

1. Introduction

A better prediction of geometry and forces during sheet forming plays an important role to realize cost reductions due to improvement of tool design and material selection in automotive industry. Using numerical models together with CAD/CAM systems will lead to this purpose. The state of the art at this moment is as follows.
-Axisymmetric and one-dimensional problems can be handled rather well [8]; the attention is now focussed on other geometries like rectangular parts.
-Constitutive equations of material behaviour are often limited to isotropic material; no effect of strain rate has been taken into account.
-In most cases a constant coefficient of friction is taken.
   In a common research program of Hoogovens Group B.V. and the University of Twente (The Netherlands) a simulation model for the sheet forming process is developed. In a first stage calculations are made with a finite difference model of axisymmetric parts in which different formulations for material behaviour and friction are available. Comparing these calculations with experiments of pressed parts will give arguments for an experimental program in order to improve constitutive equations. A finite element model is developed simultaneously for axisymmetric parts [9]; and in a next stage for rectangular parts. In this last case model calculations will be verified by experiments of pressed parts on industrial scale.

111

J. L. Chenot and E. Oñate (eds.), Modelling of Metal Forming Processes, 111–121.
© *1988 by Kluwer Academic Publishers.*

The subject of this paper is the comparison of calculations with experiments on axisymmetric parts using different formulations for material behaviour and friction. Three principal choices are made in this work:

-The finite difference model is chosen for mathematical convenience; particularly if constitutive equations are changed. This choice will limit geometries of the parts only to one-dimensional strips and axisymmetric parts.

-Axisymmetric parts are prefered because in strips only one state of stress will occur during forming. In case of forming axisymmetric parts a wide range of stress states occurs, which can give us very useful information about the yield surface.

-Only normal anisotropy is taken into account in the yield criterion; a good mathematical description for the effect of planar anisotropy has not been found until now.

2. Description of the model

2.1. GEOMETRICAL DESCRIPTION

The cup is divided in geometrical cells which are related to the original position of the blank. For the determination of the geometry the drawing process is divided into five zones as given in figure 1. A basic assumption is made for the unsupported region (zone 3) which always has a conical shape in order to get a stable solution. A second assumption is a linear relationship between thickness and surface within the cell.

During normal cupdrawing (figure 2a) the new geometry is determined after a certain time step from the punch radius to the edge of the blank. The computation at every cell point starts with the initial thickness. Using invariancy of volume the hoop strain increment, $\Delta\varepsilon_\varphi$, can be determined. From theory of plasticity the thickness strain increment, $\Delta\varepsilon_z$, is determined and a new thickness is used for the next iteration in this cell point.

In case of stretch forming from the flat bottom (figure 2b) the calculations of the radial position of the cell points starts at the die radius and is continued just before the flat bottom. The flat bottom is considered as one cell in which uniform equibiaxial stretching occurs and completes the calculation because of invariancy of the total volume.

When stretch forming occurs only in the conical unsupported region (figure 2c) a stretch cell point of localized necking is determined based on a stress criterion. The geometrical calculation for the other cell points is performed from the two radii to the neighbouring cell points. Localized thinning in the stretch cell point is determined to complete the geometrical calculations.

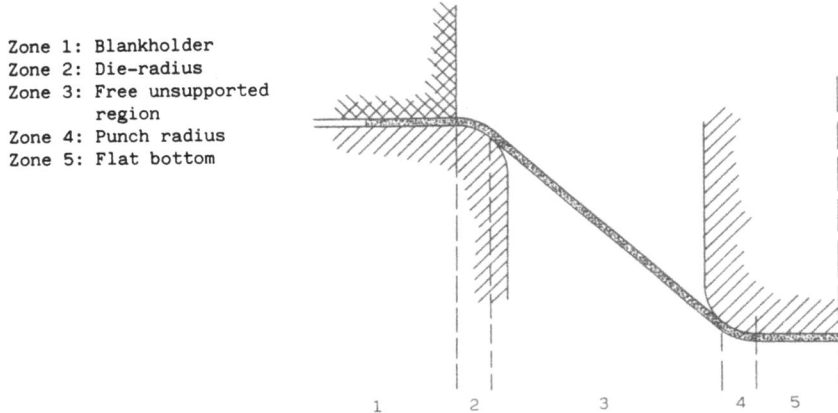

Zone 1: Blankholder
Zone 2: Die-radius
Zone 3: Free unsupported
        region
Zone 4: Punch radius
Zone 5: Flat bottom

1   2   3   4   5

Figure 1: Geometry of axisymmetric drawing

2.2. CALCULATION OF STRESS AND STRAIN

The meridional stress component over the profile of the deformed
axisymmetric part can be determined on the following force equilibrium in
the stress element (figure 3).

$$\frac{d\sigma_x}{dx} = -\left\{ \pm (\tau_a + \tau_b) + \sigma_x \frac{dh}{dx} \right\} \frac{1}{h} - \frac{\sigma_x - \sigma_\varphi}{r} \cos\alpha \qquad (1)$$

σ_x = meridional stress
σ_φ = hoop stress
τ_a = shear stress at the underside of the profile
τ_b = shear stress at the upperside of the profile
h = thickness
r = radial coordinate
x = length coordinate over the profile
α = angle between profile and radial coordinate
$dx = dr/\cos\alpha$
The shear stresses τ_a and τ_b can be determined from the normal stress and
coefficient of friction.
The normal force can be determined from the applied blank
holder-force, zone 1, (figure 1), or using normal force equilibrium
of the membrane forces in zone 2 and 4.

$$\sigma_N = -h \left(\frac{\sigma_x}{r_a} \pm \frac{\sigma_\varphi \sin\alpha}{r} \right) \qquad\qquad (2)$$

r_a = radius (die or punch)

The sign of the second term is positive in zone 4 and negative in zone 2. The blankholder force (zone 1) is usually not uniformly distributed over the surface but is dependent on the material thickness and the elastic deformation within the blankholder and die. This behaviour is not simulated here, but in the computer program an option is available to define the part of the tool, which is in contact with the material. In this contact area the normal stress can be linearly distributed over the radial position.

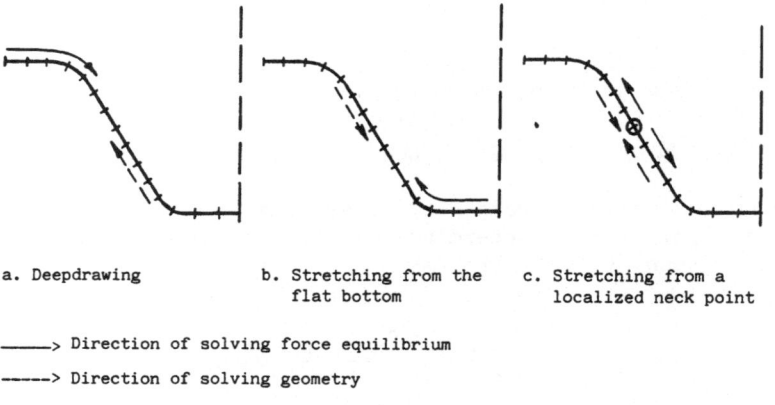

a. Deepdrawing   b. Stretching from the   c. Stretching from a
                         flat bottom   localized neck point

——⟶ Direction of solving force equilibrium

- - - -⟶ Direction of solving geometry

Figure 2: Three ways of solving force equilibrium and geometry.

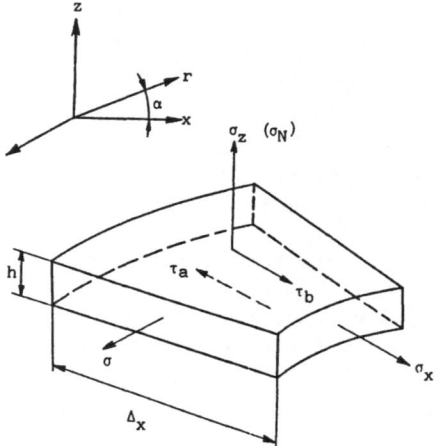

Figure 3: Stress components and coordinates
of force equilibrium cell

The hoop stress σ_φ can be determined from σ_x and σ_z using a yield criterion. The stress σ_z is equal to the normal force in case of the blankholder area (zone 1), in zones 2-5 we take $\sigma_z = 0$. From the yield criterion the deformation state is fixed because the deformation vector is perpendicular to the yield surface. Furthermore the effective plastic strain increment can be determined from theory of plasticity; at every cell point the total effective strain and effective strain rate is known. Three cases can occur concerning the boundary condition (figure 2) from which the calculation of the forces starts.
-normal deepdrawing: $\sigma_x = 0$ at the edge of the blank
-stretching from the flat bottom: $\sigma_x = \sigma_\varphi$ = yield stress in the equibiaxial point
-stretching from a localized necking point in the unsupported region: σ_x = yield stress in the plane strain point.

2.3. CONSTITUTIVE EQUATIONS

2.3.1. Description of the material behaviour

Several possibilities are available for the description of the yield surface in which the effect of normal anisotropy is taken into account. As a first approach the conventional Hill-criterion [2] is used:

$$R(\sigma_x - \sigma_\varphi)^2 + (\sigma_\varphi - \sigma_z)^2 + (\sigma_z - \sigma_x)^2 = (R+1)\ \sigma_v^2 \qquad \text{Hill-0 (3)}$$

with: σ_v = yield stress

In many cases this criterion is unsatisfactory, because it is completely fixed by performing uniaxial tensile tests from which the unknown parameters σ_v and R can be determined. The extrapolation for other stress conditions is rather speculative. Because of this reason a third parameter is used by introduction of real powers from which two variants are given here [3, 4].

$$(2R+1)\ |\ \sigma_x - \sigma_\varphi\ |^q + |\ \sigma_x + \sigma_\varphi - 2\sigma_z\ |^q = 2(R+1)\ \sigma_v^q \qquad \text{Hill-1 (4)}$$

$$R\ |\ \sigma_x - \sigma_\varphi\ |^p + |\ \sigma_\varphi - \sigma_z\ |^p + |\ \sigma_z - \sigma_x\ |^p = (R+1)\ \sigma_v^p \qquad \text{Hill-2 (5)}$$

Both criterions give the same results as in case of the conventional Hill-criterion using powers of 2. The difference is relatively high in the plane-strain point and the equibiaxial point in formula (4) in particular. By using powers which are greater than 2, lower plane strain yield stresses and equibiaxial yield stresses are achieved for a material with an R-value 1.5 - 2 like steel.
The yield stress can be determined as a function of strain using a Nadai relation:

$$\sigma_v = C\bar{\varepsilon}^n \tag{6}$$

$\bar{\varepsilon}$ = total accumulated effective strain
C = constant (N/mm²)
n = strain hardening exponent

A more advanced model is available too, based on a Voce-like relation [6] including the influence of strain rate[7].

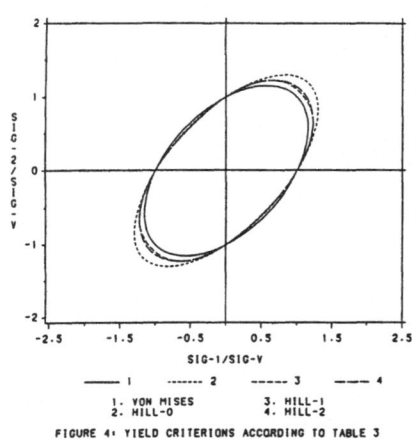

FIGURE 4: YIELD CRITERIONS ACCORDING TO TABLE 3

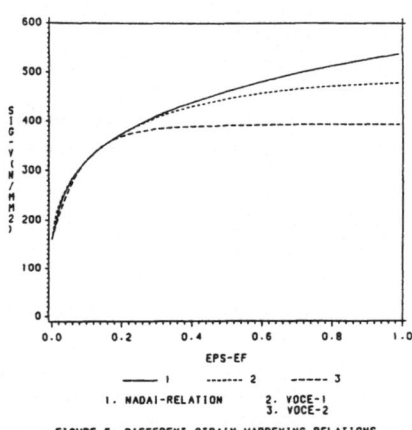

FIGURE 5: DIFFERENT STRAIN HARDENING RELATIONS

2.3.2. Friction between material and tool

For the calculation of the shear stress on the material surface which is in contact with tool, a constant coefficient of friction is usually taken. In this model friction is described based on a theory of mixed lubrication by Emmens [5]. The coefficient of friction is a function of the following parameters:
-viscosity of the lubricant
-relative speed of the material
-normal pressure
-surface roughness of the material

3. Experiments of pressed parts

Three geometries are used for the verification of the model (table 1):
1. Deep drawn   part   Ø 75 mm
2. Stretch   formed part Ø 75 mm
3. Deep drawn   part   Ø 293 mm

The punch force is measured during pressing of the deep drawn parts. On the blanks of the parts with a diameter of 75 mm a grid of circles has been put, so strains could be determined afterwards. Thickness is measured too by means of an ultrasonic method. In all cases preserving oil is used as a lubricant, viscosity at 20°C η = 0.19 N/m²s.

The experiments are performed on a cold rolled Al-killed low carbon steel which has mechanical properties and surface roughness as given in table 2.

Part no.	D_p (mm)	D_d (mm)	r_{ap} (mm)	r_{ad} (mm)	D_{bl} (mm)	h_{bl} (mm)	Speed mm/s	F_{bl} kN
1	75	77.4	10	8	162	0.8	5	20
2	75	77.4	22.5	3	158	0.8	5	200
3	293	300	15	20	600	0.8	30	260

D_p = punchdiameter $\qquad$ D_{bl} = blankdiameter

D_d = die hole diameter $\qquad$ h_{bl} = blank thickness

r_{ap} = radius of the punch $\qquad$ F_{bl} = blank holder force

r_{ad} = radius of the die

Table 1: Geometry and process variables of the pressed parts

dir (°)	R_p (N/mm²)	R_m (N/mm²)	A %	A_{80} %	C N/mm²	n-value	R-value	ho (mm)	R_{pm} (µm)
0	158	313	27.8	28.1	534	0.231	2.01	-	-
45	164	325	25.3	27.2	549	0.222	1.43	-	-
90	157	310	27.0	27.8	525	0.226	2.16	-	-
$\bar{x}$	161	318	26.3	27.6	539	0.225	1.76	0.8	2.4

$$x = (x_0 + 2X_{45} + X_{90}) / 4$$

Table 2: Mechanical properties of Al-killed steel

4. Determination of the constants of the constitutive equations

Preliminary results of plane strain tensile tests give a plane strain yield stress between the Von Mises criterion (R=1) and the conventional Hill criterion (R=1.76). The values of the exponents p and q are given in table 3. The differences in both criterions exist in the stretch forming region between the plane strain point and the equibiaxial point (figure 4).

Criterion	Exponent	R-value	$\dfrac{\sigma_1\text{-ps}}{\sigma_v}$	$\dfrac{\sigma_1\text{-bi}}{\sigma_v}$
Mises	2	1.00	1.155	1.000
Hill-0	2	1.76	1.298	1.174
Hill-1	2.3	1.76	1.226	1.052
Hill-2	4.4	1.76	1.226	1.076

Table 3: Constants of the different yield criterions

The results of the tensile tests at constant strain rate (10^{-2} s^{-1}) have been averaged over the different directions and have been fitted using the Voce-relation. Two variants are chosen; in the second case a big difference with the Nadai-relation (6) exists at effective strains larger than 0.3 (figure 5).

The influence of the strain rate has not been determined. Results of tensile tests at different strain rates and temperatures on pre-cold rolled specimens are used in this work. These are carried out for the determination of the deformation resistance of several steels during cold rolling at Hoogovens Group [7].

The constants of the friction model are determined by drawing flat strips between two flat platens. In this way a wide range of conditions of speed and pressure can be achieved. A complete test program using lubricants of different viscosity on materials with different surface roughness.

5. Comparison of model calculations with experiments

As a starting point for the calculations the conventional constitutive equations are taken.
-Strain hardening behaviour is described by the Nadai relation;
-No influence of strain rate;
-The conventional Hill criterion is taken (Hill-0);
-Constant coefficient of friction (μ=0.2)

For understanding the effect of the proposed alternative description of material behaviour and friction, every new equation is tested separately in respect of the basic model. The results of all these calculations are given here qualitatively.

Concerning strain distribution and instability in the stretched part we mention the following results:
-Strain hardening behaviour after the instability strain has a very
 strong influence; from the three mentioned variants the Nadai relation
 will give the most optimistic prediction.
-An increase is found if the effect of strain rate is taken into account.
-A yield criterion with a large difference between the plane strain stress
 and equibiaxial stress (Hill-1 criterion; table 3) has the most
 favourable effect in comparison with the conventional Hill-0 criterion.
-A low coefficient of friction on the punch radius is favourable; in our
 case only boundary lubrication will occur according to the proposed
 friction model, because a low speed difference exists between material
 and tool.

In case of the deep drawn parts the influence on strain distribution can be summarized as follows:
-Strain hardening behaviour has no effect.
-The minimal thickness at the punch radius will decrease if strain rate
 influence is taken into account because the necessary force for drawing
 the material out of the blankholder area will increase by this effect.

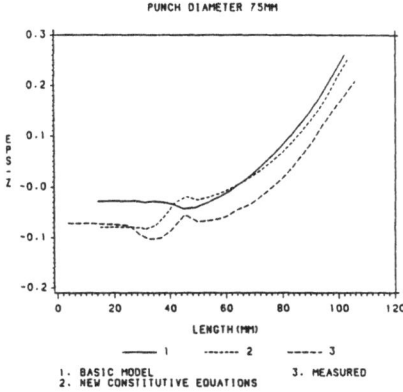

PUNCH DIAMETER 75MM

1. BASIC MODEL
2. NEW CONSTITUTIVE EQUATIONS
3. MEASURED

FIGURE 6: STRAIN DISTRIBUTION AFTER DEEPDRAWING

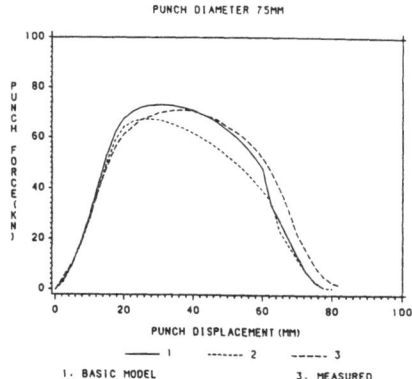

PUNCH DIAMETER 75MM

1. BASIC MODEL
2. NEW CONSTITUTIVE EQUATIONS
3. MEASURED

FIGURE 7: FORCE DISPLACEMENT CURVE DURING DEEP DRAWING

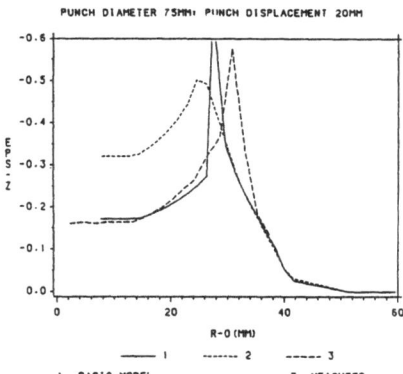

PUNCH DIAMETER 75MM: PUNCH DISPLACEMENT 20MM

1. BASIC MODEL
2. NEW CONSTITUTIVE EQUATIONS
3. MEASURED

FIGURE 8: STRAIN DISTRIBUTION AFTER STRETCH FORMING

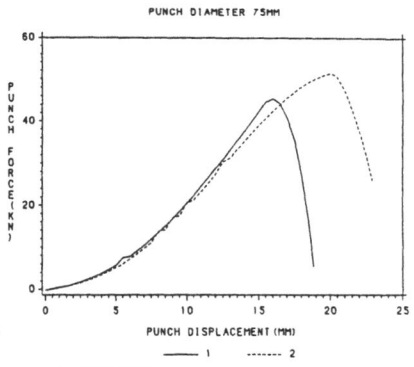

PUNCH DIAMETER 75MM

1. BASIC MODEL
2. NEW CONSTITUTIVE EQUATIONS

FIGURE 9: FORCE DISPLACEMENT CURVE DURING STRETCH FORMING

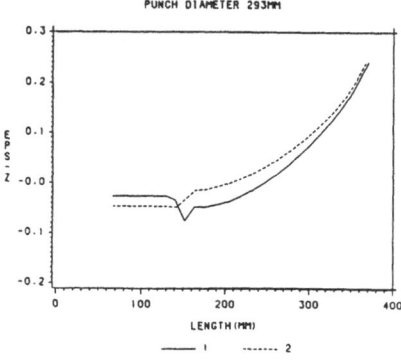

PUNCH DIAMETER 293MM

1. BASIC MODEL
2. NEW CONSTITUTIVE EQUATIONS

FIGURE 10: STRAIN DISTRIBUTION AFTER DEEPDRAWING

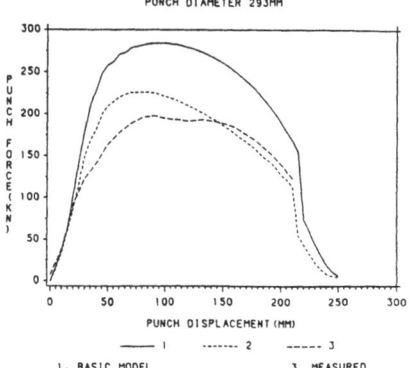

PUNCH DIAMETER 293MM

1. BASIC MODEL
2. NEW CONSTITUTIVE EQUATIONS
3. MEASURED

FIGURE 11: FORCE DISPLACEMENT CURVE DURING DEEP DRAWING

120

-The choice of yield criterion effects strain in the stretch forming
 region as well as in the deep drawing region. Because the chosen parts
 are not too critical, it is very hard to judge which one is most
 favourable.
-A low coefficient of friction is favourable. In case of the small part
 only boundary lubrication occurs, but in the large part mixed lubrication
 occurs according to our friction model.

The improvement of using the proposed equations in the model in
respect of the basic model is shown in figures 7 to 12 and in some cases
experimental data are available. The basic model has been taken and one
model with another set of constitutive equations:
-Strain hardening behaviour is described by Voce-2 relation (figure 5).
-Influence of strain rate has been taken into account.
-The yield criterion Hill-2 is used from table 3.
-Emmens' model for the description of friction is used.

In case of the stretched formed part this alternative gives no real
improvement in predicting strain distribution, but it gives a better
prediction of the displacement at which instability occurs, as we can see
in the force displacement curve (figure 9). The strain distribution in the
deep drawn part is described in a better way with the alternative model,
particularly if we look at the two minima in the measured strain
distributions (figure 6). The calculated force displacement curves of the
deep drawn parts show a more pronounced maximum at lower displacements
(figures 7 and 11). A possible explanation is the increase in the
coefficient of friction during the experiments which is also found in the
strip tests used for the determination of the constants in the friction
model. This effect could not be taken into account in a physically based
model.

6. Conclusions

-Differences in strain hardening behaviour after equivalent strains which
 exceed the uniform strain of the tensile tests, affect strain
 distribution of stretched formed parts considerably; in case of deep
 drawing of not too critical products this effect is very small.

-Strain distribution during stretch forming is affected favourably, by
 taking into account the influence of strain rate, because stretching
 from the flat bottom will be increased.In a deep drawing process, a high
 influence of strain rate is unfavourable, because it increases the
 necessary force of the draw-in process in the blankholder area.

-The type of yield criterion affects the strain distribution of stretched
 formed parts, which is related to the difference between the plane strain
 stress and the equibiaxial stress. In case of the deep drawing process
 the minimal thickness is not affected -only some differences in the
 stretch form region and draw-in region are found.

-The proposed model for the description of the friction coefficient gives
 better predictions in case of the large deep drawn part
 (diameter = 293 mm).

7. Future work

For a good description of the sheet forming process the material behaviour
has to be investigated:
-at large effective strains
-at different strain rates
-at multi-axial stress states
This requires a good balanced research program in which choices have to be
made to realize a good simulation model within reasonable time limits. In
a next publication the attention is focussed on this subject. As a start
of this future work plane-strain tensile tests are being performed to get
more information on material behaviour under multi-axial stress state.

The description of friction which is proposed here describes the
lubricant phenomena under steady-state-like conditions. Moreover, changing
of the surface roughness during the deformation process will influence
friction. This subject has to be investigated too.

8. Acknowledgements

The author wishes to thank Prof.Ir. P. Jongenburger, Dr.Ir. J. Huetink
(University of Twente) and Ir. P. Vreede (University of Twente) for
discussion.

9. References

[1]   Woo, D.M. - Analysis of the cup drawing process;
        J. Mech.Eng.Mech. (1964) 6, 2, p.116

[2]   Hill, R. - The mathematical theory of plasticity

[3]   Hill, R. - Theoretical plasticity of textured aggregates;
        Math.Proc.Camb.Phil.Soc. 85 (1979), p. 179

[4]   Wagoner, R.H. - Constitutive equations for sheet forming
        analyses; Computer Modelling of Sheet Metal Forming Process,
        p. 77, AIME 1985

[5]   Emmens, W.C. - The influence of surface roughness on friction -
        accepted for publication at the 15th IDDRG Biennial Congress,
        Dearborn, Michigan, USA, May 16-18, 1988

[6]   Bergstrom, Y. - A dislocation model for the stress strain
        behaviour of polycrystalline α-Fe with special emphasis of the
        densities of mobile and immobile dislocations; Mat.Sci. En. 5
        (1969), p. 1983

[7]   Van Liempt, P. - Internal report Hoogovens 1985

[8]   Massoni, E., et.al. - A finite element modelling for deep
        drawing of thin sheet in automotive industry; Advanced
        Technology of Plasticity, Stuttgart 1987, p. 615

[9]   Vreede, P. & Huetink, J. - To be published yet

INFLUENCE OF ANISOTROPY IN SHEET METAL FORMING

E. Doege & M. Seydel
Institut für Umformtechnik und Umformmaschinen
University of Hannover, W. Germany

ABSTRACT: Numerical analysis is carried out in order to
investigate the influence of plastic anisotropy on sheet
metal forming processes such as stretching and deep-drawing.
The analysis is based on the deformation theory and Hill's
theory of plastic anisotropy. The numerical results are
compared to experiments. It is shown that the neglect of the
anisotropy in the simulations leads to inaccurate results.
Using the yield stresses and anisotropic parameters from
tensile and bulge tests, the earing in deep-drawing of a
circular blank is calculated correctly.

1. Introduction

The deformation behaviour of sheet metal under press forming
is affected by the properties of the material. The anisotropy of the sheet generally induced by rolling is a desired
feature improving the plastic formability. In the mathematical analysis of plasticity taking into account the anisotropy, the fundamental theory was postulated by Hill /1/. The
application of this theory to sheet material presumes the
principal axes of anisotropy to coincide with the local
Cartesian axes of the sheet with the rolling and the thickness direction as references. The texture due to the rolling
justifies this reduction to the orthotropic case. As Hill's
yield function is quadratic in stress components without
linear terms, it is easy to deal with and in preference
implemented in computer programs /2,3/. But it can only be
applied to materials forming two or four ears in an axisymmetrical deep-drawing process and having the ratio of
width-to-thickness strain, labelled as r-value, greater than
unity. In order to overcome this latter restriction, Hill
proposed a new yield criterion in 1978 /4/.
    Press forming operations can be distinguished between
stretching and deep-drawing which differ entirely in defor-

J. L. Chenot and E. Oñate (eds.), Modelling of Metal Forming Processes, 123–130.
© *1988 by Kluwer Academic Publishers.*

mation mechanism. The role played by anisotropy in stretch
forming has been discussed by Hill himself theoretically
/5/, whereas several workers, among others, more recently
treated the stretching /2,6,7,8/ and deep-drawing /8,9,10/
under the numerical and experimental aspects of the anisotropy.
     The r-value gives a useful measure of a materials
thinning resistance. In an approximation equating the in-
plane properties of the sheet such as the axisymmetric one,
an average value is adopted for the normal anisotropy.
An evaluation of Hill's first yield criterion for anisotropic materials provides the tensile yield stress in the
thickness direction. Hill also has given an expression for
the effective stress determined from the fact that equal
amounts of plastic work obtained during an uniaxial test in
different directions must give rise to equal effective stress
levels. In this way, a test on a tensile specimen cut at 45°
with respect to the rolling direction also supplies the
ratio of the yield shear stress to the reference yield
stress /2/. Thus, all required yield stress ratios can be
obtained by three tensile tests on strips cut at 0°, 45° and
90° with respect to the rolling direction. This method
assumes these ratios to be constant during plastic deformation, i. e. the state of anisotropy remains effectively the
same.
     In this investigation, Hill's original yield function
is applied to study the effects of the anisotropy upon the
deformation behaviour of stretching and deep-drawing in
numerical simulations which are compared to experiments.

2. Stretching

It has become common to obtain work-hardening characteristics over simple strain paths and attempt correlation
between them by means of the anisotropic theory /8/. The
most widely used uniaxial tension test provides the flow
curve and the anisotropy parameter r. However, this method
gives pairs of corresponding values of yield stress versus
plastic strain only up to the limit of uniform deformation.
This range has to be extended because strain during sheet
metal forming reaches higher levels. Stretching like the
hydraulic bulge test with a balanced biaxial stress state is
suitable for this purpose.
     From the measured data of the bulge test, the flow
stress is evaluated by the equilibrium conditions of an
axisymmetric shell. By reason of the equivalence between a
through-thickness compression and a balanced biaxial tension, the following relationship is valid:

$$(Z/X)^2 = (B/X)^2 = (1+r)/2 \tag{1}$$

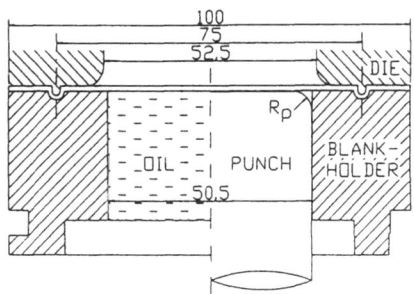

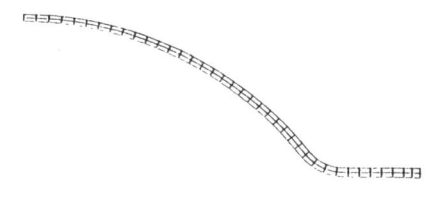

Fig. 1: Set-up for stretching   Fig. 2: Deformed mesh of
      by oil pressure or punch   hydraulic bulging

with the through-thickness yield stress Z, the uniaxial and
the biaxial in-plane yield stress, X and B, resp..
     Without regard to the anisotropy, the flow curve from
the bulge test rises to nearly 20 % higher levels than that
which results from the tensile test. The mechanical properties of the investigated deep-drawing quality steel are
listed in Table 1. Based on (1), the flow curves can be
transposed that they approach each other /8/. Fig. 1 shows
the experimental set-up which serves also for stretch-for-
ming by interchangeable punchs with different shapes.
     For the calculation, the ABAQUS program has been used.
It requires pairs of plastic strain and stress values to
define the plastic material behaviour. The data input are
given in Table 2. These data were identified numerically in
order to fit the measured curve of pressure versus pole
displacement as shown in Fig. 3. Fig. 2 shows a deformed
mesh composed of axisymmetric bilinear continuum elements.
An alternative calculation using axisymmetric shell elements
revealed no significant deviation. The bulge test simulation
without consideration of the anisotropy does not reach the

Table 1: Mechanical properties of the investigated deep-dra-
        wing quality steel St 1303, thickness 1mm, from
        tensile tests.

Tensile test angle with respect to the rolling direction	Proof stress $R_{p0.2}$ (MPa)	Ultimate tensile strength R_m (MPa)	$r(\varepsilon=20\%)$
$0°$	182	291	1.699
$45°$	196	301	1.215
$90°$	191	290	2.098

126

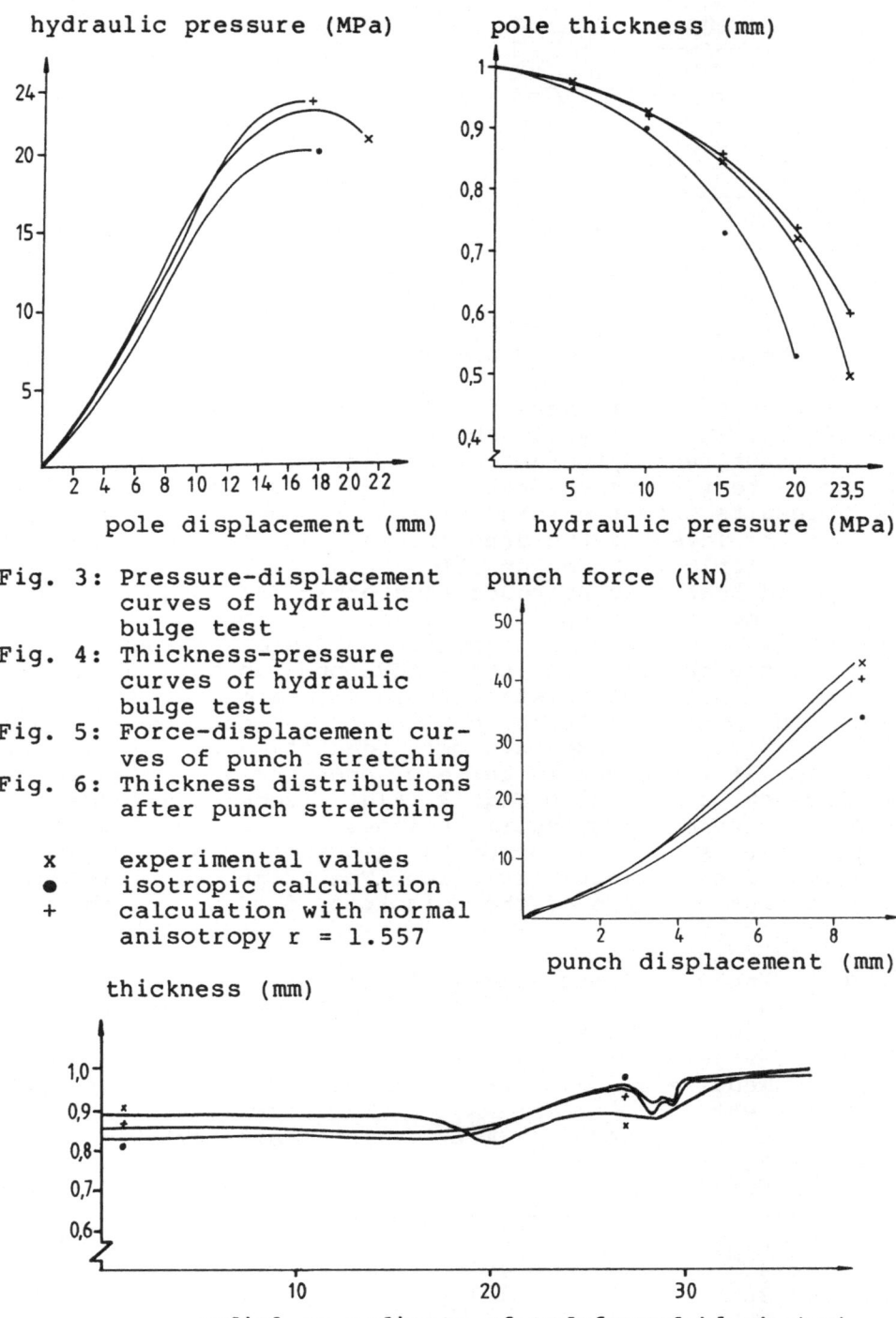

hydraulic pressure (MPa)

pole displacement (mm)

pole thickness (mm)

hydraulic pressure (MPa)

Fig. 3: Pressure-displacement curves of hydraulic bulge test

Fig. 4: Thickness-pressure curves of hydraulic bulge test

Fig. 5: Force-displacement curves of punch stretching

Fig. 6: Thickness distributions after punch stretching

x   experimental values
●   isotropic calculation
+   calculation with normal anisotropy r = 1.557

punch force (kN)

punch displacement (mm)

thickness (mm)

radial co-ordinate of undeformed blank (mm)

Table 2: Material definition input into ABAQUS:

Young's modulus: 206 GPa   Poisson's ratio: 0.3

plastic behaviour: plastic strain (%) versus stress (MPa)

ε	0	3	6	8	10	13	19	25	32	45	60	80
σ	237	269	299	319	339	379	422	446	467	470	472	476

yield stress ratios for anisotropic calculation:

Calculation	X/X_o	Y/X_o	Z/X_o	T_{xy}/T_o
only normal anisotropy	1	1	1.13	1
2D deep-drawing 0°	1	1	1.13	1
2D deep-drawing 45°	1.08	1	1.13	1
3D deep-drawing	1	1	1.13	1.07

measured maximum pressure, in contrast to the anisotropic
calculation (Fig. 3). The same discrepancy between isotropic
and anisotropic calculation is found in the thickness
decrease (Fig. 4). Moreover, all strains in the isotropic
calculation exceed the measured values. Figs. 5 and 6 show
the thickness distributions and the force-displacement-plots
measured and calculated in stretching with a punch with 9mm
edge radius. In this case, stresses and strains are strongly
influenced by friction.

3. Deep-Drawing

The deep-drawability of the sheet measured by the limiting
drawing ratio increases in a significant correlation with
the normal anisotropy /8/. Besides, compared to the stretching process, the planar anisotropy has some influence on
deep-drawing operations. Drawing a circular blank of the
material described in Table 1, ears form in the rolling
direction and its transverse. In these directions, the proof
stress and the ultimate tensile strength have minimal
values. By this reason, the strains reach maximal values in
these directions. Ears and hollows form at those points of
the rim where the radial and circumferential directions are
the principal axes of strain increments. Assuming plane
stress state, the only non-zero stress component at the
margin is the hoop stress. This fact entails that the

principal axes of stresses and strains coincide at the
points of ears and hollows. Hence, taking advantage of these
symmetry conditions, it is sufficient to model a 45° sector
of the circular blank for a three-dimensional calculation.
    Simulations of a deep-drawing process have been carried
out and compared to experiments of which the set-up is shown
in Fig. 7. In axisymmetric calculations the cross-sections
in the rolling direction and in 45° are described by the
different tensile yield stress ratios ensuing from Table 1
and listed in Table 2. The results are also shown in Fig. 7.
Obviously, this method cannot replace a full three-dimensional analysis. Thus, a 45° sector of the blank is idealized
using 3D-shell elements with full integration and five
integration points in thickness direction. In order to
reduce the computing time, the element mesh is rather
coarse. Nevertheless, the calculated results visible in
Figs. 7 to 9 show a good conformity to the reality. In the
normal view (Fig. 8), the deformed mesh is superimposed on
the original one. It may be easily seen that an ear forms in
0° direction and that the circumferential displacement
components in the flange are directed towards this ear. The
original mesh has a symmetry at the 22.5° line in order to
exclude all mesh influences on the result.

4. Conclusion

The numerical simulations of stretching and deep-drawing
show the important influence of plastic anisotropy on sheet
metal forming. The neglect of the anisotropy in the calculation leads to inaccurate results in comparison with

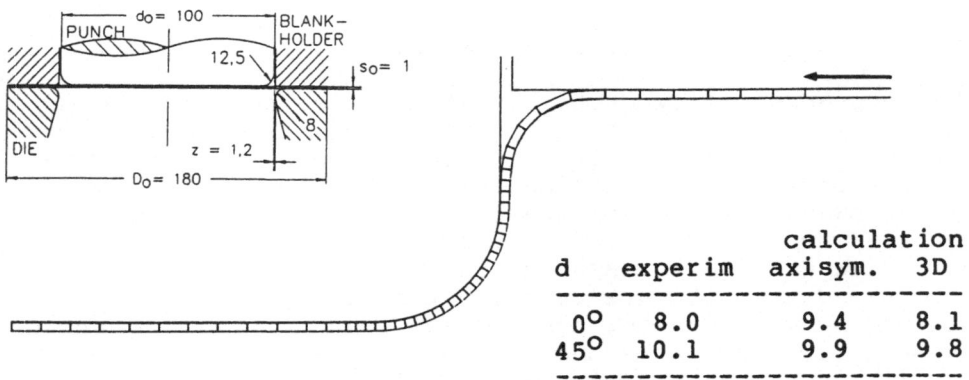

d	experim	calculation	
		axisym.	3D
0°	8.0	9.4	8.1
45°	10.1	9.9	9.8

Fig. 7: Set-up for deep-drawing; deformed mesh of axisym-
        metric calculation with listed values of rim dis-
        placements d, indicated by an arrow in the figure.

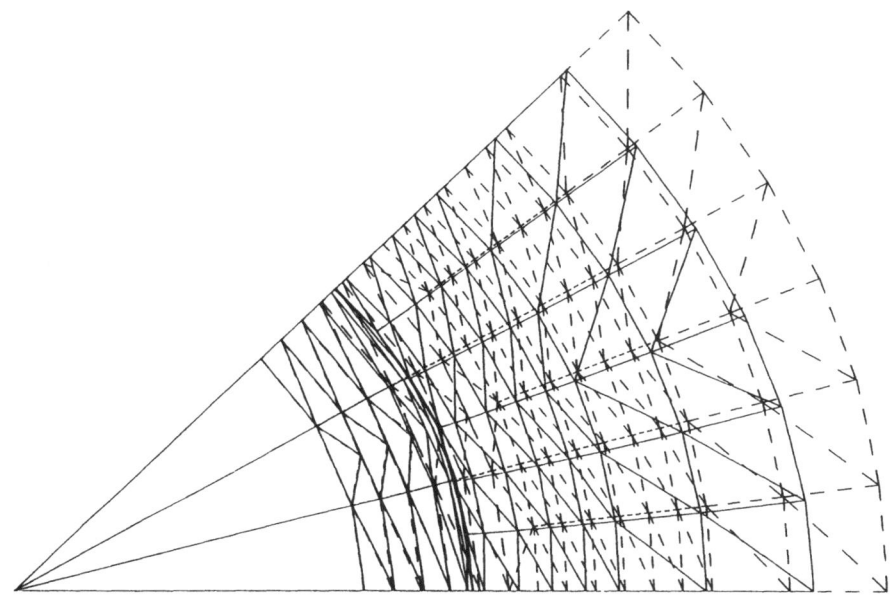

Fig. 8: Nodal displacements showing the ear at the 0° line
(dashed lines = original mesh)

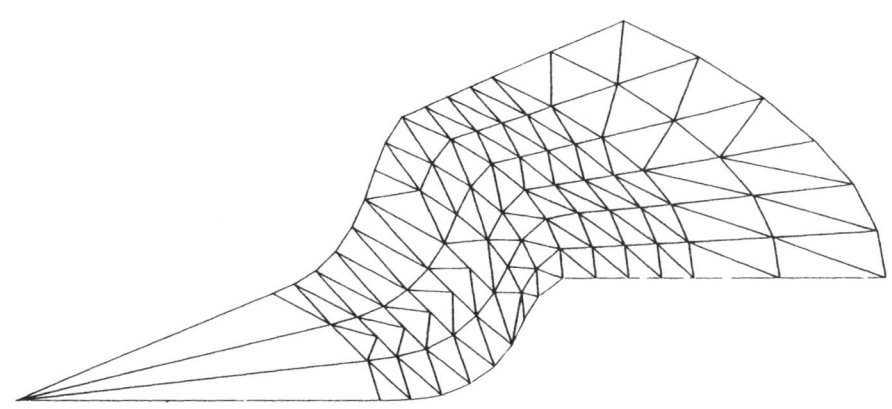

Fig. 9: Deformed mesh of the deep-drawn cup,
modelled by a 45° sector

experiments. The material data can be identified by the hydraulic bulge test because tensile tests are not sufficient. The obtained yield stress values are used for two- and three-dimensional calculations. This way is a first step to a complete simulation of deep-drawing processes involving anisotropic phenomena such as earing.

Acknowledgements

This study is part of a research project supported by Deutsche Forschungsgemeinschaft under grant of SFB 300. We also acknowledge gratefully the provision of ABAQUS to HKS and the valuable co-operation with its German agency ABACOM.

References

/1/ Hill, R.: The Mathemetical Theory of Plasticity. Oxford Univ. Press, 1950

/2/ Ekmark, B.: On Large Strain Theories in Sheet Metal Forming. Doctoral Thesis Lulea University, 1983

/3/ Hibbitt, H.D. et al.: ABAQUS Theory Manual. Hibbitt, Karlsson and Sorensen Inc., Providence R.I., 1985

/4/ Hill, R.: Theoretical Plasticity of Textured Aggregates. Math. Proc. Camb. Phil. Soc. 85 (1979), pp. 179-191

/5/ Hill, R.: A Theory of the Plastic Bulging of a Metal Diaphragm by Lateral Pressure. Philosophical Magazine 41 (1950), pp. 1133-1142

/6/ Fuchizawa, S.: Influence of Plastic Anisotropy on Deformation of Thin-walled Tubes in Bulge Forming. In: Advanced Technology of Plasticity 1987, Vol. 2 (Ed. K. Lange) Springer 1987, pp. 727-733

/7/ Park, J. J.; Oh, S. I.; Altan, T.; Analysis of Axisymmetric Sheet Forming Processes by Rigid-Viscoplastic Finite Element Method. Trans. ASME J. Eng. Ind. 109 (1987) pp. 347-354

/8/ Mellor, P. B.: Experimental Studies of Plastic Anisotropy in Sheet Metal. In: Mechanics of Solids, The Rodney Hill 60th Anniversary Volume (Eds. H. G. Hopkins, M. J. Sewell), Pergamon Press 1982, pp. 383-415

/9/ Kawai, N. et al.: Effects of Punch Cross-Section on Deep-Drawability of Square Shell of Aluminium Sheet. Trans. ASME J. Eng. Ind. 109 (1987) pp. 355-361

/10/ Alworth, H. M. et. al.: The Effects of Second Cold Reduction on the Plastic Anisotropy, Crystallographic Texture and Earing Behaviour. J. Applied Metalworking 4 (1987) pp. 327-330

EXPERIENCE WITH EXPLICIT AND IMPLICIT FINITE ELEMENT PROGRAMS FOR SHEET FORMING ANALYSIS

R. H. Wagoner, E. Nakamachi, and Y. Germain[*]

Department of Metallurgical Engineering
The Ohio State University
Columbus, Ohio, 43210, U. S. A.

[*]Centre de Mise en Forme des Materiaux
Ecole des Mines de Paris
06560 Valbonne, France

1. INTRODUCTION

Two finite–element modeling (FEM) programs have been developed to simulate three–dimensional sheet forming operations. The rigid–visco–plastic (R–P) [1] and elastic–plastic (E–P) [2] programs make use of constant–strain triangular elements, membrane mechanics, Hill's yield surfaces [3,4], updated Lagrangian formulations, and a Coulomb model of friction at tool interfaces. The R–P program is based on a fully nonlinear treatment (geometry, material, contact, friction) at each time step while the E–P program is completely linearized at each time step, with all updating taking place only at the next increment. Operational experience illustrates the advantages and disadvantages of the two approaches.

2. RIGID–PLASTIC (R–P) PROGRAM FEATURES

The R–P program is formulated in terms of nodal displacements and is nonlinear and implicit. Convergent solutions are consistent with material laws, contact conditions, and geometric nonlinearities. Several unusual features are introduced.

2.1. Formulation

The R–P program utilizes Hill's new theory of anisotropy [3] and assumes a special, proportional plastic path over each time increment. The path is characterized by constant ratios of eigenvalues of the deformation rate tensor (D) and by the eigendirections remaining parallel to fixed material lines in the interval. This assumption allows the effective strain increment to be calculated very simply because the path represents the minimum of plastic work to reach the new state. Numerical integrations are therefore avoided while the material hardening behavior is obeyed precisely.

The effective strain rate is assumed to be constant within each time step, such that $\bar{\epsilon}$ is simply equal to $\Delta\bar{\epsilon}/\Delta t$. The simplicity of the strain path makes incorporation of rate effects straight–forward. Because elastic effects are neglected, the total stress (not stress rate) may be formulated directly as a function of strain increment. This approach has the advantage that equilibrium is automatically satisfied at each time step (rather than the weaker, rate–of–equilibrium condition) but also demands that nonlinear equations be solved. The nonlinear equations are generated with the incremental nodal displacements as the unknowns and are

J. L. Chenot and E. Oñate (eds.), Modelling of Metal Forming Processes, 131–138.
© *1988 by Kluwer Academic Publishers.*

solved using a special implementation of the Newton–Raphson (N–R) technique starting from a trial solution equal to the solution of the previous time step.

The actual formulation in terms of total stress (or force) is made on the basis of equating internal virtual work increments with external work increments:

$$\delta W_i = d(\Delta W_i) = \int_{A_0} \bar{\sigma} \frac{\partial \Delta \bar{\epsilon}}{\partial \Delta u} d(\Delta u) \, dA = F_i d(\Delta u) = F_e d(\Delta u) \tag{1}$$

2.2 Contact Algorithm

At the beginning of each solution step, a punch travel increment is applied and a trial solution for all nodal displacements is assumed. The nodal positions are then checked for any penetration into the tools. All penetrating nodes are projected onto the closest tool position before the first N–R iteration. After each iteration, newly–penetrating nodes are re–projected in the same manner. Therefore, subsequent trial displacements are not obtained entirely from the N–R scheme but their positions involve contact conditions as well.

Coulomb friction conditions are imposed in a similar manner. For each trial set of nodal displacements, a consistent internal force vector at each node may be found . The component of this force normal to the mesh surface represents the external force (from tool contact) required to maintain local equilibrium. Further, the tangential component of the contact force acts opposite to the direction of nodal displacement relative to the punch Δu_t with a magnitude proportional to the

normal component. Each of these quantities may be evaluated using the trial solution. Following Oden et al. [5], a modified version of Coulomb friction has been introduced that provides better numerical stability without serious reduction of accuracy:

$$F_t = -\mu \, \phi(\Delta u_t) \frac{\Delta u_t}{\|\Delta u_t\|} F_n \tag{2}$$

The function ϕ is unity at displacements greater than some tolerance d, but drops linearly to zero as $\Delta u_t \rightarrow 0$. The modification produces small slip for all nodes at all times, replacing the usual slip–stick condition.

3. ELASTIC–PLASTIC (E–P) PROGRAM FEATURES

The E–P program is formulated step–wise linearly in a convected coordinate system. Time discretization is based on simple use of nodal velocities at time t to update nodal displacements at time $t + \Delta t$. Strain rate effects are not considered.

3.1 Formulation

The E–P program utilizes Hill's old theory of normal anisotropy and solves for a velocity field at each discrete time. All terms are linearized at an instant in time and nodal positions are updated by assuming a constant–velocity path (i.e. $\Delta u = v\Delta t$). Since the linearization is complete, all of the equations become homogeneous in time (within one time step) and either v or Δu may be considered the principal unknown.

Assuming a constant velocity within a time step leads to an incremental variational theorem of the following form:

$$\int_{V_0}(\Delta\tau^{\alpha\beta}+\tau^{\alpha\beta}\Delta u^k|_k+\tau^{\alpha\gamma}\Delta u^\beta|_\alpha)\ \delta(\Delta u)_\beta|_\alpha dV = \int_{A_0}\Delta T^\alpha\delta(\Delta u)_\alpha dA \qquad (3)$$

where ΔT^α is the component of the incremental external surface force. This equation may be written in linear, FEM form as follows:

$$k\ \Delta u =(\ k^{im}_{jn} + k^{im}_{\sigma jn} + k^{im}_{g jn} + r^{im}_{s jn})\Delta u^j_m = \Delta p^i_n \qquad (4)$$

where, k^{im}_{jn} is the incremental stiffness matrix, $k^{im}_{\sigma jn}$ the initial stiffness matrix, $k^{im}_{g jn}$ the initial–rotation matrix, and $r^{im}_{s jn}$ is the nonsymmetric incremental loading matrix.

The total incremental displacement Δu is given by the sum of the known punch travel increment ΔU and an unknown increment Δu^* tangent to the tool surface. For non–contacting nodes, $\Delta U = 0$. Therefore the stiffness equation may be written as follows:

$$K\ \Delta u = K\ (\Delta U+\Delta u^*) = 0, \quad K\ \Delta u^* = -K\ \Delta U \qquad (5)$$

3.2 Contact Algorithm

The incremental contact force is expressed as the sum of a normal force and tangential (friction) force. The normal force can be evaluated explicitly by employing local equilibrium conditions in a direction normal to the membrane defined by the tool normal at the centroid of each element. Only elements which have all three nodes in contact with the tools are considered contacting. The increment of normal force depends only on tool curvature, tool normal, and internal stress increment.
The tangential component of the contact force is proportional to the normal component if the node is slipping (Coulomb friction). The direction of the tangential component is taken opposite to the velocity of each node during the previous time step and contact and slip/stick conditions are only updated at the end of the current time step.

The explicit nature of the contact pressure scheme depends on the current stress (and previous nodal velocity), so any accumulated inaccuracy in current stress can lead to a large variation of pressure, and ultimately, node position. The elemental nature of the pressure calculation (with subsequent assignment to nodes) requires that all three nodes be in contact for stability.

4. HEMISPHERICAL PUNCH–STRETCH (HPS) SIMULATIONS

The tool and FEM mesh geometries for the hemispherical punch–stretch (HPS) simulation are shown in Figs. 1 and 2. Although the problem is axisymmetric when using circular blanks, it was simulated using four–fold symmetry to investigate the general formulation of each program.

The formulation of the R–P program has been tested extensively (previously in 2–D form) [6] and its accuracy verified for analytically solvable problems as well some simple experimental geometries [7]. Fig. 3 compares experimental and R–P FEM strain distributions for HPS of circular blanks and verifies the formulation with an uncertainty associated only with an arbitrary friction coefficient.
The dashed lines (Case 2) in Fig. 4a show strain distributions originally calculated using the E–P program and corresponding to the HPS simulation with circular blanks. These are in marked disagreement with both experiment and R–P simulations. Careful investigation revealed that a minute change in constitutive equation disrupted the smooth elastic–plastic transition (the plastic modulus was larger than the elastic one at the transition), resulting in large accumulated deviations from the proper constitutive law. Fig. 4b shows the material hardening law that resulted by this accumulated error and the proper one after a small correction at the elastic–plastic transition (solid lines). The non–iterative nature of the E–P code allowed no direct method for revealing this major inaccuracy.

Fig. 5 shows typical comparisons for circular–blank HPS between the E–P and R–P programs after smooth elastic–plastic transitions were assured. Computation times for various step sizes and friction coefficients are shown in Table I the dependence of solution on step size for the E–P program is illustrated in Fig. 6. The rough appearance of strain distributions is a result primarily of the explicit contact algorithm. The medium step size was used for comparing with the R–P program. The E–P program generally requires about eight times the computer time to achieve similar accuracy.

Figs. 5a and 5b show good agreement for circular blanks but the difference between R–P and E–P results for the 25mm–wide strip is significant. Analysis reveals that the main source of the difference is in the contact algorithm. The narrow strips, similar to a tensile test, are much more sensitive to contact than the axisymmetric case.

Fig. 7 shows E–P results for an intermediate blank width of 76mm. The R–P program was unable to obtain convergent solutions beyond very small punch heights. The problem lies in the longitudinal edge regions which are unconstrained and out of contact with the punch. At an early punch height, a bifurcation of solution occurs, similar to "oil–can" or "snap–through" phenomena and the membrane program cannot simulate this bending–based event. Curling–up of these edge regions is seen in experiments with this blank width. The E–P program marches forward through this instability, obtaining unrealistic solutions at the edges, Fig. 8. In spite of the membrane indeterminacy at the edges, the center region can be simulated, yielding what appear to be reasonable results. This is a clear benefit to the E–P formulation, namely that it continues to obtain solutions even in the face of physical indeterminacy.

5. CONCLUSIONS

Explicit E–P calculations require small time steps and smoothly–varying boundary conditions to obtain accurate results. The implicit R–P results are relatively insensitive to most numerical parameters but convergence is greatly influenced. The E–P program is robust in the sense that a solution, of unknown accuracy, is always obtained. The R–P is accurate in the sense that any solution is consistent with material, equilibrium, and boundary constraints, within the time and spatial discretizations employed. Each approach may have benefits depending on the trade–off desired between robustness and accuracy.

ACKNOWLEDGEMENT

The authors would like to thank the Center for Net Shape Manufacturing at The Ohio State University and the Chrysler Corporation for supporting this work.

REFERENCES

1. Y. Germain, K. Chung, and R. H. Wagoner, "A Rigid–Visco–Plastic Finite Element Program for Sheet Metal Forming Analysis", *Int. J. Mech. Sci.* accepted for publication)
2. E. Nakamachi, NUMIFORM 86 (Edited by K. Mattiasson, A. Samuelsson, R.D. Wood and O.C. Zienkiewicz, Belkma, Rotterdam), p.333 (1986)
3. R. Hill, *Math. Proc. Camb. Phil. Soc.*, 85, p.179 (1979)
4. R. Hill, *The Mathematical Theory of Plasticity*, Oxford University Press. London.(1950)
5. J. T. Oden and E. B. Pires, *J. Appl. Mech.*, 50, p.67 (1983)
6. K. Chung and R. H. Wagoner, *Metall. Trans. A.*, 19A, p.293 (1988)
7. J. R. Knibloe, Master's Thesis, The Ohio State University (1988)

Table I   **Step numbers and computation times(DEC VAX 8550)**

	Friction coefficients μ	R–P FEM			E–P FEM	
		number of steps	number of iteration (average)	CPU (sec)	number of steps	CPU (sec)
Disk	0.0	20	4–5	428	(s) 4576 (m) 1222 (l) 251	16460 4623 1196
	0.25	18	4–6	395	(s) 4113 (m) 1018 (l) 211	14826 3903 1055
	0.50	16	4–7	360	(s) 3950 (m) 1005 (l) 201	14251 3857 1020
Strip 1 inch	0.0 0.25 0.50	30 28 26	4–9 4–10 4–13	744 680 610	(m) 1217 (m) 1216 (m) 1216	4144 4140 4140
Strip 3 inch	0.0 0.25 0.50				(m) 1301 (m) 1171 (m) 928	13290 11964 9486

* (s) = small step, (m) = medium step, (l) = large step

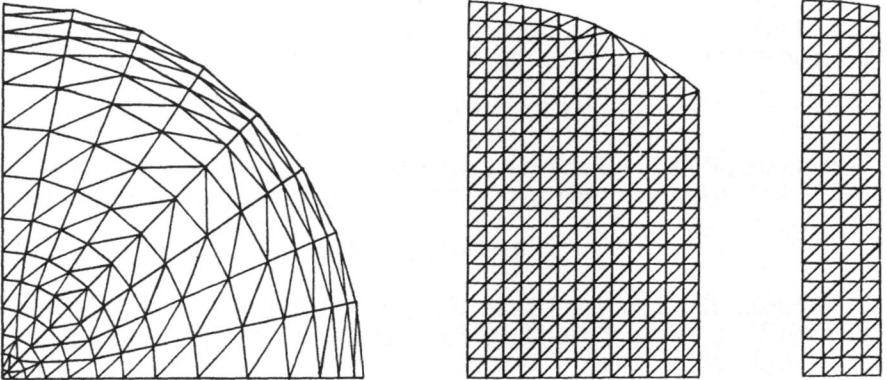

Fig.1 —   FEM mesh for three blanks, hemispherical punch—stretch simulation.

59.18 mm

52.83 mm

Die R=6.35mm

Sheet

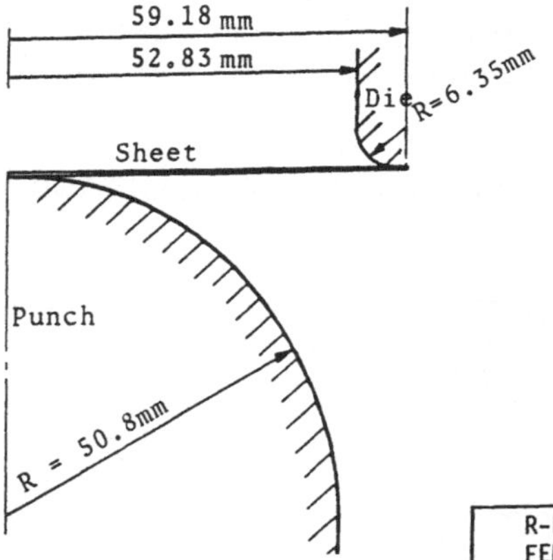

Fig.2 —   Tool geometry for hemispherical
            punch—stretch simulation.

Punch

R = 50.8mm

Fig.3 —   Comparison of experimental and FEM
            strain distributions for hemispherical
            punch stretching of disks.

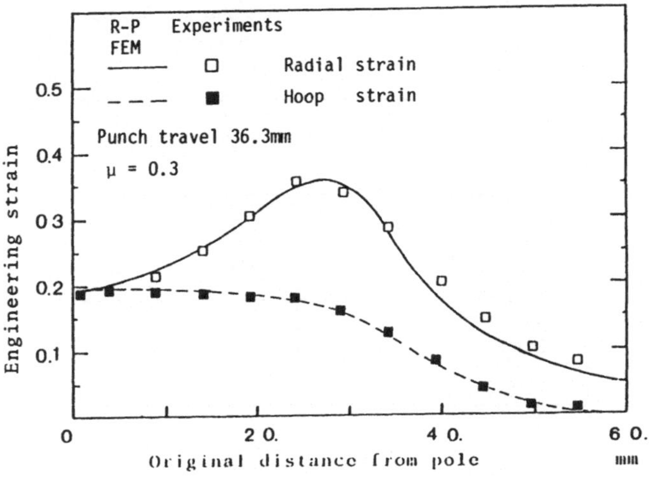

R-P   Experiments
FEM
                □   Radial strain
        - - -   ■   Hoop   strain

0.5

Punch travel 36.3mm

0.4 μ = 0.3

0 3

0.2

0.1

Engineering strain

0   2 0.   4 0.   6 0.
            Original distance from pole   mm

137

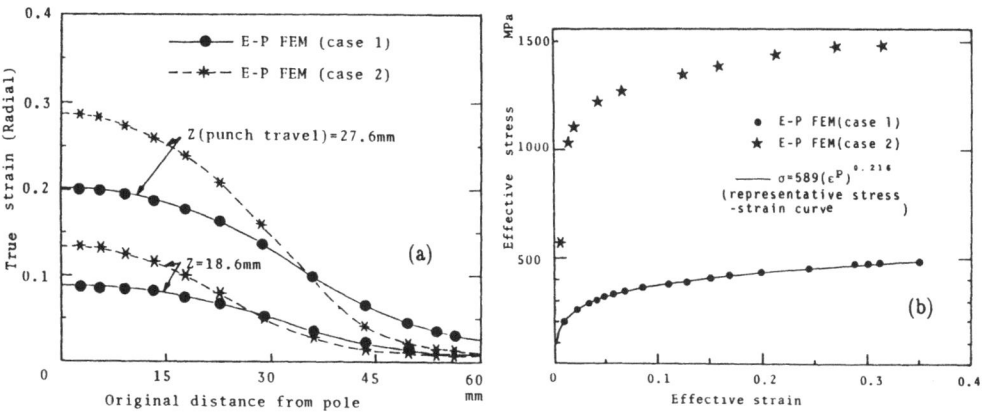

Fig.4 — Comparison of E–P FEM results before (Case 1) and after (Case 1) small change in constitutive equation.
(a) strain distributions
(b) adherence to material hardening law.

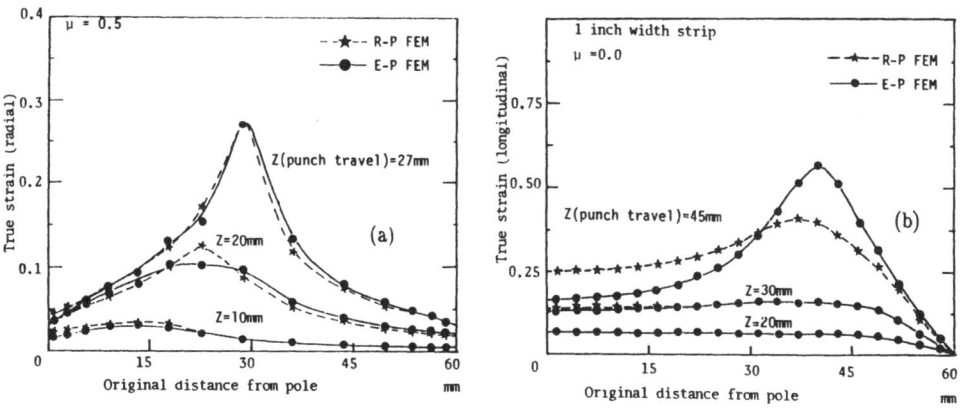

Fig.5 — Comparison of E–P and R–P strain distributions for hemispherical punch stretching.
(a) circular blank
(b) 25mm–wide blank

138

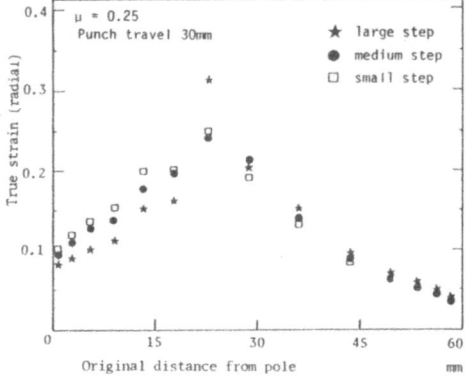

Fig.6 — Dependence of calculated strain distribution on step size, E–P FEM.

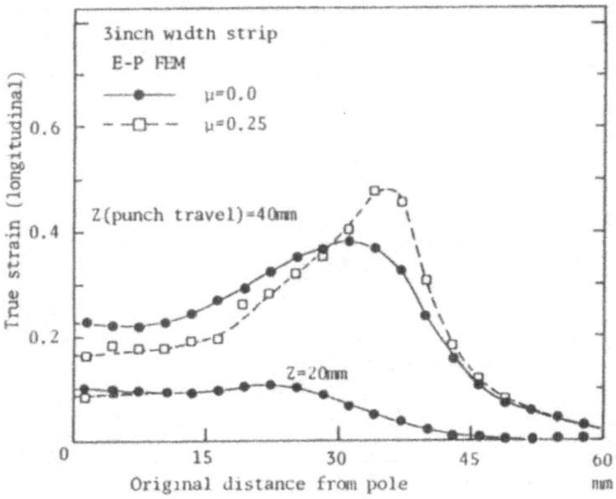

Fig.7 — Strain distribution along axis of 76mm–wide blank, E–P FEM.

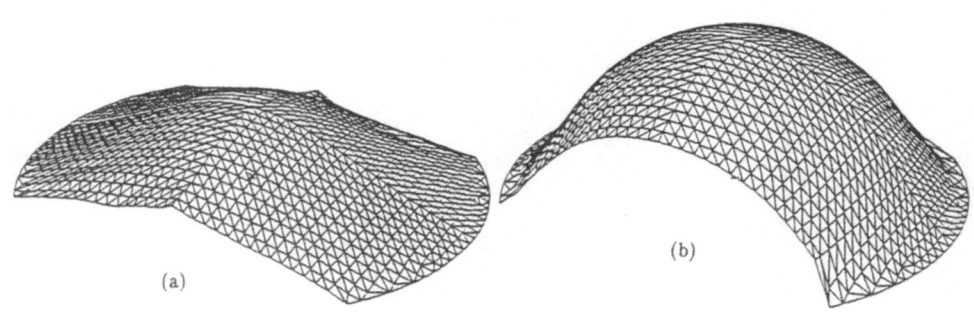

Fig.8 — Deformed FEM mesh for 76mm–wide blank, E–P FEM.
    (a) 20mm punch height
    (b) 40mm punch height

INFLUENCE OF THICKNESS AND CURVATURE
ON THE FORMABILITY OF METAL SHEETS

Philippe PIERRE - Jean-Pierre CORDEBOIS
Laboratoire de Mécanique et Technologie
E.N.S. de Cachan / Université Paris VI / C.N.R.S.
61, avenue du Président Wilson 94230 Cachan - France

Abstract : An overall theory modelizing the onset of necking is presented. The first developments were made on the assumption that the fields considered were uniform (stresses, strains...). But this didn't allow us to take into account the influence of the thickness and curvature of sheets on necking. The same theory is used again but we introduce non-uniform fields. To validate this approach, numerical simulations are presented, and the results are compared with the relevant experiments.

I. Introduction.

In sheet metal forming, we know that a localization phenomenon appears at the point where strains are the largest. This is unlikely to ensure the good quality of pieces manufactured by deep drawing. This is the necking phenomenon.

The studies about necking are numerous and are not recent. They explain the onset of necking in different ways, and generally lead to a local function of stresses which is equal to zero when necking appears.

Jean-Pierre Cordebois and Pierre Ladevèze (L.M.T. Cachan) have propounded an approach in which necking is treated as a structural instability problem.

If we assume uniform fields on the volume considered, we obtain a local criterion which is only a function of stresses. In this case, we can't show the effects of thickness and curvature. The aim of this study is to propose a new local necking criterion which is a function of stresses and their gradients.

II. Necking criterion.

2.1. FORMULATION OF THE PROBLEM.

The problem of instability is treated for large strain elastoplasticity in the case of an orthotropic material without evolutive anisotropy, i.e. with isotropic hardening.

The structure will be considered unstable if, when we stop the loading, $\{ \dot{\mathcal{F}}_e \} = \{0\}$, the velocity field $\vec{v}_{(M)}$, the solution of the problem, is different from $\vec{v} \equiv \vec{0}$.

When the unique solution is $\vec{v} \equiv \vec{0}$, the structure is stable.

139

The $\{\dot{\mathcal{F}}_e\} = \{0\}$ condition is introduced in the time derivative of the principle of the virtual work rate. This leads to the following equation :

$$(1) \quad \forall \, \vec{v}^* \in \mathcal{V} \qquad \mathcal{A}(\vec{v}, \vec{v}^*, \sigma) = 0$$

$\mathcal{V}$ is the space of virtual admissible velocity fields.

$\mathcal{A}$ is an integral over the considered volume.

It has been shown that the problem can be written in the following way :

$$(2) \quad \text{The } \vec{v}_{(M)} \text{ solution minimizes } \mathcal{A}(\vec{v}, \vec{v}, \sigma) \text{ over the space } \mathcal{V}.$$

Indeed we remark that :

$\mathcal{A}(\vec{v}, \vec{v}, \sigma) \geq 0$ for small stresses σ and then is equal to zero only for $\vec{v} \equiv \vec{0}$.

$\mathcal{A}(\vec{v}, \vec{v}, \sigma)$ is strictly convex in the neighbourhood of $\vec{v} \equiv \vec{0}$.

$\mathcal{A}(\vec{v}, \vec{v}, \sigma)$ is no longer strictly convex in the neighbourhood of $\vec{v} \equiv \vec{0}$ as soon as the stresses become sufficiently high.

If we call $\vec{v}_m$ the velocity field which minimizes $\mathcal{A}(\vec{v}, \vec{v}, \sigma)$, then $\mathcal{A}(\vec{v}_m, \vec{v}_m, \sigma) = 0$ is the local criterion.

We can notice that each space $\mathcal{V}$ gives a different criterion.

If no boundary condition is prescribed, we obtain the lower mode.
If a maximum of velocity conditions is prescribed, we have a high mode.
Then we suppose that the solution to any other problems is situated between these two modes.

2.2. LOOKING FOR SOLUTIONS.

For convenience sake, to carry out the minimization, we introduce new variables which depend on the velocity field $\vec{v}_{(M)}$.

$$\beta = \text{Tr} [\, \sigma_D . H . K [\, D \,] \,]$$

$\vec{\omega}$ the vector associated with the antisymmetric tensor W.

$$\tilde{D} = D - \frac{\beta}{X} H [\, \sigma_D \,]$$

With : $\quad \sigma_D$ the deviatoric stress tensor

$\quad\quad$ H $\quad$ the operator of anisotropy which is associated with Hill's criterion of plasticity

$\quad\quad$ K $\quad$ the operator of elasticity

$\quad\quad$ W $\quad$ the spin rate tensor

$$X = \text{Tr} [\, \sigma_D . H . K . H [\, \sigma_D \,] \,]$$

$\quad\quad$ D $\quad$ the strain rate tensor

After some simplifications, we obtain the following form :

$$(3) \quad \mathcal{A}(\vec{V}, \vec{V}, \sigma) = \int_{\mathcal{D}} \{ \mathrm{Tr} \, [\, \mathbb{K}[\tilde{D}].\tilde{D}\,] + \vec{\omega} . A \, [\vec{\omega}] + \frac{2\beta}{X} \vec{\omega} . \vec{y}$$

$$+ \frac{\beta^2}{X(1+gX)} - \frac{\beta^2}{X^2} \, \mathrm{Tr} \, [\, \sigma . H[\sigma].H[\sigma]\,] \} \, d\mathcal{D}$$

The constitutive equations employed are : $\begin{cases} D_e = \mathbb{K}[\sigma^J] \\ D_p = g(p) < \dot{\sigma}_{II} > H[\sigma] \end{cases}$

and $A = \mathrm{Tr} \, [\, \sigma] \, 1 - \sigma$

$\vec{y}$ is the vector associated with the antisymmetric tensor $\sigma . H[\sigma] - H[\sigma] . \sigma$

The equation (2) is solved on the assumption of uniform fields. We minimize the expression with respect to β, $\vec{\omega}$ and $\tilde{D}$ which are considered as independent variables.

The low and high mode are represented, respectively, by the following equations :

$$\begin{cases} D(\sigma) = 0 \\ D(\sigma) + L(\sigma) = 0 \end{cases}$$

All the developments and Forming Limit Diagrams drawn with these criteria can be found in <1>.

The constitutive equations are written in <2>.

III. Influence of thickness and curvature.

The study summed up in the above paragraph doesn't allow us to show the influence of thickness and curvatures on necking. But in <3>, <4> and <5>, we can see that, experimentally, these parameters have an effect on necking (fig.1).

We then use again the same approach as in the previous paragraph. This time, the fields depend on the z variable in the thickness.

The variable z varies between $-\frac{h}{2}$ and $\frac{h}{2}$.

We can show that $d\mathcal{D} = \det (\mu) . d\Sigma_m . dz$, where $\mu = 1 + z . \frac{\partial \vec{N}}{\partial m}$, and $\frac{\partial \vec{N}}{\partial m}$ is the curvature operator of the middle surface Σ_m. Then,

$$\mathcal{A}(\vec{V}, \vec{V}, \sigma) = S(\Sigma_m) \int_{-\frac{h}{2}}^{\frac{h}{2}} \det (\mu) \, \{ \mathrm{Tr} \, [\, \mathbb{K}[\tilde{D}].\tilde{D}\,] + \alpha (\beta, \vec{\omega}, \sigma) + \beta^2 . f(\sigma) \} \, dz$$

With respect to $\tilde{D}$, the minimization gives us $\tilde{D} = 0$ because $\det (\mu) > 0$, $\forall \, \Sigma_m$.

First let's show the isotropic case.

$\alpha(\beta, \vec{\omega}, \sigma) = \vec{\omega} . \mathbf{A} [\vec{\omega}]$ and the minimization with respect to $\vec{\omega}$ gives us $\vec{\omega} = \vec{0}$ (same reason as before).

To minimize with respect to β, we develop β around $z = 0$ then $\beta = \sum_{i=0}^{n} \beta_i . z^i$

The minimization with respect to β is written $\dfrac{\partial \mathcal{A}(\vec{v}, \vec{v}, \sigma)}{\partial \beta_i} = 0$ for $i = 0, 1, ..., n$

We can write these equations in the following way $[R].[\beta] = [0]$ with

$$[R] = \begin{pmatrix} \int_e \det(\mu) \, f(\sigma) \, dz & .... & \int_e z^n \det(\mu) \, f(\sigma) \, dz \\ \int_e z \det(\mu) \, f(\sigma) \, dz & ..... & \int_e z^{n+1} \det(\mu) \, f(\sigma) \, dz \\ .... & & \\ \int_e z^n \det(\mu) \, f(\sigma) \, dz & .... & \int_e z^{2n} \det(\mu) \, f(\sigma) \, dz \end{pmatrix} \quad \text{and} \quad [\beta] = \begin{pmatrix} \beta_0 \\ \beta_1 \\ .... \\ \beta_n \end{pmatrix}$$

The solution $[\beta]$ musn't be equal to zero, then $\det [R] = 0$

Remarks : . $\det(\mu) = 1 - 2.H.z + k.z^2$ where $2.H$ and k are respectively the mean
curvature and the total curvature of Σ_m
. With each degree n, we obtain more or less simplified criteria.

We can show that in the orthotropic case, the same process can be used. The minimization with respect to $\vec{\omega}$ leads us to a linear system of equations depending on the coefficients β_i and the unknowns of which are the coefficients $\vec{\omega}_i$ of the expansion of $\vec{\omega}$. Then the minimization with respect to β leads us to an other linear system of equations which is homogeneous. Then its determinant must be equal to zero which is a new criterion.

IV. Application.

4.1. TEST.

To test this criterion, we imagine an experiment verifying, as closely as possible, the hypothesis of the previous theory. This experiment should allow us to use a calculation of stresses thanks to a finite elements method, and it should be reasonably easy to apply. We have chosen the case of a long rectangular sheet (length = 200 mm), bent by a cylindrical punch. The sheet's ends are held in order that the bent part comes under the punch only, so as to localize necking there.

For some values of thickness (h = 2 or 4 mm), and some values of curvature (R = 25 or 50 mm), we obtain, by a finite elements analysis made with the package Abaqus, the stresses for different positions of the punch. Thanks to the criteria described, we can determine the position of the punch which corresponds with the onset of necking.

4.2. RESULTS AND CONCLUSION.

For these examples the hardening of the steel is modelized by $\bar{\sigma} = 200 + 500.\bar{\epsilon}_p^{0.33}$. The strained mesh obtained, and the graphs representing, for the different approximations (n = 0, 1, 2, 3 and 4), the value of the staving possible for each abscissa (origin on the middle of the sheet) are presented on figures 2, 3 and 4.

We can see that the higher the order n, the lower the solution. Moreover, in view of the results, we can think that the orders 3 or 4 give us reasonable approximations of the exact solution.

We note that necking is always localized under the punch (but not always on the middle).

We can sum up the influence of h and R through these three calculations by : The higher h is or the smaller R is, the earlier necking appears. Quantitatively, we note almost the same possibility of staving of the punch in both cases with the same ratios h/R.

We note that the finite elements analysis cannot be easily developed for higher punch stroke because the numerical instability is obtained. The difficulty in reaching the stage where the criterion in uniform fields (n = 0) shows that there is necking, proves that this criterion is too optimistic and justifies the determination of more accurate criteria.

In order to locate this instability with regard to the external loading, we can represent (fig. 5) the deep drawing force F_p versus the staving of the punch. We then note that the material's instability doesn't correspond to a maximum of F_p. On the other hand, we have drawn a dotted line representing the tractive effort F_t in the non-bent part, and we can see that the maximum of F_t is rather near. Which means that instability with uniform fields happens a little later than that with non-uniform fields.

Quantitatively, for example, for h = 4 mm and R = 25 mm, the criteria with n = 4 and n = 0 lead respectively to $\bar{\sigma} = 464$ MPa, $\bar{\epsilon}_p = 14.76 \%$ and $\bar{\sigma} = 499$ MPa, $\bar{\epsilon}_p = 21.55 \%$ when instability happens. The maximum stroke is respectively 43.5 mm and 47.7 mm.

REFERENCES :

<1> J.P. Cordebois, P. Ladevèze : "Sur la prévision des courbes limites de formage" Journal de Mécanique Théorique et Appliquée. Vol 5, n°3, pp. 341-370.

<2> P. Ladevèze : "Sur la théorie de la plasticité en grandes déformations" Laboratoire de Mécanique et Technologie - E.N.S. de Cachan - Paris VI Rapport interne n°9 - 1981.

<3> J.M. Jalinier : Endommagement, instabilité plastique et courbes limites de formage - Thèse de Docteur Ingénieur - Université de Metz - 1978.

<4> A propos du formage des tôles - Cetim Informations - 1973

<5> M.Y. Demeri : Strain analysis of the Hemispherical Stretch Bend Test J. Applied metal Working. Vol 4, n°2, pp. 183-187.

144

Fig.1

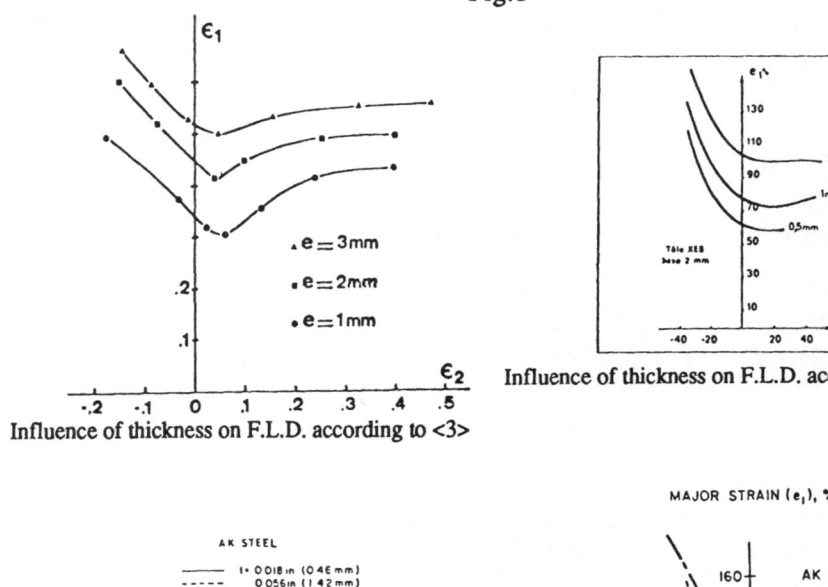

Influence of thickness on F.L.D. according to <3>

Influence of thickness on F.L.D. according to <4>

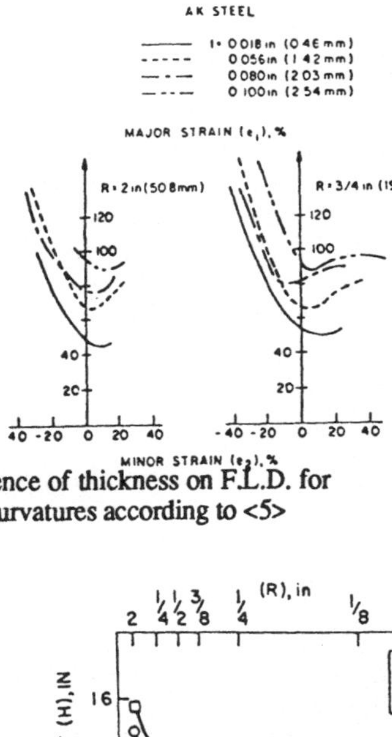

Influence of thickness on F.L.D. for
two curvatures according to <5>

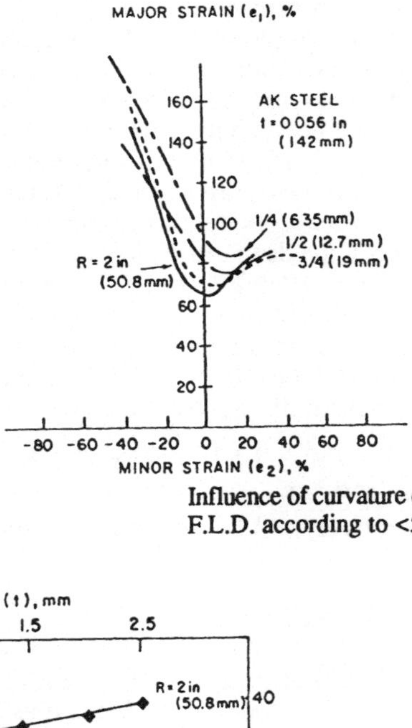

Influence of curvature on
F.L.D. according to <5>

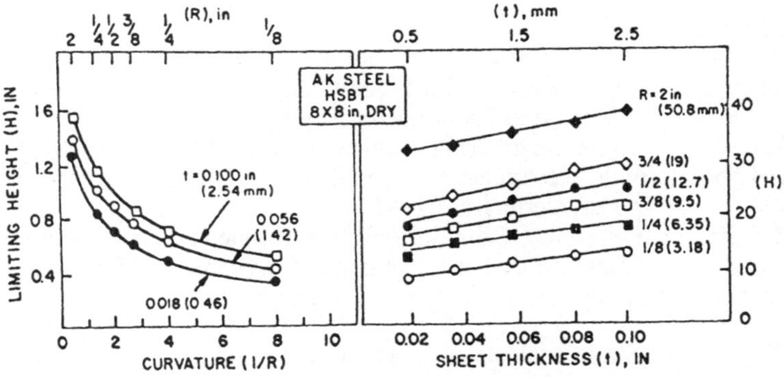

Influence of thickness and curvature on the maximum staving of a   hemispherical punch according to <5>

Fig.2

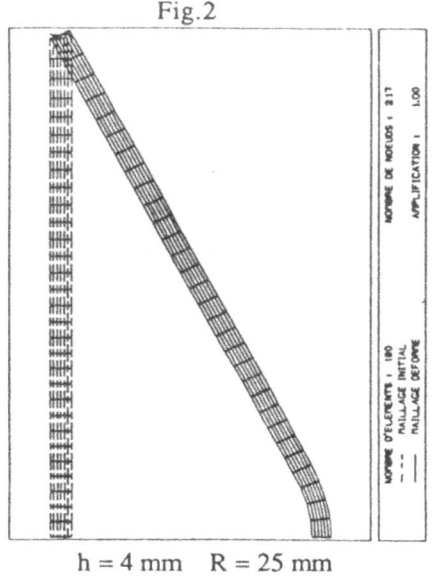

h = 4 mm   R = 25 mm

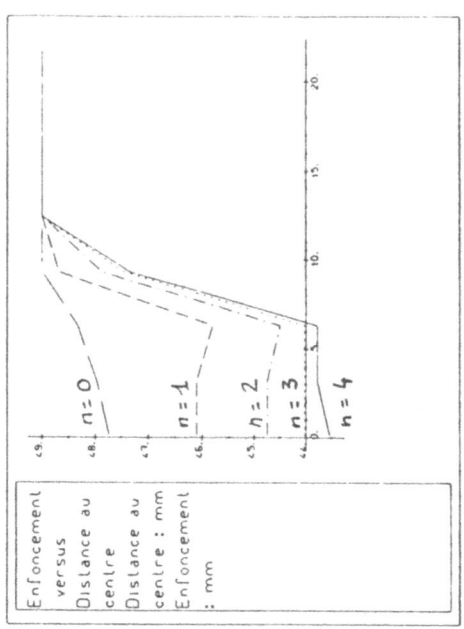

Fig.3

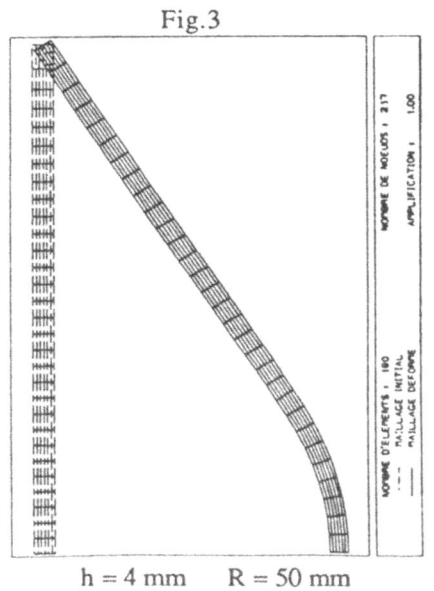

h = 4 mm   R = 50 mm

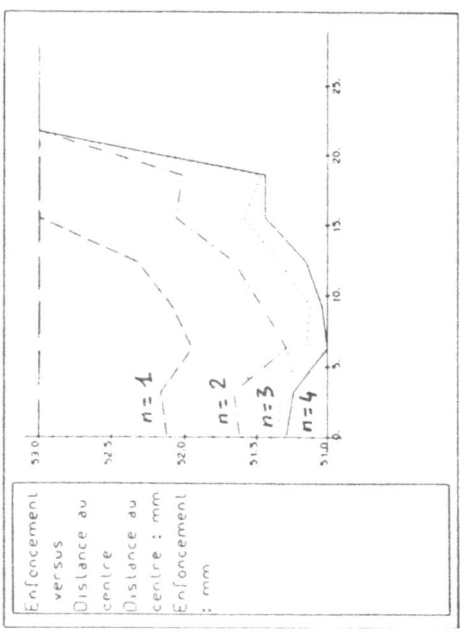

Fig.4

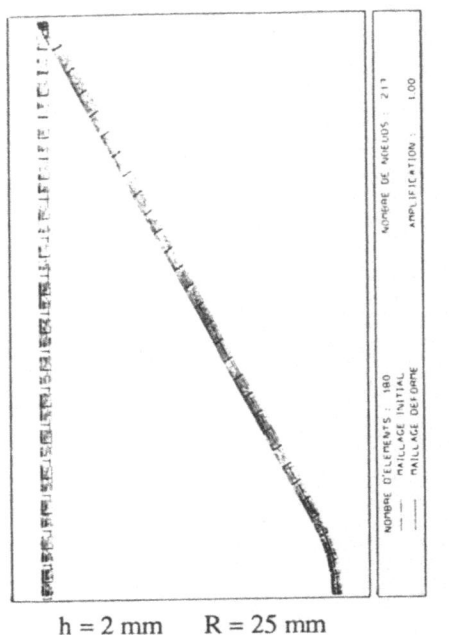
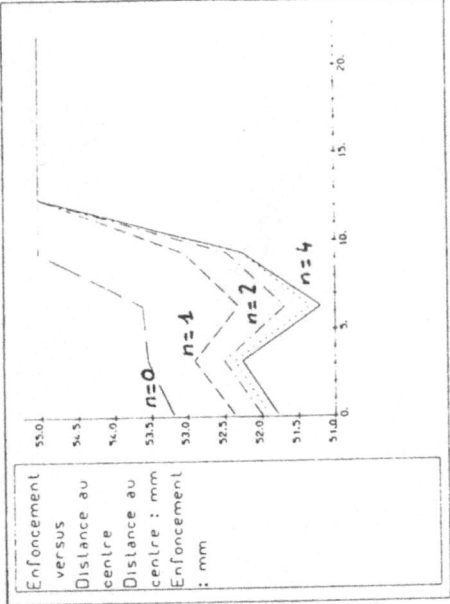

h = 2 mm   R = 25 mm

Fig.5

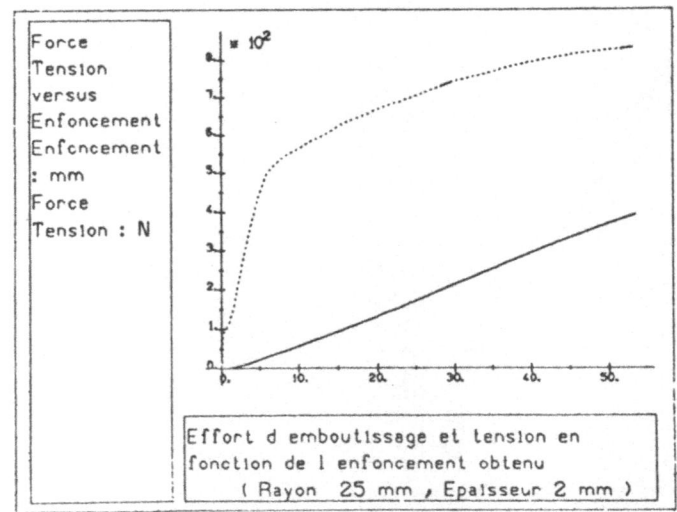

EXPERIMENTAL ANALYSIS OF AN AXISYMMETRIC DEEP DRAWING

Giovanni PEROTTI^, Giovanni BELINGARDI^^, Pier Luigi BELLARDI^^^,
Michele GUIDA^^^^, Angelo MESSINA^, Filippo ZINGARIELLO^^

^   Politecnico di Torino, Dip. di Tecnologia e Sistemi di Produzione
^^   Politecnico di Torino, Dipartimento di Meccanica
^^^   FIAT Auto, Direzione Produzione Presse, Torino
^^^^ Centro Ricerche FIAT, Orbassano

ABSTRACT.

A cooperative research program was established in order to verify the
present possibility of a computer simulation of sheet metal forming.
A first simulation was performed applying a Finite Element code to a
test case: the deep drawing of an axisymmetric cup characterized by a
double curvature profile.
The obtained computational results need a validation, through a
comparison with experimental measurements. It is also important to know
if the simplification adopted for the mathematical model are too
drastic. This paper presents the adopted experimental procedure and
a discussion of the experimental results.
A low carbon steel was used for the experimental evaluations; some tests
were also performed using a high strength steel. The local strains and
the punch force were measured and compared with computational results.
The measurements were planned in some sets, each characterized by a
different test condition, in order to explore a grid of variables such
as punch travel speed, surface lubrication, blankholder force.

1. INTRODUCTION.

In recent years the problem of sheet metal forming has been widely
investigated from the computational point of view, because of its great
interest for manufacturers, such as automotive companies, where a lot of
parts are made by those technologies.
Stamping of sheet metal parts by means of punches and dies is a standard
manufacturing process. However, despite its broad application in
industry, the design of the forming process is largely based on
experimental techniques, such as the use of circular grid system or
forging limit diagrams.
The development of reliable analytical procedures to predict the
behavior of sheet metal deformation processes has encountered many
serious obstacles. Together with the non linearity of the material
properties, the large strains and displacements, the discontinuity of

147

J. L. Chenot and E. Oñate (eds.), Modelling of Metal Forming Processes, 147–154.
© *1988 by Kluwer Academic Publishers.*

the contact between the blank and the punch or the die, and the friction effects make the study of such a complexity that it requires a large-scale computer [1,2,3,4].

A cooperative research work was begun between the Politecnico di Torino, the FIAT Auto Manufacturing Division and the FIAT Research Center, in order to demonstrate the present feasibility of a computer simulation of sheet metal forming.

The well known MARC finite element program (which allows elasto-plastic behavior of the material, large deformations through the constitutive equations of the element, large displacements and lagrangian updating of the geometry) was used to execute the calculations necessary for the simulation of the deep forming of an axisymmetric part [5].

Axisymmetric stamping processes are obviously easier to treat than the more general problem, but the complexity of the mechanics is still considerable and therefore they are suitable for initial testing.

The finite element simulation needs an experimental validation [6,7] therefore with this paper we want to present some experimental results.

2. THE FINITE ELEMENT SIMULATION.

To make a meaningful test, a part, manufactured in FIAT Auto for a motor car, has been considered. At present this part is obtained through a three step stamping process. In the following sections we will present results only for the first drawing step.

Figure 1 shows the geometry of the blank, of the punch and of the die for the first stamping step. It is an axisymmetric cup with a moderately complex profile characterized by a double curvature.

The mathematical model of the blank was built with 50 four-node axisymmetric toroidal elements arranged in one row.

In order to model mathematically the above situation, the MARC updated Lagrange procedure was utilized together with the finite strain plasticity, the large displacement and the isotropic hardening options.

The problem of modeling the punch, the die and the blank-holder was solved writing some

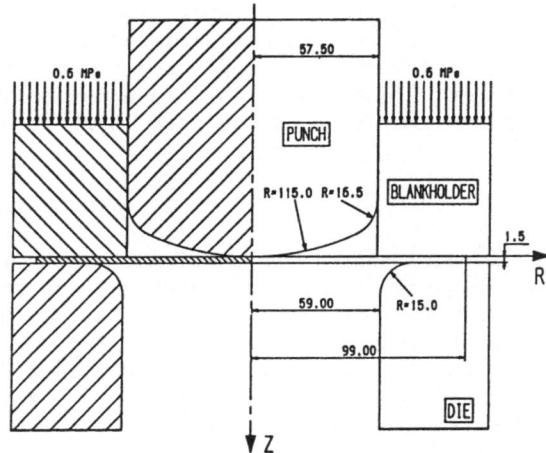

Fig. 1 Geometry of the blank, of the punch and of the die.

subroutines, linked with the main MARC. These subroutines describe the geometry of the 102 gap elements, check step by step the position of the nodes on the sheet relative to the reference surfaces and if necessary change the linkage, thus modeling the changes in contact between surfaces.

The blankholder is simulated as a constant force one.

The whole calculation was executed stating 40 increments of the punch displacements.

To perform the first computational test some simplifications were assumed being aware that they did not lead to a completely wrong mathematical model of the physical phenomenon. Neither the friction of the blank sliding over the punch and the die, the thermal field connected with the friction mechanism, the anisotropy of the mechanical properties of the material, the dynamic conditions of the operation, nor the viscous behavior of the material are taken into account at this stage of the research work.

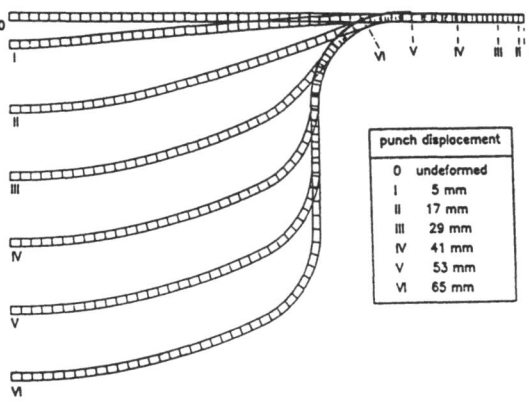

Fig. 2 Superimposition of six progressive deformed shapes.

Figure 2 shows the superimposition of the deformed shapes of the metal sheet at six different values of the punch displacement.

3. EXPERIMENTAL PROCEDURE.

The true stress-strain curves of the steels used in this analysis were determined through traction tests. Three sets of standard specimens were used, each set characterized by an angle of 0° or 45° or 90° between the specimen axis and the rolling direction. Figure 3 shows the experimental true stress-strain curves; they are the results of averaging within each set and then between the three directions. A second set of tests was executed in order to determine the anisotropy coefficient (ASTM norm E 517-81), in this case a maximum value of 20% of the strain was fixed.

The experimental work was executed with the 160 ton Oleotronic drawing press installed in the FIAT Research Center Laboratories.

The press is of the oleodynamic type with double effect cylinders, the blankholder is pushing upward.

The press is guided by a programmable electronic controller, which allows the modulation of the blankholder pressure and of the punch speed

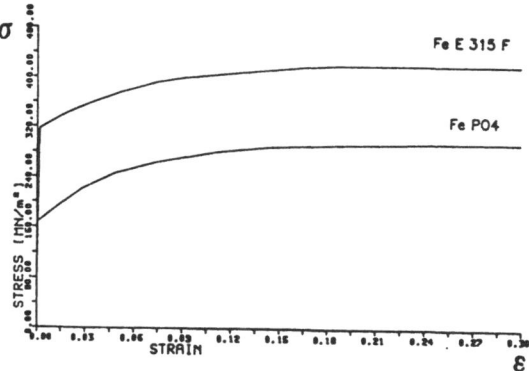

Fig. 3 Experimental true stress-strain curves.

during the stamping process.
The press is instrumented to measure the actual pressure in the pushing cylinder and the displacement of the main slide.
The measure of the stamping force as a function of the punch stroke was obtained with the press instrumentation, plotting directly the punch force versus the punch displacement. From that diagram it is also possible to evaluate the real force exerted by the blankholder. This procedure saves from possible errors coming from an indirect evaluation of the stamping force, based on a previous calibration test for the obtainment of the oil pressure-punch force relationship.
The measure of the local strains of the metal sheet was performed using the circle grid technique. Each sheet is grided with 5.0 mm diameter circles. During the drawing the circles follow the sheet surface deformation and become ellipses. After the drawing, it is possible to measure, in two previously stated directions, the values of the ellipses axis. The difference between the original diameter and the ellipses axis is proportional to the local strain.
The same grid was also used as a reference frame for the measurement of the sheet thickness before and after the drawing in order to obtain the thickness strain, along a cup diametral plane.
The measurement of the ellipses axis was executed with an optical instrument (0.05 mm precision), while the thickness measures were executed with an electronic high precision instrument (0.005 mm precision).
With both steels used , in the formed cup four ears and four hollows were observed, related to the 0° and 90° and the ±45° directions with respect to the lamination direction. This is in agreement with the anisotropy of the materials, as pointed out from the measured anisotropy coefficients. The phenomenon is more evident with the Fe PO4 steel ($r_{m,20\%}$ = 1.69) than with the Fe E315F steel ($r_{m,20\%}$ = 1.24).
For this reason, for each blank the strain measures were executed in the above specified four directions, but, to show the results along a diametral plane of the cup, we prefer to use an averaged value for each strain and for each radial location.
In the following figures the strain values are plotted against the radii of the circle centers in their original positions (undeformed sheet).

4. EXPERIMENTAL RESULTS.

The first aim of this experimental work was to obtain the local strains and the punch force values to make a comparison with the corresponding

parameter	experimental	computational
punch speed lubrication blank holder force	2.3 mm/s double teflon sheet 8000. N	quasi static no friction 8000. N

Table I. Experimental and computational parameters.

values available [5] from the finite element computations. In order to reduce the effects of casual differences in the parameters of the test process and of the unavoidable errors in making the strain measures, for each testing conditions a certain number of drawing were executed. Again the strain values are here reported after their averages.

A first set of 10 tests were performed with Fe PO4 steel, the drawing conditions were chosen with the aim of reproducing the con-

Test number	Lubrication state				Deep-drawing speed		Blankholder force	
	Teflon 2 sheets	Teflon 1 sheet	Butex 276/29	Dry	2.3 [mm/s]	63.4 [mm/s]	8800 [N]	19700 [N]
1	X				X		X	
2		X			X		X	
3			X		X		X	
4				X	X		X	
5	X				X			X
6		X			X			X
7			X		X			X
8				X	X			X
9	X					X		X
10		X				X		X
11			X			X		X
12				X		X		X

Table II. Plane of experimental variation of the process parameters.

ditions assumed for the finite element computations (see table I).
The results shown in the figures 4, 5, and 6 are the averaged values, obtained from all the ten tests, respectively, of the punch force versus punch displacement, of the radial and hoop strains versus the radial position. Figure 7 shows the thickness strains versus the radial position, after having averaged results from five tests.
In order to study the influence of each single parameter, some other sets of tests were then performed keeping fixed all the specified parameters, except one. This second part of the experimental analysis also aims at producing information about the real consistency of the simplifications used for the numerical computations. Moreover it is of interest to begin a data base.
In table II the process parameters are summarized for the different testing conditions explored. For each testing conditions now only three tests were executed.
Finally the same testing procedure was completely repeated also for the high strength steel Fe E315F.
Figure 8 shows the influence of the lubrication parameter on the punch force displacement diagrams. The eight curves show the results of deep drawing operations performed in the same experimental conditions (of punch speed and blankholder pressure), on the two above quoted materials. For each material there are four different lubrication conditions: three with different lubricants and one with dry surfaces. This experimental work program requested a total amount of about 200 tests.

152

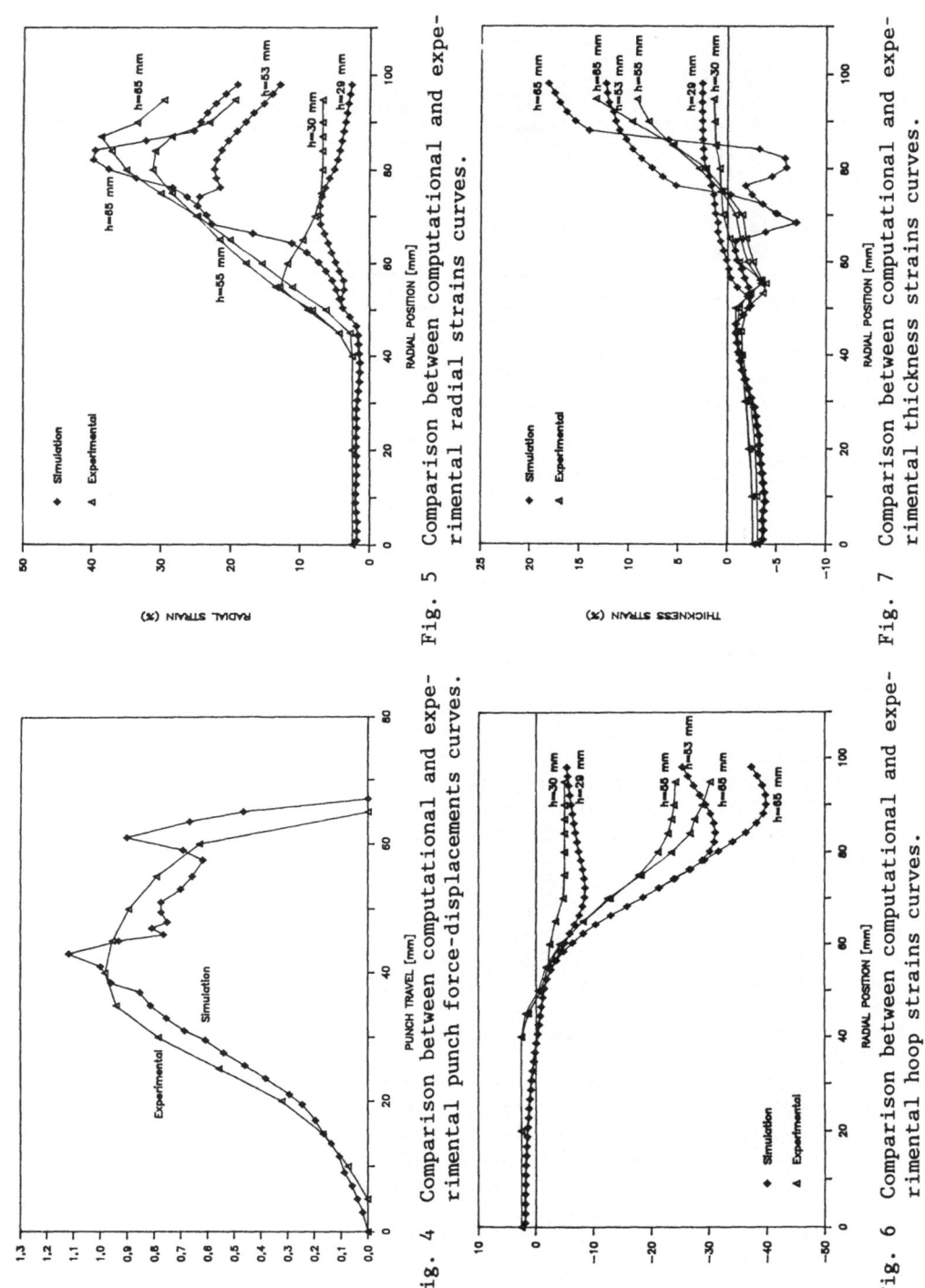

Fig. 4   Comparison between computational and experimental punch force-displacements curves.

Fig. 5   Comparison between computational and experimental radial strains curves.

Fig. 6   Comparison between computational and experimental hoop strains curves.

Fig. 7   Comparison between computational and experimental thickness strains curves.

5. DISCUSSION AND CONCLUSIONS.

The deep drawing of a steel axisymmetric cup has been analyzed experimentally with the aim of validating the numerical simulation of that stamping operation, previously performed with a finite element computation. Careful examination of the obtained results allows to present some conclusions:
- punch force displacement diagrams (figure 4): the shapes of the computational curve and of the experimental one are in good agreement. The computational maximum value of the force is slightly over the experimental one. The areas under the curves, and therefore the forming energies, are quite the same.
- local strain diagrams: considering the radial strains (figure 5) there is a good agreement between computational and experimental results, both from the curve shape and from the numerical values point of view.
Considering the hoop strains (figure 6) there is a good agreement of the curve shapes, but there is a slight difference between the computational and the experimental numerical values, moderately increasing in correspondence with the outer edge of the cup.
Considering now the thickness strains (figure 7) there is a good agreement between the experimental and numerical values, except for the zone where the wall of the cup is cylindrical. The computational results are affected by the concentrated forces exerted on the deformed sheet at the top (from the die) and the bottom (from the punch) of that zone. The effects of those concentrated forces are so great as to hide the minimum thickness location in the computational results.
The global result of these comparisons allows to formulate a favourable judgment about the degree of reliability of the computational simulation and about the possibility of representing the considered axisymmetric deep drawing process with that finite element model.

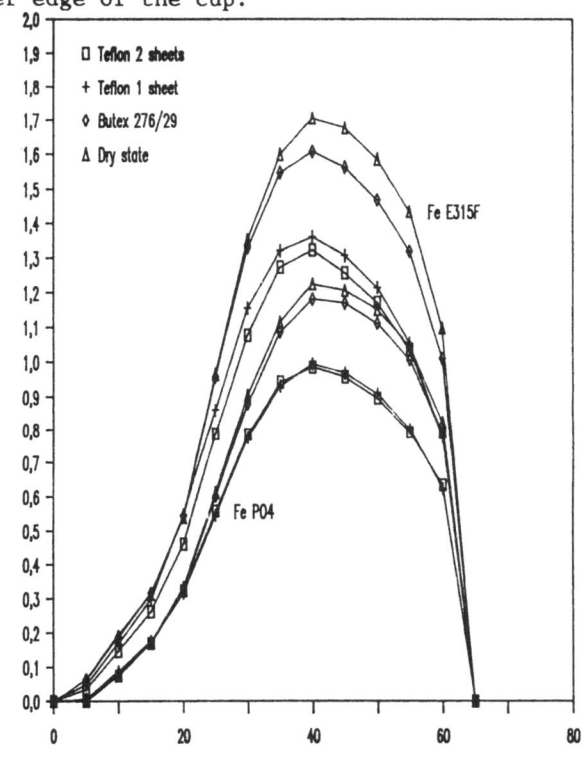

Fig. 8 Influence of the lubrication parameter on the punch force-displacement curves.

154

The pointed out differences are at present under examination in order to identify the aspects that request an improvement, for example a simulation that performs a better mathematical model of the sheet bending characteristics [8] could yield better results.
The experimental work was completed with some sets of measurements as planned in table II, changing one process parameter for each set in order to explore the influence of those variables. These results are not shown here for sake of brevity, except for what concerns the lubrication conditions. Figure 8 shows the punch force-displacement diagrams for the two considered steels with four different surface lubrication.

6. ACKNOWLEDGMENTS.

The authors wish to acknowledge Prof. P.M. Calderale of the Mechanical Engineering Department, Dr. G. Cozzari head of the FIAT Auto Stamping Division and Prof. A. Garro of the FIAT Auto Technical Center for their constant encouragement during the development of the research work.

7. REFERENCES.

[1] Vemuri K.R. et alii -"A computer aided design system for axisymmetric deep drawing process", proc. CIRP, vol. 36/1, 1987.
[2] Perotti G., Maggiorano E., Spirito F., Tornincasa S. -"A calculation program for deep drawing forms", proc. CIRP, vol. 34/1, 1985.
[3] Reissner J., Ehrismann R. -"Computer aided deep drawing of two part cans", proc. CIRP, vol. 36/1, 1987.
[4] Zienkiewicz O.C. -"Flow formulation for numerical solution of forming processes", in Numerical Analysis of Forming Processes, J.Wiley, 1984.
[5] Belingardi G., Calderale P.M., Cozzari G., Zingariello F. - "Experience in approaching sheet metal forming by finite element technique", proc. of the Int. Conf. Computational Plasticity, Barcelona (Spain), 6th-10th April, 1987.
[6] Mattiasson K., Saran M., Melander A., Schedin E., Gustafsson C. - "Finite element simulation of deep drawing of low and high strength steel", proc. 2nd Int. Conf. Technology of Plasticity, Stuttgart (West Germany), 24th-28th August, 1987.
[7] Massoni E., Bellet M., Chenot J.L., Detraux J.M., de Baynast C. - "A finite element modeling for deep drawing of thin sheet in automotive industry", proc. 2nd Int. Conf. Technology Plasticity, Stuttgart (West Germany), 24th-28th August, 1987.
[8] Belingardi G., Zingariello F. - "Sul ruolo della flessione nella simulazione di operazioni di formatura di particolari in lamiera col metodo degli elementi finiti", proc. AIAS National Conf., Pisa (Italy), 15th-19th September, 1987.

COMPUTER AIDED DESIGN OF A PROGRESSIVE DIE

K. Bergström, S. Kivivuori, S. Osenius and A. Korhonen

Helsinki University of Technology
Dept. of Materials Science and Engineering
Lab. of Metal Working and Heat Treatment
Vuorimiehentie 2 A, 02150 ESPOO
FINLAND

Abstract

The mathematical simulation of even the simplest metal forming operations is often tedious and time consuming. Despite a considerable amount of activity especially in the sheet metal forming field, complete programs which would take into account all the complex dimensional changes during plastic forming do not exist. More simple methods have to be used for designing the tools and dies on practice.

The present paper discusses the recent experiences of the authors on the use of CAD-techniques for the design of a progressive die. The ultimate goal of any CAD/CAM system is to improve productivity and to reduce the time spent for design and manufacture. Sheet metal forming and blanking dies are complicated tools and a progressive die may contain several hundred parts. This makes the conventional design extremely time-consuming.

In this work a commercially available CAD-package called Auto-trol formed the basis of the design system. The system was installed into an Apollo workstation computer. The original program contained the flat pattern development operation and the calculation of bend allowances as standard functions.

The commercial program was expanded to a die design expert system by adding routines for the evaluation of spring-back, proper die clearance and the minimum bending radius and for calculation of the blanking and bending forces and their points of action.

Practical design examples on the use of the system to design sheet metal forming and blanking tools will be given and both merits and limitations of the present CAD-technique will be discussed.

1. Introduction

The competitiveness of the metal forming industry can be improved using Computer Aided Design (CAD) which makes it possible to considerably reduce the time spent on tool design. The versatile and rapid manipulation of the geometry by a CAD-system has also made computational analyses of metal forming processes economically viable /1/.

155

J. L. Chenot and E. Oñate (eds.), Modelling of Metal Forming Processes, 155–162.
© *1988 by Kluwer Academic Publishers.*

The effective use of the designer's skill and experience in designing processes and manufacturing tools with the aid of Numerically Controlled machining is important. Modern theoretical and experimental techniques for solving complex engineering problems using numerical computer methods and NC machining have been applied successfully to metal forming processes and tooling /1,2,3/.

In sheet metal industry a very large number of products can be manufactured by using only bending and blanking processes. Very often a progressive die is the most suitable and economical alternative when the tooling for these products is selected. However, progressive dies are complicated tools involving functions such as lifting and positioning of the strip, as well as punching and bending operations /4/. The tooling may involve several hundred individual parts which make the conventional design of a progressive die extremely time-consuming /5/.

The present paper discusses experiences with the use of CAD-techniques for the design of a progressive die. In this work a commercially available CAD-package called Auto-trol forms the basis of the design system. The system was installed into an Apollo workstation computer.

2. Fortran integrated CAD/CAM

The design work using CAD can be accelerated by automating several design-operations and stages. The most flexible solution is to connect user-created fortran-programs with a CAD-system.

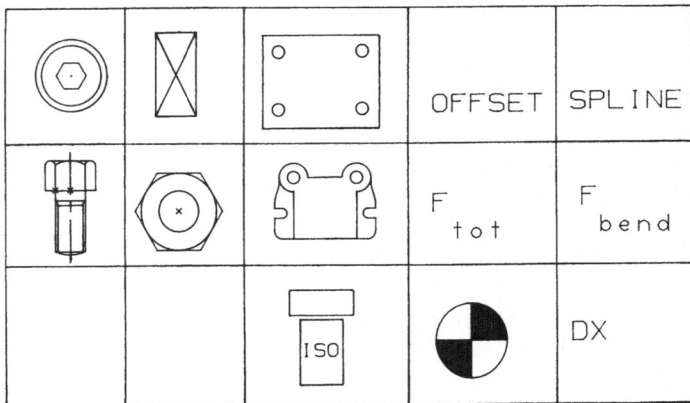

Fig. 1. A tailored CAD-menu for designing punching and bending tools.

A CAD-system is called fortran-integrated if its interactive
commands can be called from an user-created fortran program. Such a
CAD-system allows the user to master the data-bases of his geometrical
models automatically. Further, it is also possible to create new
menu-buttons, which will perform arbitrary user-created applications
as seen in Fig. 1.

A fortran program created by the CAD-user communicates with the
core of the CAD-system using normal fortran-based calls. The possible
subroutines and their parameter lists are listed in the user manual
for appropriate use. For successful application, the formats of the
parameter lists of individual calls must be strictly followed. When
the whole core of the CAD-system -i.e. all graphic commands- is
available, it is possible to perform normally interactive CAD-commands
automatically. This means commands such as creating elements,
selecting elements, performing calculations based on the selected
elements and drawing on the screen a new model based on the
calculations.

For executing one's own fortran applications in connection with
CAD the user must first compile them with the host-computer's
fortran-compiler and then bind the resulting binary files. After this
the CAD-system is startable with the new user-created applications. In
an advanced CAD-system the procedure requires time of only from a few
seconds to some minutes.

When a single element on the screen is selected, only its
pointer is transferred to the external user-created fortran programs.
Based on the pointer, the topology, geometry, attributes and hierarchy
of an individual element can be read from the model data-base and used
for all kinds of fortran operations.

3. Progressive die design process

The effectiveness of Computer Aided Design (CAD) in the construction
of a progressive die (Fig. 2) is based on the use of a real-time model
including all geometrical information on the product. The most
essential feature of the designing is interactivity, which means that
all adding or modifying operations can be seen instantaneously on the
screen after performing them.

When creating a sheet metal product for the first time, it must
be generated onto the screen using elements, such as coordinate
points, lines and arcs. The resulting 3-dimensional wire-frame model
is then flattened to a 2-dimensional part, when the bend allowances
are automatically compensated.

Design of a progressive die follows the workflow shown in Fig 3.
The work starts using the flattened part and the material utilization
is estimated comparing various combinations of multiple copies. The
optimum nesting unit is selected and all the punches, bend stations,
pilots, dowels and springs as well as plates are added. During all
these steps the same geometry of the unfolded part is used.

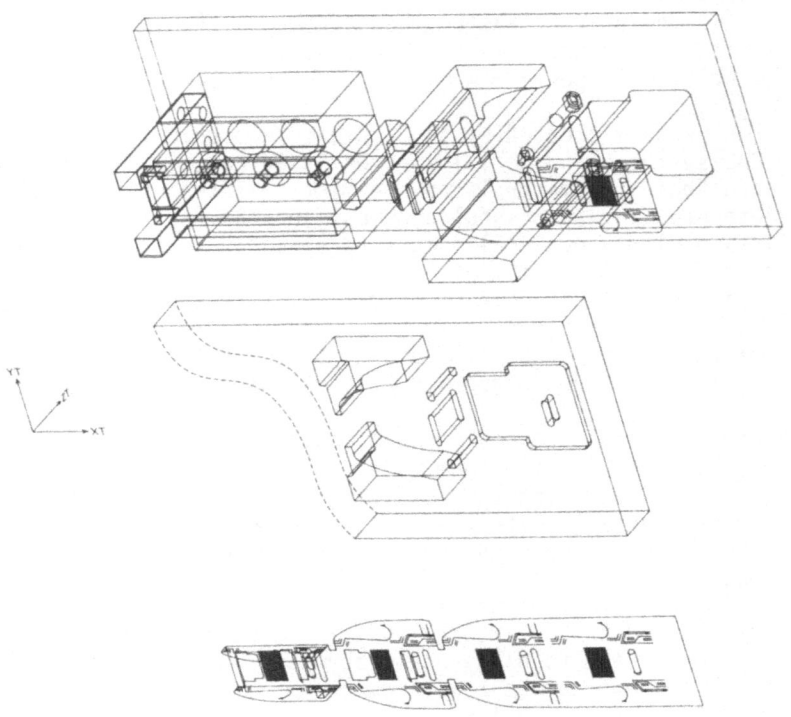

Fig. 2. The punches, the punch plate and the die of a progressive
tool. The buckle below is produced from a continuous steel strip in
four consequent pressing stations. In every stroke of the tool the
strip is transferred corresponding to the length of a single forming
station.

The designing of progressive dies was improved by using the special
expert operations. The present system was developed by adding the
following routines to the program:
- Calculating the minimum diameter and the height of the punch
  to prevent cracking and buckling. These values depend on the
  strength and thickness of the punched material.
- Calculating the stripper force.
- Determining the punching forces and their centre. The value of
  the punching force gives information on the press needed for
  punching. By determining the centre of the forces, the uneven
  loading leading to press failures can be avoided.
- Calculating bending forces and their centre.
- Calculating the minimum radius of the material to be bent.
- Calculating the waste material percentage.

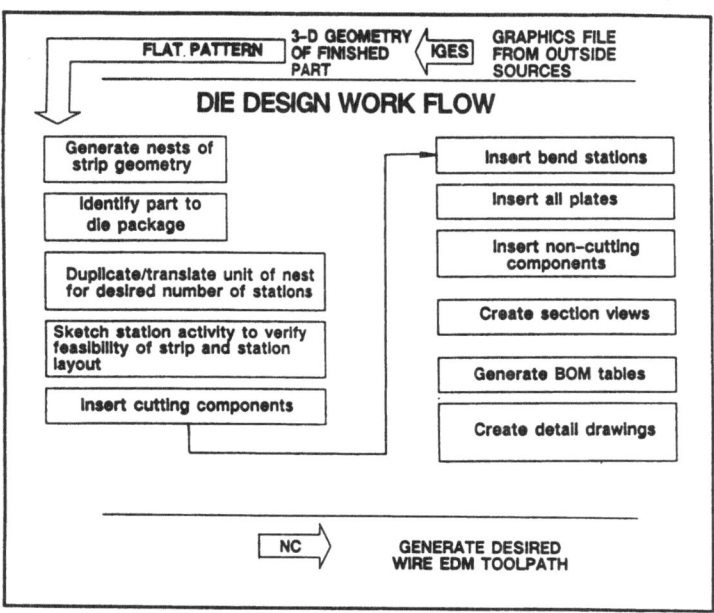

Fig. 3. The workflow of the design procedure of a progressive die /6/.

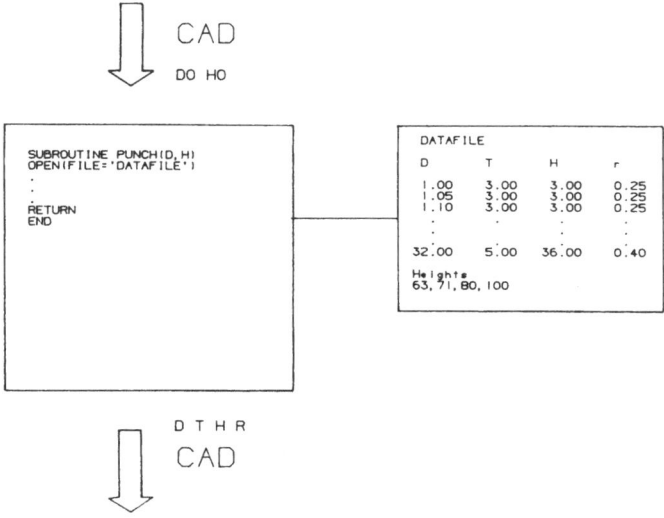

Fig. 4. Building a standard part library using fortran-programming.

It is shown in Fig. 4, how for example a usual cylindrical punch can be programmed with fortran into the standard part library of the CAD/CAM system. The user-created fortran-program is surrounded with the dashed line in Fig. 4. As will be noted, the created program searches for a suitable punch based on numerical standard values stored in a datafile. For the searching the program needs only the diameter of the hole to be punched and the height of the punch. A similar program has been created for all kinds of standard tool parts. The fortran-programming can be done outside the CAD-system.

For complete connecting to CAD the input/output routine of the user-created fortran-program must be replaced by that of the CAD-system. This makes it possible for the user to select elements on the screen and transfer their parameter values into the fortran-program.

Consequently, to execute the program the user only selects a hole on the screen and the program automatically creates a punch for it. The user is not required to be skilled in programming fortran, and usually he does not notice that he is using a user-created function.

Conventional engineering drawings of forming tools are being replaced by equivalent mathematical representations of geometric models constructed by a CAD/CAM system and stored on a CAD/CAM database for subsequent retrieval. An important use of the part's geometric model is as input to programs designed to automatically generate instructions for manufacturing the part. There are many APT-based programming systems available.

Die machining part programs are generated by the function. The basic steps generating tool paths are NC-set identification, initializing technological parameters such as the machine tools axis system and definition of the NC pass. The tool paths are generated from line and arc representations of the model. Some APT statements are added to the tool paths as operations to control the postprocessor execution. The cap ability of tool path verification is used to prevent machining errors.

4. Discussion

The flow chart of the design of a progressive die using CAD does not differ from that of the conventional design. Nevertheless, the CAD procedure reduces the time spent on designing and manufacturing the progressive die. Errors can be identified and corrected before the incorrect data causes costs and difficulties in manufacturing.

The designing of a progressive tool, the choice of forming equipment and the mounting of the dies on a certain machine essentially requires the determination of several parameters. There have been difficulties in making these estimations when the conventional planning of a forming operation is used. The calculation of these parameters has become possible when using a fortran-integrated CAD-system. The programming effort needed to build up a program for the computing parameters mentioned above has decreased dramatically - thanks to the existence of fortran-integrated CAD/CAM.

As a consequence of the above, with a fortran-integrated
CAD-system it is possible for the user to build up and execute
computer applications fast and easily for the following design
purposes:
- To carry out the parametric design of families of parts. This
  means, that in creating a part model only the numerical
  dimensions of the parts are needed as input.
- To construct a die-design package including the standard
  tool-parts and the empirical rules for designing.
- To automate tool parts lists generation and tool dimensioning.
- To optimize the forming operation planned.

The CAD/CAM programs based on a wire-frame model are not very
sophisticated, as can be understood from Fig. 2. With a volume-based
folding and unfolding CAD system, such as the German PROREN-system
developed by ISYKON Software GmbH /7/ Bochum, which has recently been
installed in Helsinki University of Technology, the designing of sheet
metal parts and tools becomes much more sophisticated, Fig. 5.

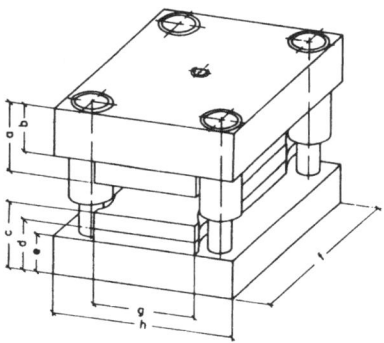

Fig. 5. A die frame created using a volume model based CAD-system /4/.

The generating of a tool side view from a 3D wire-frame model
requires effort in interactive trimming. In Auto-trol's case this can
be avoided by using the system-dealer-created side 2D-view generating
program.

5. Conclusions

The use of the CAD/CAM technique considerably speeds up the design
work of a progressive tool. This is based on the CAD/CAM system's
ability to unfold a 3D sheet part into a plane and the idea of using
this geometry for all the process and tool design phases. Tool
standard part libraries and stored expert data speed up the designing
process further. However, the design and manufacture demand a
considerable number of highly-skilled personnel and the effective use
of the designer's skill and experience in designing metal-forming
processes is still required.

Acknowledgment

The authors are grateful to the Technology Development Centre (TEKES)
for the financial support in this work.

References

1. Mattiasson, K. et al, Numerical Methods in Industrial Forming
Processes. Balkema, Boston 1986.

2. Altan, T. et al, Metal Forming, Fundamentals and applications.
Metals Park, Ohio 1986.

3. v. Finckenstein, E. & Kleiner, M., Process Simulation and Adaptive
Controlling in Sheet Metal Forming. Advanced Technology of Plasticity,
Berlin 1987 p. 1187-1194.

4. Stracke, J., Methodische Grundlagen für die rechnerunterstützte
Bearbeitung von Anpassungskonstruktionen. Ruhr-Universität Bochum,
Bochum 1980.

5. VDI Bildungswerk, Leistungssteigerung bei Werkzeugen der
Stanztechnik. VDI, Düsseldorf 1987.

6. User's manual, Auto-trol CAD-system. Auto-trol Corporation, Denver
1986.

7. Seifert, H., Rechnerunterstütztes Konstruieren mit PROREN.
Ruhr-Universität Bochum, Bochum 1986.

FINITE ELEMENT ANALYSIS OF SHEET METAL FORMING PROBLEMS USING A VISCOUS VOIDED SHELL FORMULATION

Eugenio Oñate
and
Carlos Agelet de Saracibar

Escuela Técnica Superior de Ingenieros de
Caminos, Canales y Puertos
Universidad Politécnica de Cataluña
08034 Barcelona, Spain

Summary

A formal analogy between the equations of pure plastic and viscoplastic flow theory for void containing metals and those of standard non linear elasticity is presented. It is shown how by direct simplifications of the general equations, the standard incompressible flow expressions for non voided metals are obtained. The general formulation is particularized for the analysis of sheet metal forming problems and details of the viscous voided shell and membrane formulations for dealing with the axisymmetric case are given. Finally, some examples of applications of pure and hemispherical stretching and deep drawing of a circular sheet are presented.

1. Introduction

It is well known that an effective way of treating the continuous deformation of metals is to use a rigid plastic flow model in which elastic effects are neglected. The simplest and perhaps most widely used model, uses the Von Mises yield criterion, which results in the incompressibility of the material flow. The governing equations in this case are entirely analogous to those of standard elasticity with a single material parameter, the shear modulus, playing the role of the non linear strain rate dependent viscosity, and the displacements and strains that of the velocities and strain rates in the analogous flow model respectively [1]. This analogy has allowed the solution of complex metal forming problems with standard finite element programs originaly written for 2D and 3D elasticity [2]-[4]. Applications of this approach in the context of sheet metal forming analysis leaded to the derivation of the so called *viscous shell model*. This is based on a simple modification of standard small displacement elastic shell theory using the mentioned flow-elasticity analogy [5].

163

J. L. Chenot and E. Oñate (eds.), Modelling of Metal Forming Processes, 163–178.
© *1988 by Kluwer Academic Publishers.*

Recently, Oñate *et al.* [6] have extended the viscous shell model to deal with *material degradation* effects by taking into account nucleation, growth and coalescence of microscopic voids in the deforming metal sheet. The resulting *viscous voided shell model* introduces the effect of material compressibility in the form of a two parameters constitutive model which can be simply identified as the equivalent shear modulus and Poisson's ratio of an analogous non linear elasticity material. This allows finite element solutions to be obtained for such potentially more difficult problems by directly using computer programs written for standard —compressible— elasticity.

The objective of this paper is to present in an unified form the basic concepts of the viscous shell model for plastic/viscoplastic materials including the effect of material degradation due to the development of microscopic voids, for the finite element analysis of sheet metal forming problems. The formulation will be particularized for the axisymmetric case using simple linear axisymmetric shell and membrane elements. Details of the treatment of the contact and friction effects are also briefly given. Finally, some examples of application of the general formulation to three examples of pure and hemispherical stretching and deep drawing of a circular sheet are presented.

2. Basic concepts

The basis of the plastic/viscoplastic flow approach is to neglect elastic stresses and strains in the deforming material [1],[3]. This assumption allows to write the following rate equation

$$\dot{\epsilon}_{ij} = \dot{\epsilon}_{ij}^{NL} = f(\sigma_{ij}) \tag{1}$$

where $\dot{\epsilon}_{ij}$ and $\dot{\epsilon}_{ij}^{NL}$ account for the total and non linear —plastic/viscoplastic— strain rate tensors, respectively. The form of function f depends on the type of plastic/viscoplastic constitutive model used. In any case (1) describes the behaviour of an equivalent *fluid* in which strain rates and velocities u_i are simply related by

$$\dot{\epsilon}_{ij} = \frac{1}{2}\left(\frac{\partial u_i}{\partial x_j} + \frac{\partial u_j}{\partial x_i}\right) \qquad \text{or} \qquad \dot{\epsilon} = Lu \tag{2}$$

and the stresses satisfy the standard equilibrium conditions [7]

$$L^T \sigma + b = 0 \qquad \text{in the volume } V$$
$$M^T \sigma + t = 0 \qquad \text{in the boundary } \Gamma \tag{3}$$

where b and t are body force and surface load vectors, respectively, and M is a matrix containing the components of the unit normal to the boundary Γ.

In the following sections particular forms of (1) for different types of material behaviour will be presented.

2.1. RIGID PLASTIC FLOW OF VOID CONTAINING METALS

The yield condition for a randomly voided material with spherical —for 3D problems— or circular cylindrical voids —for plane stress problems— may be assumed following Gurson [8] as

$$\Phi = \frac{3}{2}\frac{s_{ij}s_{ij}}{\sigma_M^2} - \omega = 0 \tag{4}$$

where σ_{ij} and s_{ij} are the macroscopic Cauchy stress and stress deviator, respectively, σ_M is the tensile yield limit of the matrix material (assumed incompressible), f is the void volume fraction and

$$\omega = 1 - 2f\cosh\left(\frac{\sigma_{hh}}{2\sigma_M}\right) + f^2 \tag{5}$$

Note that for porosity parameter $f = 0$, $\omega = 1$ and (4) reduces to the classical Von Mises yield condition [2]. The change of the void volume fraction during the deformation increment is taken as [8]-[11]

$$\dot{f} = \dot{f}_g + \dot{f}_n + \dot{f}_c \tag{6}$$

where subscrits g, n and c stand for growth, nucleation and coalescence of voids. Also, it can be assumed that [9]-[10]

$$\dot{f}_g = (1-f)\dot{\epsilon}_{kk}^{(p)} \qquad \dot{f}_n = \frac{k}{\sigma_M}\left(\dot{\sigma}_M + \frac{\dot{\sigma}_{kk}}{3}\right) \tag{7}$$

the material parameter k being the volume fraction of particles converted to voids per unit fractional increase in stress. Nucleation is assumed to occur only if the approximate value of the maximum normal stress $\sigma_M + \frac{\sigma_{kk}}{3}$ exceeds in the current time increment   its previous maximum. Finally, the term $\dot{f}_c$ can be numerically accounted for in the following way. According to [10], coalescence takes place at $f \simeq 0.2$. Thus a simple and effective numerical scheme can be used to reproduce this phenomenon if, at points for which $f = 0.2$, a proportional increase in f is assumed for a number of fixed incremental steps ($\simeq 5$), up to $f = 1$, for which the material carrying capacity in that point is effectively zero.

Using standard plasticity theory [8], it is possible to arrive at the following expression for the non linear —plastic— strain rates of eq. (1)

$$\dot{\epsilon}_{ij}^{(p)} = \dot{\epsilon}_{ij} = \frac{1}{2\bar{G}}\left(\sigma_{ij} - \frac{\bar{\nu}}{1+\bar{\nu}}\sigma_{kk}\delta_{ij}\right) \tag{8}$$

where

$$\bar{G} = \frac{\sigma_M}{3\bar{\epsilon}}\left(\frac{\omega + fAS}{1-f}\right) \tag{9}$$

$$\bar{\nu} = \frac{1-B}{2+B} \tag{10}$$

with $B = \frac{LS}{2A}$, $A = \frac{\sigma_{hh}}{2\sigma_M}$, $S = \sinh A$ and $\bar{\dot{\epsilon}} = \sqrt{\frac{2}{3}\dot{\epsilon}_{ij}\dot{\epsilon}_{ij}}$. After some manipulations a simpler expression for $\bar{G}$ can be found as

$$\bar{G} = \frac{\sigma_M \sqrt{\omega}}{3\sqrt{\bar{\dot{\epsilon}}^2 - \frac{2}{9}\dot{\epsilon}_{hh}^2}} \tag{11}$$

Comparing (8) with the constitutive equation for classical elasticity [7], and taking into account (2) and (3) it can be easily be concluded that *there is a perfect analogy between the equations of plastic flow of a voided metal and those of standard elasticity.*

Therefore, displacements and strains in the analogous elastic model, can be identified with the velocities and strain rates, respectively, and the elastic shear modulus and Poisson's ratio with parameters $\bar{G}$ and $\bar{\nu}$ given by (8)-(10). Note the stress/strain rate dependence of $\bar{G}$ and $\bar{\nu}$ which makes the analogous elastic problem non linear and the numerical solution must therefore be found iteratively.

2.2. PLASTIC FLOW OF NON VOIDED METALS

For *classical plastic materials* $f = 0$ and $\omega = 1$ and therefore, from (8)-(10) we have

$$\bar{\nu} = \frac{1}{2} \qquad \bar{G} = \frac{\sigma_M}{3\bar{\dot{\epsilon}}} \tag{12}$$

and

$$\dot{\epsilon}_{ij} = \frac{1}{2\bar{G}}\left(\sigma_{ij} - \frac{\sigma_{hh}}{3}\delta_{ij}\right) = \frac{s_{ij}}{2\bar{G}} \tag{13}$$

Thus the *incompressible* form of the deformation is recovered and the expression for the equivalent shear modulus $\bar{G}$ coincides with that of the non-Newtonian viscosity of the standard plastic flow problem [2].

2.3. INCLUSION OF VISCOPLASTIC EFFECTS

The expression for the viscoplastic strain rate can be postulated as [7]

$$\begin{aligned} \dot{\epsilon}_{ij} = \gamma\chi^n\frac{\partial\phi}{\partial\sigma_{ij}} \qquad &\text{for} \quad \chi > 0 \\ \dot{\epsilon}_{ij} = 0 \qquad &\text{for} \quad \chi = 0 \end{aligned} \tag{14}$$

where γ is the material fluidity and χ is the overstress parameter defined as

$$\chi = \sqrt{\frac{3}{2}\frac{s_{ij}s_{ij}}{\omega}} - \sigma_M \tag{15}$$

Note that for the non viscous case $\chi = 0$ (see eq. (4)).

Oñate *et al.* [6] have shown that accounting for the viscous properties of a voided metal in the manner described above corresponds to replacing in (9) or (11) the matrix yield value σ_M by

$$\sigma_M\left[1+\left(\frac{\dot{\bar{\epsilon}}}{\gamma\sigma_M^n\sqrt{\omega+2\beta_1^2}}\right)^{\frac{1}{n}}\right] \tag{16}$$

with $\beta_1 = \frac{\dot{t}s}{2}$. It is easy to see that for the particular case of non voided material $(f = 0, \omega = 1)$ eqs. (9) and (16) lead to

$$\bar{G} = \frac{\sigma_M + (\frac{\dot{\bar{\epsilon}}}{\gamma})^{\frac{1}{n}}}{3\dot{\bar{\epsilon}}} \tag{17}$$

which again exactly coincides with the expression of the fluid viscosity derived in [2].

3. Application to thin sheet metal forming problems

The analogy presented in previous sections allows the treatment of large plastic/viscoplastic deformations of thin sheets of metal making direct use of classical small displacement elastic shell theory. The solution scheme is thus as follows

1) Identify an elastic shell formulation. If standard finite element techniques [7] are used, a discrete system of equations is obtained, upon discretization, of the form

$$K(G,\nu)a = f \tag{18}$$

where K is the shell stiffness matrix and a and f are the displacement and nodal force vectors, respectively. The equivalent *viscous voided shell* is formulated by simply identifying displacements and strains with velocities and strain rates respectively, and the shear modulus and Poisson's ratio with parameters $\bar{G}$ and $\bar{\nu}$ given in previous section. For the simpler non voided case $\bar{\nu} = \frac{1}{2}$ and $\bar{G}$ is given by (12) or (17). Eq. (18) becomes a system of non linear equations which must be solved iteratively. In the initial solution values of velocities a^0, and void volume fraction f^0 must be specified.

2) Solve for a^1. If direct iteration is used the first iteration becomes

$$a^1 = [K(a^0)]^{-1}f \tag{19}$$

3) Check for convergence of velocities. If desired convergence is not achieved go back to 2.

4) Once convergence has been achieved the geometry is updated by $a\Delta t$, where Δt is an appropriate time step size, which can be taken as constant or equal to the time increment for which the first node of the non contacting region comes into contact with the indenting punch [5],[6]. Also, the boundary conditions must be changed if new points have come into contact with the tool surface. Finally, the values of the sheet thickness and void volume fraction are updated according to

the values of the thickness and volumetric strain respectively.

5) The process is restarted with the new values of the sheet geometry and void volume fraction.

The algorithm is thus very simple and it also allows other effects like strain hardening and friction conditions to be included in a straight- forward manner [6] (see also Section 5).

It is worth nothing that in well developed sheet forming stages the spatial velocity field does not change much between two consecutive solutions. Thus, significant savings in computer time can be obtained by updating the sheet geometry using the constant spatial velocity field for a number of incremental steps [6].

We have also to point out that direct iteration usually yields convergence of the velocity field after a small number of iterations. This is due to the well posed boundary value nature of the problem in which velocities are prescribed at the tool–blank contact nodes, and forces (reactions) are obtained 'a posteriori' from the converged velocity field. Thus for each solution the initial velocities can be guessed to be not too far from their correct values and convergence is rapidly achieved. Special care, however, must be taken to define the cut-off value of the equivalent shear modulus in zones of the sheet where rigid deformations are expected, to prevent ill conditioning of the stiffness matrix.

In the next sections we present details of the finite element viscous voided shell formulation for axisymmetric sheet metal forming problems.

4. Axisymmetric formulation

4.1. AXISYMMETRIC SHELL FORMULATION

The basis of the success of the *viscous shell approach* described in previous sections lies in the efficiency of the analogous elastic shell formulation. We will briefly present here the relevant expressions of the finite element axisymmetric formulation developed by Oñate *et al.* [5],[6] for thin sheet metal forming analysis. The formulation is based on Reissner-Mindlin shell theory and uses the simple two node linear element. Details of the shell theory can be found in [6],[13].

The velocity field, after discretization of the shell in axisymmetric linear elements, can be expressed as

$$\boldsymbol{u} = \left\{ \begin{array}{c} u \\ w \\ \theta \end{array} \right\} = \sum_{i=1}^{2} N_i \boldsymbol{a}_i \quad \text{with} \quad N_i = N_i I_3 \quad \text{and} \quad \boldsymbol{a}_i = \left\{ \begin{array}{c} u_i \\ w_i \\ \theta_i \end{array} \right\} \tag{20}$$

with u_i, w_i and θ_i being the two global velocities and the angular velocity of node i, respectively and N_i the linear shape function of node i. (*Figure 1*).

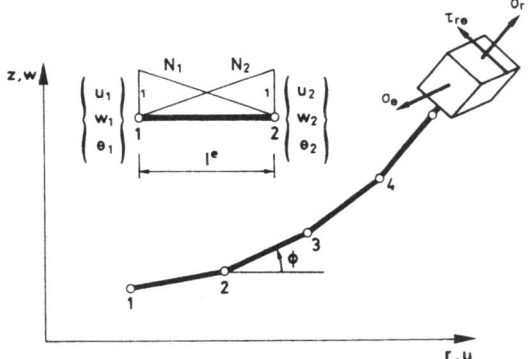

Figure 1.- Axisymmetric shell. Discretization in axisymmetric linear elements

The generalized strain rate and stress vectors can be expressed as

$$\dot{\epsilon} = [\dot{\epsilon}_r, \dot{\epsilon}_\theta, \dot{\gamma}]^T = \sum_{i=1}^{2} B_i a_i \qquad (21)$$

$$\sigma = [\sigma_r, \sigma_\theta, \tau]^T = D\dot{\epsilon} = D\sum_{i=1}^{2} B_i a_i \qquad (22)$$

where for an isotropic material

$$D = \begin{bmatrix} d_{11} & d_{12} & 0 \\ d_{21} & d_{22} & 0 \\ 0 & 0 & d_{33} \end{bmatrix} \qquad (23)$$

with

$$d_{11} = d_{22} = 2\bar{G}\frac{1}{1-\bar{\nu}}, \qquad d_{12} = d_{21} = 2\bar{G}\frac{\bar{\nu}}{1-\bar{\nu}}, \qquad d_{33} = \bar{G} \qquad (24)$$

where $\bar{G}$ and $\bar{\nu}$ are given by (9)-(11), and the expression of the strain rate matrix B_i is given in the Appendix.

The element contributions to the stiffness matrix K, and the nodal force vector f are

$$K_{ij}^{(e)} = 2\pi \int_{l^{(e)}} B_i^T \hat{D} B_j r ds \qquad (25)$$

$$f_i^{(e)} = 2\pi \int_{l^{(e)}} N_i t r ds + 2\pi r_i p_i \qquad (26)$$

where $l^{(e)}$ is the element length, r the radial distance, t and p, surface and point load vectors, respectively and the expression of matrix $\hat{D}$ is given in the Appendix. It has been shown that for a successfull use of this formulation the integral of (25) must be numerically computed using a single Gaussian integration point. This allows to obtain an explicit form of $K_{ij}^{(e)}$ as

$$K_{ij}^{(e)} = 2\pi \bar{B}_i^T \hat{\bar{D}} \bar{B}_j \bar{r} l^{(e)} \qquad (27)$$

where $(\bar{\cdot})$ denotes values at the element midpoint. The expression of $\bar{B}_i$ is readily obtained by substituting the terms N_i and $\frac{\partial N_i}{\partial s}$ in eq. (A.1) by $\frac{1}{2}$ and $\frac{(-1)^i}{l(e)}$, respectively.

4.2. AXISYMMETRIC MEMBRANE FORMULATION

The membrane formulation can be easily derived from the general case presented in previous sections by simply neglecting in all expressions the flexural and shear terms. The relevant matrices and vectors are now defined as

Velocity field:
$$\mathbf{u} = \left\{ \begin{matrix} u \\ w \end{matrix} \right\} = \sum_{i=1}^{2} N_i I_2 \mathbf{a}_i \qquad \mathbf{a}_i = \left\{ \begin{matrix} u_i \\ w_i \end{matrix} \right\} \tag{28}$$

Generalized strain rate field:
$$\hat{\dot{\epsilon}} = \left\{ \begin{matrix} \dot{\epsilon}_r \\ \dot{\epsilon}_\theta \end{matrix} \right\} = \sum_{i=1}^{2} B_{m_i} \mathbf{a}_i \tag{29}$$

Generalized stress field:
$$\hat{\sigma} = \left\{ \begin{matrix} \hat{\sigma}_r \\ \hat{\sigma}_\theta \end{matrix} \right\} = D_m \hat{\dot{\epsilon}} = D_m \sum_{i=1}^{2} B_{m_i} \mathbf{a}_i \tag{30}$$

where
$$D_m = t \begin{bmatrix} d_{11} & d_{12} \\ d_{21} & d_{22} \end{bmatrix} \tag{31}$$

where t is the thickness, d_{ij} are given in (24) and

$$B_{m_i} = \begin{bmatrix} \cos\phi \frac{\partial N_i}{\partial s} & \sin\phi \frac{\partial N_i}{\partial s} \\ \frac{N_i}{r} & 0 \end{bmatrix} \tag{32}$$

Finally the explicit expression of the stiffness matrix for the linear element in this case is identical to (27) with B_{m_i} and D_m instead of B_i and $\hat{D}$, respectively.

5. Treatment of friction

An algorithm to simulate friction effects between the contact interfaces can be based on an simple adjustment of nodal reactions at contact nodes after each iterative solution, until they satisfy a Coulomb type of friction law. This method has been succesfully implemented by the authors [5],[6] and it is used in the examples presented in this paper.

A more consistent procedure to model contact and friction can be developed by imposing the contact conditions via a penalty type approach, using the total potential of the contact forces with the geometric compatibility conditions. Generally the contact constraints can be written as

$$r = Ca - s \tag{33}$$

where a is the velocity field at a particular deformed configuration and C and s are obtained from the appropiate sticking or sliding contact rule [14].

The expression of the total potential energy penalized using (33) is expressed as

$$\Pi^* = \frac{1}{2}a^T K a - a^T f + \frac{1}{2}r^T \alpha r \tag{34}$$

where α is a diagonal penalization matrix. The stationarity of (34) gives the modified system of equations to be solved at each iteration as

$$[K + C^T \alpha C]a = f + C^T \alpha s \tag{35}$$

The friction conditions can now be taken into account at each iteration by imposing a Coulomb type of relationship between the total resultant normal and tangential forces acting on each element belonging to the contact surface. This procedure is currently under development by the authors. For more general information see [14].

6. Examples

6.1. STUDY OF THE INFLUENCE OF POROSITY IN THICKNESS CHANGE

In the first example the effect of porosity development in the change of thickness in a circular sheet, of radius = 2.20 in. and an initial thickness of 0.035 in., under uniform stretching has been studied. The uniaxial stress–effective strain curve of the matrix material is given by

$$\sigma_Y = 5.4 + 27.8\bar{\epsilon}^{0.504} \quad \text{tn/in}^2 \qquad \bar{\epsilon} < 0.36$$
$$\sigma_Y = 5.4 + 24.4\bar{\epsilon}^{0.375} \leq 22.0 \quad \text{tn/in}^2 \qquad \bar{\epsilon} \geq 0.36$$

Fifty axisymmetric linear elements have been used in the analysis. The thickness strain and porosity distributions obtained for two cases with $k = 0.005$ and $k = 0.01$ respectively, in all the sheet and $f^0 = 0.0$ in all elements except $f^0 = 0.01$ in element 20 ($r \simeq 1.0$ in.) for both cases, are shown in *Figure 2*. It can be clearly seen in these figures that porosity development affects very strongly the otherwise uniform thickness strain field and causes strain localization as expected. Also from *Figure 2* it is clear that the inclusion of the void nucleation parameter k amplifies the mentioned localization effect.

6.2. HEMISPHERICAL STRETCHING OF CIRCULAR ISOTROPIC SHEET

The geometrical configuration of the problem is shown in *Figure 3*. Fifty axisymmetric linear elements have been used in the analysis. The uniaxial stress–effective strain curve of the matrix material is the same as in previous example. A friction coefficient of 0.04 has been used, as suggested in [15]. The problem has

172

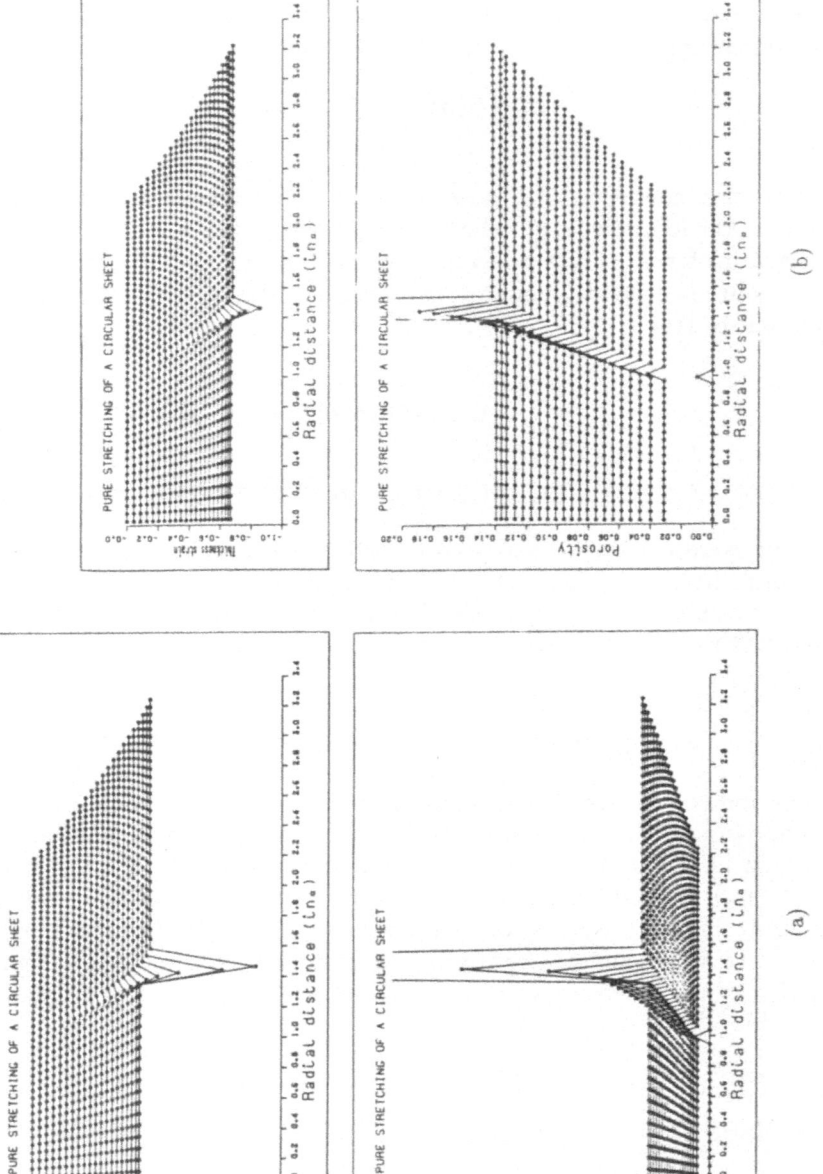

Figure 2.- Pure stretching of a circular sheet. Effect of porosity in thickness strain. $f^0 = 0$ in all the sheet except $f^0 = 0.01$ in element 20 ($r \simeq 1$ in.). (a) $k = 0.005$ (b) $k = 0.01$

173

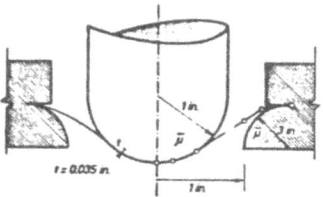

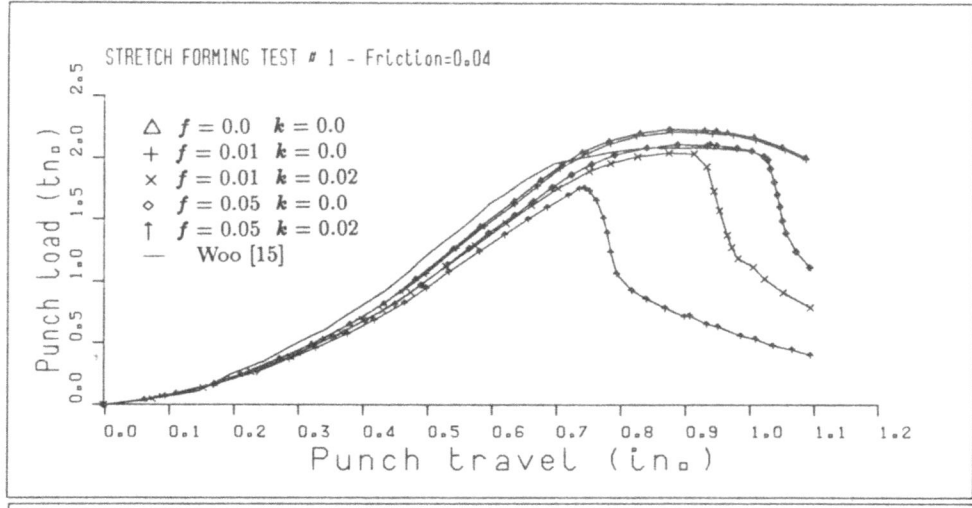

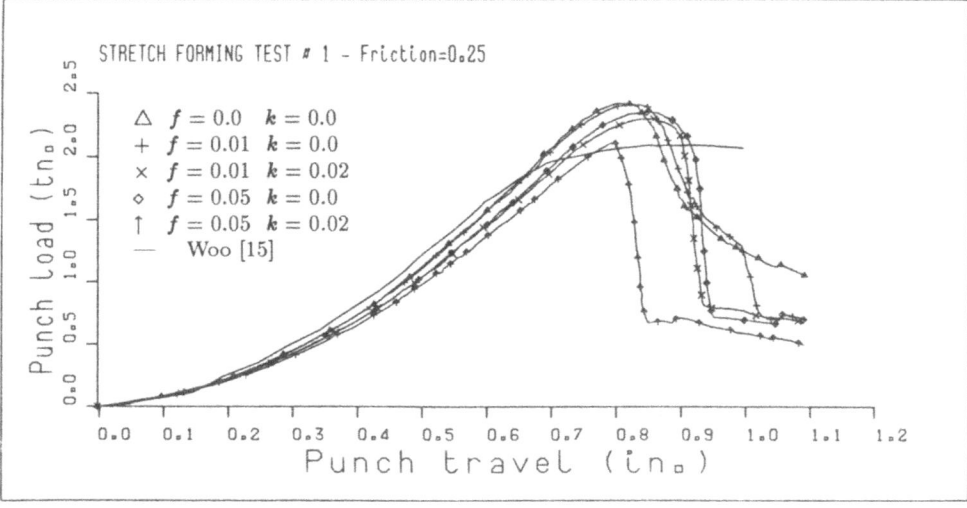

Figure 3.- Hemispherical sheet forming.
Punch load-displacement curve for different porosity conditions.

been analysed for different initial void volume fractions of $f^0 = 0.0$, 0.01, and 0.05, and nucleation parameters $k = 0.0$ and 0.02 in all elements. Numerical results for the punch load–displacement curves for various values of f^0 and k are shown in *Figure 3*. From these results it can be deduced that:

(a) An increase of the initial void porosity and nucleation parameters causes a progressive reduction of the load carrying capacity of the sheet. Values of the maximum punch load obtained for $f^0 = 0.0$ and 0.01 with $k = 0.0$ and 0.02, and $f^0 = 0.05$ with $k = 0.0$, are in agreement with the experimental results reported in [15]. However, for $f^0 = 0.05$ and $k = 0.02$ a reduction of the maximum punch load of 35 per cent is obtained.

(b) Inclusion of void porosity induces localized failure with a rapid loss of rigidity which causes an almost vertical descent of the load–displacement curve.

6.3. HEMISPHERICAL DEEP DRAWING OF CIRCULAR ISOTROPIC SHEET

The geometrical configuration of the problem is shown in *Figure 4*. Fifty axisymmetric linear elements have been used for the analysis. The uniaxial stress–strain curve of the material is the same as in previous example. A friction coefficient of 0.04 has been used [16]. Numerical results of the punch load–displacement curve and hoop and thickness strain distributions for $f = 0$ and $k = 0$ have been plotted in *Figure 4*. Numerical results compare reasonably well with experimental ones [16]. Finally a 3D perspective of the sheet geometry at different deformation stages is shown in *Figure 5*.

7. Conclusions

The equations describing plastic and viscoplastic flow of metals including the effects of nucleation, growth and coalescence of voids, are analogous to those of classical non linear elasticity. The formulation for non voided materials is directly obtained from the general case simply by neglecting the effect of voids, thus yielding the classical form analogous to incompressible elasticity. This allows standard finite element methods developed for elastic shell problems to be directly used for the analysis of complex sheet metal forming processes, including material degradation effects by development of microscopic voids.

The examples analysed show that by adjusting parameters such as the value and distribution of the initial void volume fraction and of the fraction of particles converted to voids per unit increase of stress, the model should be able to predict development of voids and localized material failure.

8. References

1. Zienkiewicz, O.C. and Godbole, P.N., 'Flow of plastic and viscoplastic solids with special reference to extrusion and forming processes', *Int. J. Num. Meth. Engng.*, **8**, 3–16 (1979).

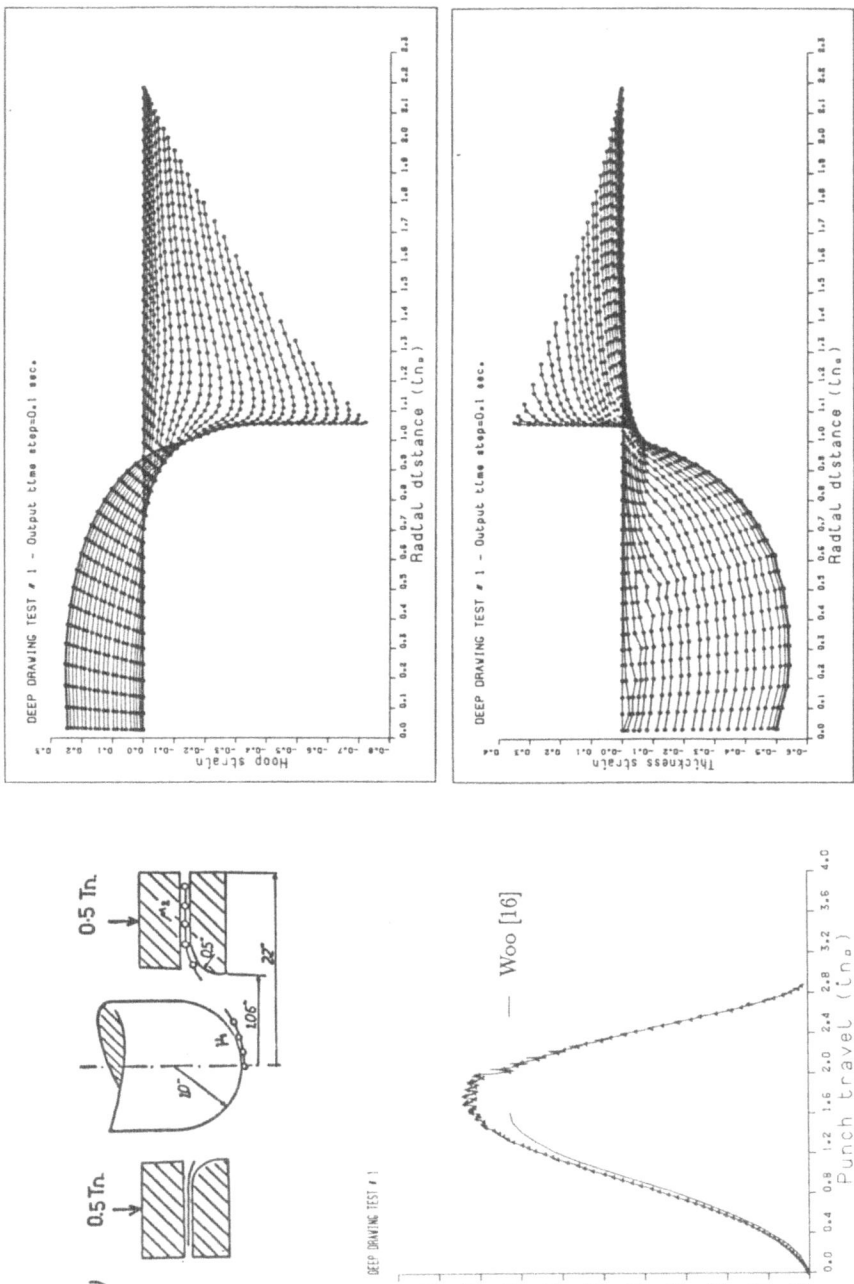

Figure 4.- Hemispherical deep drawing. (a) Punch load-displacement curves. (b) Hoop and thickness strain distributions for various deforming configurations.

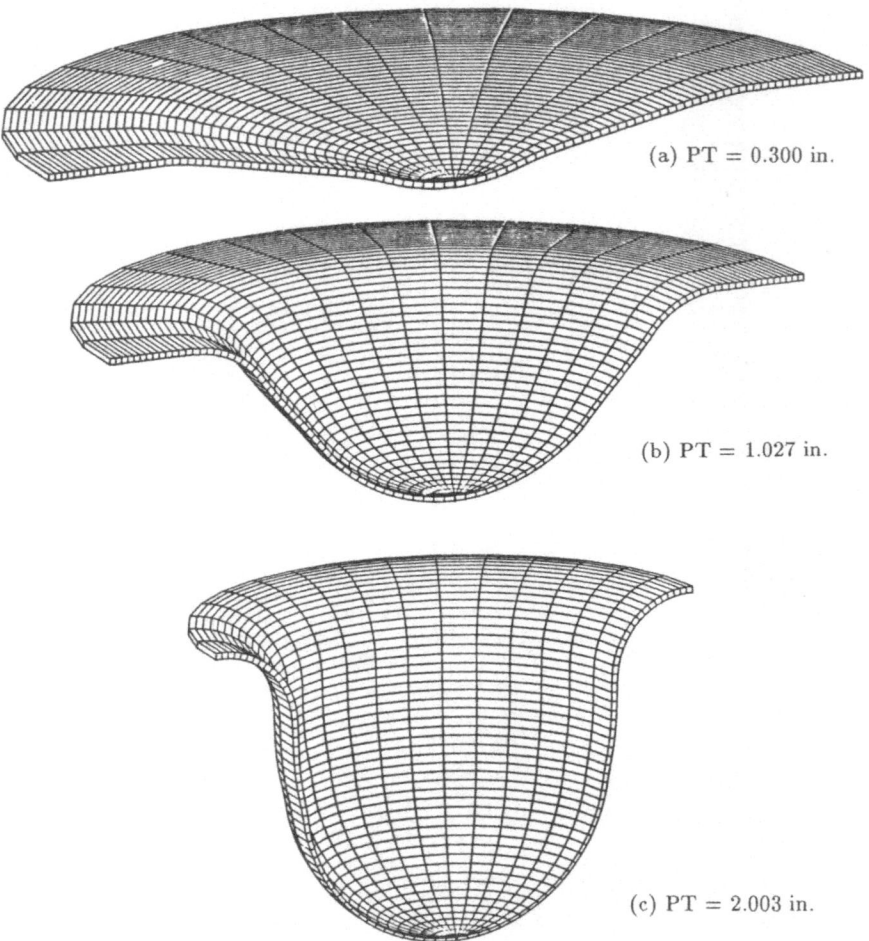

(a) PT = 0.300 in.

(b) PT = 1.027 in.

(c) PT = 2.003 in.

Figure 5.- Hemispherical deep drawing.
Geometry of the sheet for various deforming configurations.
(PT = Punch travel)

2. Zienkiewicz, O.C., Jain, P.C. and Oñate, E., 'Flow of solids during forming and extrusion. Some aspects of numerical solutions', *Int. J. Solids Struct.*, **14**, 15–38 (1978).

3. Zienkiewicz, O.C., Oñate, E. and Heinrich, J.C., 'A general formulation for coupled thermal flow of metals using finite elements', *Int. J. Num. Meth. Engng.*, **17**, 1497–1514 (1981).

4. Pittman, J.F.T., Zienkiewicz, O.C., Wood, R.D. and Alexander, J.M. (eds.), *Numerical Analysis of Forming Processes*, Wiley, New York, 1984.

5. Oñate, E. and Zienkiewicz, O.C., 'A viscous shell formulation for the analysis of thin sheet metal forming', *Int. J. Mech. Sc.*, **25**, 305–335 (1983).

6. Oñate, E., Kleiber, M. and Agelet de Saracibar, C., 'Plastic and viscoplastic flow of void containing metals. Applications to axisymmetric sheet forming problems', *Int. J. Num. Meth. Engng.*, **25**, 225–251 (1988).

7. Zienkiewicz, O.C., *The Finite Element Method*, McGraw–Hill, 1979.

8. Gurson, A.L., 'Continuum theory of ductile rupture by void nucleation and growth. I. Yield criteria and flow rules for porous ductile media', *J. Eng. Mater. Tech.*, **99**, 2–15 (1977).

9. Needleman, A. and Rice, J.R., 'Limits to ductility set by plastic flow localization', *Mechanics of Sheet Metal Forming*. D.P. Koistinen and N.-M. Wang (eds.), 237–266, Plenum, N.Y., 1978.

10. Tvergaard, V., 'On localization in ductile materials containing spherical voids', *Int. J. Fracture*, **18**, 237–252 (1982).

11. Kleiber, M., 'Numerical study on necking-type bifurcations in void-containing elastic-plastic material', *Int. J. Solids Struct.*, **20**, 191–210 (1984).

12. Baynham, J.M.W. and Zienkiewicz, O.C., 'Developments in the finite element analysis of thin sheet drawing and direct redrawing processes using the rigid plastic approach', *Proceedings of International Conference on Numerical Methods in Industrial Forming Processes*, J.F.T. Pittman *et al.* (eds.), Pineridge Press, Swansea, 1982.

13. Zienkiewicz, O.C., Bauer, J., Morgan, K. and Oñate, E., 'A simple and efficient shell element for axisymmetric shells', *Int. J. Num. Meth. Engng.*, **11**, 1545–1559 (1977).

14. Bathe, K.J. and Chaudary, A., 'A solution method for planar and axisymmetric contact problems', *Int. J. Num. Meth. Engng.*, **21**, 65–88 (1985).

15. Woo, D.M., 'The stretch forming test', *The Engineer*, **220**, 876–889 (1965).

16. Woo, D.M., 'On the complete solution of the deep–drawing problem', *Int. J. Mech. Sci.*, **10**, 83–94 (1968)

Appendix

STRAIN RATE AND CONSTITUTIVE MATRIX FOR AXISYMMETRIC SHEET FORMULATION

Strain rate matrix
$$B = [B_1, B_2]$$

$$B_i = \begin{bmatrix} \cos\phi\frac{\partial N_i}{\partial s} & \sin\phi\frac{\partial N_i}{\partial s} & 0 \\ \frac{N_i}{r} & 0 & 0 \\ 0 & 0 & \frac{-\partial N_i}{\partial s} \\ 0 & 0 & -N_i\frac{\cos\phi}{r} \\ -\sin\phi\frac{\partial N_i}{\partial s} & \cos\phi\frac{\partial N_i}{\partial s} & -N_i \end{bmatrix} \qquad (A.1)$$

for definition of ϕ see *Figure 1*.

Constitutive matrix
$$\hat{D} = \int_{-\frac{t}{2}}^{\frac{t}{2}} S^T D S dz' \qquad (A.2)$$

$$S = \begin{bmatrix} 1 & 0 & z' & 0 & 0 \\ 0 & 1 & 0 & z' & 0 \\ 0 & 0 & 0 & 0 & 1 \end{bmatrix} \qquad D = \begin{bmatrix} d_{11} & d_{12} & 0 \\ d_{21} & d_{22} & 0 \\ 0 & 0 & d_{33} \end{bmatrix}$$

d_{ij} as given in (24). Note that the computation of $\hat{D}$ implies an integration across the thickness. This is in practice performed using numerical integration.

TIME STEPPING SCHEMES FOR THE NUMERICAL ANALYSIS OF SUPERPLASTIC
FORMING OF THIN SHEET

J. BONET, R.D. WOOD and O.C. ZIENKIEWICZ
Civil Engineering Department, University College of
Swansea, Swansea, SA2 8PP, Wales, U.K.

SUMMARY

The numerical simulation of the superplastic forming of thin sheet
involves the time integration of velocities in order to determine the
changing configuration of the sheet as it forms into the final component shape. During the course of developing a finite element
analysis of the problem various time stepping schemes have been
investigated. This paper discusses these schemes and reports on the
success or otherwise of their implementation.

INTRODUCTION

Superplastic forming of products made from thin sheet is a manufacturing technique whereby, typically, titanium or aluminium alloy sheet,
at sufficiently high temperature, can be blow formed, without frac-
turing, into a die to produce a very complex, light and strong
component .   In addition judicial use of the diffusion
bonding characteristics of superplastic alloys enables assemblages to
be formed which are structurally more integral than the same component
produced by traditional means.
    The crucial problem of predicting the relationship between the
forming pressure cycle and the product thickness distribution can be
ameliorated by using a finite element based numerical simulation of
the forming behaviour, [1-8]. This usually involves a solution scheme
in which velocities have to be integrated in order to ascertain the
changing shape of the product as forming progresses. Furthermore,
since the constitutive equations are nonlinear, the integration scheme
is intimately connected to the nonlinear solution procedure.
    This paper discusses explicit and implicit time stepping schemes
associated with a finite element solution to the problem of simulating
thin sheet superplastic forming. In particular, deficiencies experienced when the explicit scheme is used for general shapes are
overcome by introducing a trapezoidal implicit scheme which is further
improved by using a two step backward difference implicit scheme.

J. L. Chenot and E. Oñate (eds.), Modelling of Metal Forming Processes, 179–186.
© 1988 by Kluwer Academic Publishers.

GOVERNING EQUATIONS

The kinematic description of the deforming sheet is expressed in terms of the midsurface $\Gamma(t)$ which is defined by the convective coordinates ξ^1 and ξ^2 having associated contravariant base vectors $\underline{e}_\alpha$ given as,

$$\underline{e}_\alpha = \frac{\partial x_\alpha^i}{\partial \xi^\alpha} \underline{E}_i \; ; \qquad i = 1,2,3 \text{ and } \alpha = 1,2 \qquad (1)$$

A third coordinate ξ^3 is erected with unit normal $\underline{n}$ to the surface, see Figure 1

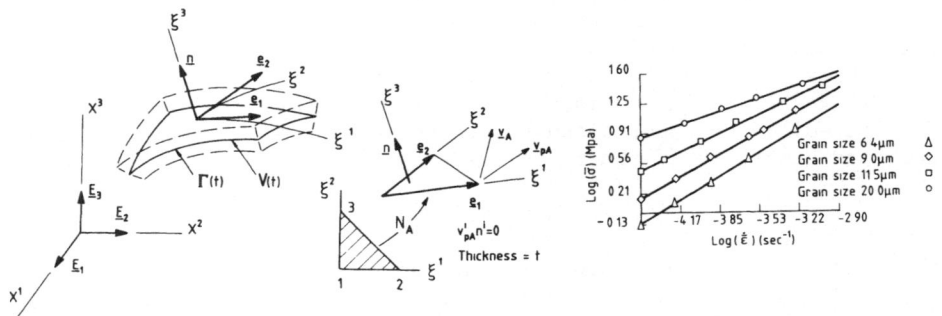

Figure 1   Coordinate definitions   Figure 2   Linearized stress-strain relation for Ti-6Aℓ-4V

The membrane rate of deformation (strain rate) is given as [6]

$$D_{\alpha\beta}^m = \frac{1}{2} (x_{\Gamma,\alpha}^i \, v_{\Gamma,\beta}^i + x_{\Gamma,\beta}^i \, v_{\Gamma,\alpha}^i) \qquad (2)$$

where v_Γ^i are the contravariant components of the midplane velocity of the sheet.   (henceforth Γ will be omitted).
    The sheet material behaviour is modelled as a non-Newtonian viscous membrane for which the constitutive relation between the Von Mises equivalent stress $\bar\sigma$ and equivalent strain rate $\dot{\bar\varepsilon}$ is given in terms of a nonlinear viscosity μ as, [9]

$$\bar\sigma = 3 \mu \, \dot{\bar\varepsilon}; \quad \mu = (K\dot{\bar\varepsilon}^{m-1})/3; \quad g = g_o(t/10)^N; \quad t > 10 \text{ min.} \qquad (1a,b,c)$$

where the constant K and the strain rate sensitivity index m are both functions of the grain size g, see Figure 2. The variable N is a function of the strain rate [10] and g_o is the initial grain size for $0 < t \le 10$ min.   Incorporating plane stress and incompressibility assumptions enables the constitutive relations for a viscous membrane shell to be developed as,

$$\sigma_m^{\alpha\beta} = 2 \ \mu t \ C^{\alpha\beta\lambda\delta} \ D_{\lambda\delta}^m \qquad (3)$$

where $\sigma_m^{\alpha\beta}$ are the membrane stress resultants over the thickness t and $C^{\alpha\beta\lambda\delta}$ is a function of the metric tensor.

The governing equilibrium expression is the virtual velocity equation,

$$\int_A \sigma_m^{\alpha\beta} \ \delta D_{\alpha\beta}^m \ dA - \int_A p\delta v_3 \ dA = 0 \qquad (4)$$

where A is the current sheet surface area, p the pressure and v_3 the velocity normal to the sheet.

FINITE ELEMENT DISCRETIZATION

The shell is discretized using the constant membrane stress triangular element for which the geometry is defined in terms of the convective coordinates ξ^α as,

$$x^i(\xi^\alpha) = x_A^i \ N_A \ (\xi^\alpha); \quad i,A = 1,2,3 \text{ and } \alpha = 1,2 \qquad (5)$$

where the shape functions N_A provide the mapping between the two and three dimensional triangles shown in Figure 1. The element midsurface inplane velocities are interpolated as,

$$v_p^i = v_{pA}^i \ N_A \quad \text{or} \quad v^\alpha = v_A^\alpha \ N_A \qquad (6)$$

where v_{pA}^i and v_A^α are the inplane components of the nodal translational velocities expressed in terms of the cartesian and convected coordinate systems respectively.

Using equations (2-6) enables the discretized equilibrium equations per element to be written as,

$$T_{mA}^i = F_A^i; \quad T_{mA}^i = \sigma_m^{\alpha\beta} \ N_{A,\beta} \ x_{,\alpha}^i \ A^e; \quad F_A^i = pA^e n^i/3 \qquad (7,a,b,c)$$

where T_{mA}^i are the components of the internal equivalent nodal forces due to membrane stresses and F_A^i the equivalent nodal forces due to the external pressure loading p. A^e is the element area.

Assembling the nodal equilibrium equations in the standard manner produces a system of differential equations in time as,

$$\underline{T}(\underline{v},\underline{x}) = \underline{F}(\underline{x},t) \qquad (8)$$

SOLUTION PROCEDURES

The assembled equilibrium equations are nonlinear with respect to both velocity $\underline{v}$ and geometry $\underline{x}$ and in order to follow the forming process

they have to be integrated in time. Consequently a solution procedure
requires an integration scheme comprising predictor and corrector
stages, the latter, due to the nonlinearity, being iterative. Here
this iterative solution is achieved using a Newton-Raphson method.
    In general the integration scheme can be expressed as,

$$x_{n+1} = x_n + \bar{V} \, \Delta t_{n+1} \tag{9}$$

where $\bar{V}$ is either an explicit or implicit function of the velocities.
Figure 3 gives expressions for $\bar{V}$ for the explicit one step forward

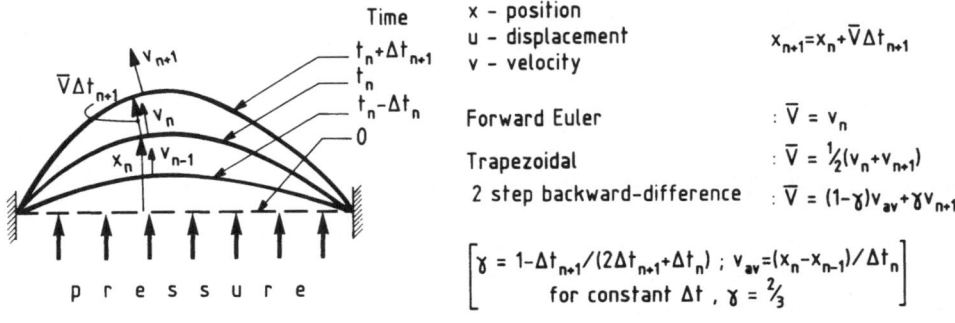

Figure 3   Time stepping schemes

Euler, the implicit one step trapezoidal and the implicit two step
backward-difference schemes. Introducing these schemes into the
equilibrium equations (8) gives,

$$\underline{T}(\underline{\sigma}[\mu(\underline{v},\underline{x}(\underline{v})), \ \underline{v},\underline{x}(\underline{v})], \ \underline{x}(\underline{v})) - \underline{F}(\underline{x}(\underline{v}),t) = \underline{0} \tag{10}$$

where $\underline{v}$ and $\underline{x}$ refer in general to values at time step $(n+1)$. To solve
equations (10) requires the development of a tangent stiffness matrix
as,

$$\underline{\underline{K}}T = \frac{D\underline{T}}{D\underline{v}} - \frac{D\underline{F}}{D\underline{v}} = \underline{\underline{K}}_{\mu v} + \underline{\underline{K}}_{\mu x} + \underline{\underline{K}}_s + \underline{\underline{K}}_x + \underline{\underline{K}}_\sigma - \underline{\underline{K}}_f \tag{11,a}$$

where,

$$\underline{\underline{K}}_{\mu v} = \frac{\partial \underline{T}}{\partial \underline{\sigma}} \frac{\partial \underline{\sigma}}{\partial \mu} \frac{\partial \mu}{\partial \underline{v}} \ ; \quad \underline{\underline{K}}_{\mu x} = \frac{\partial \underline{T}}{\partial \underline{\sigma}} \frac{\partial \underline{\sigma}}{\partial \mu} \frac{\partial \mu}{\partial \underline{x}} \frac{\partial \underline{x}}{\partial \underline{v}} \tag{11,b,c}$$

$$\underline{\underline{K}}_s = \frac{\partial \underline{T}}{\partial \underline{\sigma}} \frac{\partial \underline{\sigma}}{\partial \underline{v}} \ ; \quad \underline{\underline{K}}_x = \frac{\partial \underline{T}}{\partial \underline{\sigma}} \frac{\partial \underline{\sigma}}{\partial \underline{x}} \frac{\partial \underline{x}}{\partial \underline{v}} \tag{11,d,e}$$

$$\underline{\underline{K}}_\sigma = \frac{\partial \underline{T}}{\partial \underline{x}} \frac{\partial \underline{x}}{\partial \underline{v}} \ ; \quad \underline{\underline{K}}_f = \frac{\partial \underline{F}}{\partial \underline{x}} \frac{\partial \underline{x}}{\partial \underline{v}} \tag{11,f,g}$$

    The secant matrix $\underline{\underline{K}}_s$ and the equivalent strain rate dependent
matrix $\underline{\underline{K}}_{\mu v}$ are both symmetric and have been considered in [6]. $\underline{\underline{K}}_\sigma$
is the symmetric initial stress matrix usually occurring in geo-
metrically nonlinear analyses [11]. All the other matrix components of

$\underline{\underline{K}}T$ which depend on geometry variations, i.e. $\underline{\underline{K}}_{ux}$, $\underline{\underline{K}}_x$ and $\underline{\underline{K}}_f$ are un-symmetric and have been disregarded in the analysis.

The solution algorithms adopted herein for the integration schemes shown in Figure 3 are now given for time step n+1 as,

Predictor stage (time step n+1)

Forward Euler and
$$\underline{x}^o_{n+1} = \underline{x}_n + \underline{v}_n \Delta t_{n+1} \tag{12a}$$

trapezoidal schemes
$$\underline{v}^o_{n+1} = \underline{v}_n \tag{12b}$$

2 step B-D scheme
$$\underline{x}^o_{n+1} = \underline{x}_n + ((1-\beta)\underline{v}_{av} + \beta\underline{v}_n)\Delta t_{n+1} \tag{12c}$$

$$\underline{v}^o_{n+1} = \eta\underline{v}_n + (1-\eta)\underline{v}_{av}; \quad \eta = (2\Delta t_{n+1} + \Delta t_n)/\Delta t_n;$$

$$\beta = 1 + \Delta t_{n+1}/\Delta t_n \tag{12d,e,f}$$

Corrector stage (time step n+1, iteration k)

$$\Delta\underline{v}^k_{n+1} = (\underline{\underline{K}}T^k_{n+1})^{-1} (\underline{F}(\underline{x}^k_{n+1}, t_{n+1}) - \underline{T}(\underline{v}^k_{n+1}, \underline{x}^k_{n+1})) \tag{13a}$$

$$\underline{v}^{k+1}_{n+1} = \underline{v}^k_{n+1} + \Delta\underline{v}^k_{n+1} ; \quad \underline{x}^{k+1}_{n+1} = \underline{x}_n + \bar{\underline{v}}^{k+1}_{n+1} \Delta t_{n+1} \tag{13b,c}$$

The corrector stage continues until convergence is achieved and then the process returns to the predictor stage for the next time step. Previous papers by the authors [6,7] employed a viscous shell element that included bending in order to prevent the appearance of mechanisms when the sheet is in its initial flat start up configuration. Intro-ducing an implicit scheme obviates the need to include bending since the emergence of the initial stress matrix $\underline{\underline{K}}_\sigma$ endows the membrane sheet with additional 'out of plane' stiffness dependent upon the state of stress. Consequently a transverse stiffness can be generated when the sheet is initially flat by introducing a temporary initial fictitious stress which prevents the occurrence of mechanisms in the flat membrane.

APPLICATIONS

Experience indicates that for axisymmetric configurations the forward Euler timestepping scheme is adequate, but for a free formed ellipsoid [12] this simple technique resulted in the severe unstable velocity oscillation shown in Figure 4(a) which was prevented by introducing the trapezoidal scheme. For sheets forming with a double curvature the trapezoidal scheme is generally satisfactory despite the occurrence of small, non-propogating, velocity oscillations. But for sheets forming with a single curvature, such as a plane strain strip, the trapezoidal scheme produces significant non propogating velocity

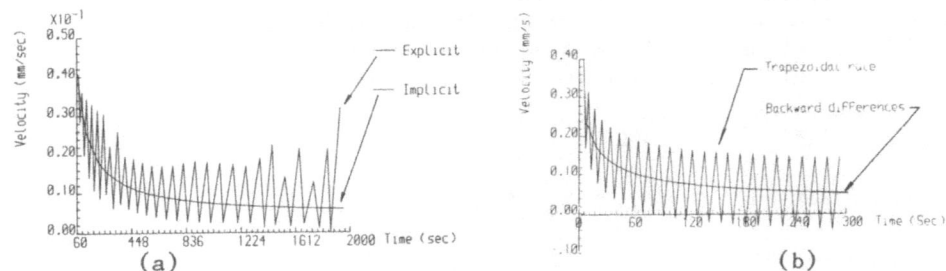

Figure 4
Velocity history for different time stepping schemes
(a) pole velocity for free formed ellipsoid, (b) centre velocity
for a plane strain strip along the minor axis
of a rectangular box (see Figure 5)

oscillations which depend, in magnitude, upon the starting procedure. This problem was overcome using the two step backward difference scheme, see Figure 4(b) which was adopted for the following rectangular box example.

Figure 5 presents information relating to the numerical simulation of the superplastic forming of Ti-6Aℓ-4V sheet into a rectangular box, which due to symmetry can be analysed using a quarter mesh. The pressure cycle is such that the maximum effective strain rate is $3*10^{-4}$, this was calculated prior to the analysis using a solution

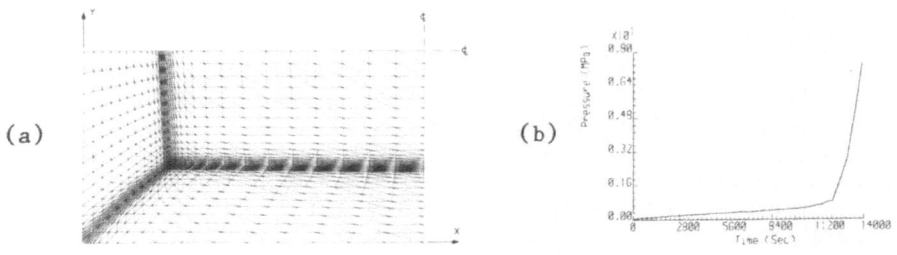

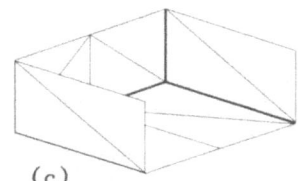

Length 280mm , width 162.6mm , height 60mm , corner radii 1mm
initial thickness 1mm , initial grain size 8μm
maximum effective strain rate $3*10^{-4} sec^{-1}$
¼ mesh , 1667 elements , 885 nodes , maximum displacement per
time step 1.5∗(current min. thickness) , material Ti-6Al-4V

Figure 5
Superplastic forming of a rectangular box,
(a) 1/4 finite element mesh, (b) pressure cycle, (c) 1/2 die mesh

technique which will be discussed in a future publication.   Figure 6
(a,b,c,d) gives final thickness and grain size distributions where
corner thinning and maximum grain growth are clearly evident. Figure
6(e) shows the deformed shape of the sheet at various forming times.
Execution on a VAX 11/750 took 8 hrs. 2 min. CPU time for 201 time-
steps with forming being completed in 3 hr. 51 mins, real time.

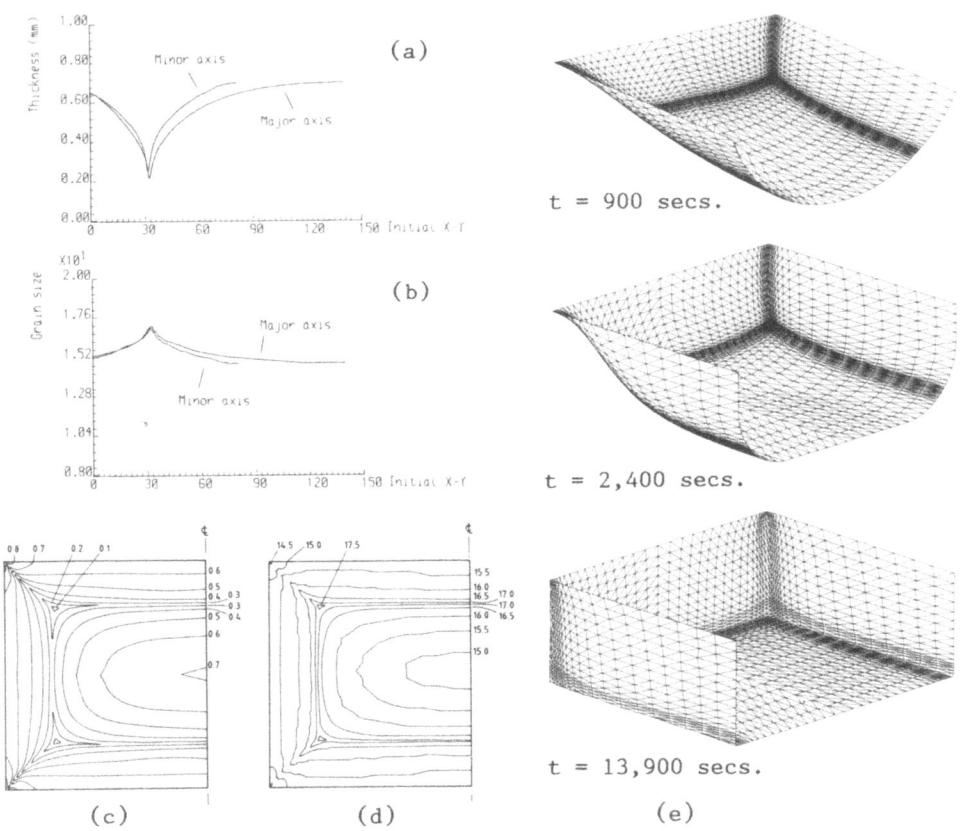

Figure 6

Superplastic forming of a rectangular box, (a,c) thickness
distribution, (b,d) grain size distribution
(e) deformed shapes at various forming times

CONCLUSIONS

Numerical simulation of the superplastic forming of thin sheet is
a viable proposition provided care is taken with regard to the time
stepping scheme employed. In particular it is shown that a two step backward difference implicit scheme is more robust than the trapezoidal scheme

186

ACKNOWLEDGEMENT

This work is associated with a contract between British Aerospace
Civil Division and the Institute for Numerical Methods in the Civil
Engineering Department at Swansea University.

REFERENCES

1. J.H. ARGYRIS and J. ST. DOLTSINIS, "A primer on superplasticity in
   natural formulation", Compt. Meths.Appl.Mech. Engrg.,16, 83-132 (1984).

2. N. CHANDRASEKARAN, R.E. GOFORTH and W.E. HAISLER, "Finite element
   formulation of superplastic metal forming processes", American
   Society for Metals, Metals/Materials Technology Series, Materials
   Week, Paper 8511-004, (1985), 1-8.

3. W.C. ZHANG, R.D. WOOD and O.C. ZIENKIEWICZ, "Superplastic forming
   analysis using a finite element viscous flow formulation",
   Aluminium Technology '86, Book 4, Session C, (1986), 111.1-111.6,
   The Institute of Metals, London.

4. J. ST. DOLTSINIS, J. LOGINSLAND and S.NOLTING, "Some developments
   in the numerical simulation of metal forming processes", Proc.Int.
   Conf. on Computational Plasticity : Models, Software and Applications, Part 2, (1987), 875-899.

5. M. BELLET, E. MASSONI and J.L. CHENOT, "A viscoplastic membrane
   formulation for 3-dimensional analysis of thin sheet metal
   forming", Ibid, 917-926.

6. J. BONET and R.D. WOOD, "Solution procedures for the finite element
   analysis of superplastic forming of thin sheet", Ibid, 927-939.

7. J.BONET, R.D.WOOD and O.C.ZIENKIEWICZ, "Finite element analysis of
   thin sheet superplastic forming", Mathematical Models for Metals &
   Materials Applications, Institute of Metals, London (1988), 2.1-2.7.

8. J. BONET, R.D.WOOD and O.C.ZIENKIEWICZ, "Finite element modelling
   of the superplastic forming of thin sheet", Int. Conf. on Superplasticity and Superplastic Forming, ASM, 1988 - to be published.

9. Superplasticity, AGARD-LS-154, Advisory Group for Aerospace Research and Development, NATO, Aug. 1987, pp. 204.

10. A.K. GHOSH and C.H. HAMILTON, "Mechanical behaviour and hardening
    characteristics of a superplastic Ti-6Aℓ-4V alloy", Metallurgical
    Trans., A, 10A, (1979), 699-706.

11. R.D. WOOD and O.C. ZIENKIEWICZ, "Geometrically nonlinear finite
    element analysis of beams, frames, arches and axisymmetric shells",
    Compt. Struct., 17, pp. 723-735, 1977

12. J. BONET, R.D. WOOD and O.C. ZIENKIEWICZ, "Numerical simulation of
    the superplastic forming of thin sheet - comparison of results
    between CEMEF and INME, Internal Report No.1, Swansea, 1988.

THIN SHEET FORMING NUMERICAL ANALYSIS
WITH A MEMBRANE APPROACH

E. MASSONI , M.BELLET , J.L.CHENOT

CEMEF ; Ecole Nationale Supérieure des Mines de Paris
SOPHIA ANTIPOLIS ; 06560 VALBONNE
FRANCE

ABSTRACT:A numerical model for solving either elastic-plastic , elastic-viscoplastic or purely viscoplastic deformation of thin sheets is presented using a membrane mechanical approach.The Finite Element Method is used associated with an incremental updated Lagrangian procedure.The mechanical equations are the principle of virtual work written in terms of plane stress at the end of each increment and an incremental implicit flow rule obtained by the time integration of the constitutive relations over the increment.The examples given here are the computation of hemispherical punch stretching for the elastic-plastic behavior and the application to superplastic forming with the viscoplastic flow rule .

1.Introduction

Thin sheet forming processes may be divided into three major classes:hydraulic bulging, punch stretching and deep-drawing . During these processes the plate material is exposed to large deformation and plastic flow.The governing equations of such a process contain therefore non linearities of both material and geometrical kind.
In this paper a Finite Element Method for the solution of non-linear elastic-plastic and viscoplastic problems is presented for the study of thin sheet subjected to hydraulic pressure or stretching processes.Due to the thinness of the sheet the membrane assumption is used for writing the governing mechanicals equations .
In order to formulate the incremental Finite Element Method we use a convected coordinates system embedded in the body and then obtain relations between incremental quantities.An updated Lagrangian formulation is used with an equilibrium equation resolution at the end of the increment.An elastic-plastic law is used for deep drawing (to take into account local unloading phenomenon) and a purely viscoplastic constitutive equation for superplastic forming process. The material is isotropic work-hardening. The constitutive law written in terms of convected coordinates is integrated according to a semi-implicit scheme.The resolution of the global non linear system is made with a Newton-Raphson method.A contact algorithm has been developed with introduction of friction terms according to a Tresca law for punch stretching and Coulomb or viscoplastic law for superplastic forming.

J. L. Chenot and E. Oñate (eds.), Modelling of Metal Forming Processes, 187–196.
© *1988 by Kluwer Academic Publishers.*

2. Membrane model

2.1. GEOMETRY

The deformed membrane is assumed to be a geometric surface.Material points are identified by two convected coordinates θ^1 and θ^2 which remain constant throughout the deformation process /2/.These material coordinates are initialized as cartesian coordinates in the undeformed initial state.At time t the position vector $\mathbf{X}$ of a material particle depends on θ^α as well as the membrane local thickness h :

$$\mathbf{X} = \mathbf{X}(\theta^\alpha, t) \qquad h = h(\theta^\alpha, t) \qquad (1)$$

The covariant base vectors tangent to θ^α lines are

$$\mathbf{g}_\alpha = \frac{\partial \mathbf{X}}{\partial \theta^\alpha} \qquad (2)$$

and we define the local normal vector $\mathbf{g}_3$ in order to get a local three-dimensional basis:

$$\mathbf{g}_3 = \frac{\mathbf{g}_1 \wedge \mathbf{g}_2}{||\mathbf{g}_1 \wedge \mathbf{g}_2||} \qquad (3)$$

The components of the metric tensor and of its inverse are:

$$g^{ij} = \mathbf{g}^i \cdot \mathbf{g}^j \text{ and } g_{ij} = \mathbf{g}_i \cdot \mathbf{g}_j \text{ with } g_{\alpha 3} = g^{3\alpha} = 0 \qquad (4)$$

2.2 NUMERICAL RESOLUTION

2.2.1. Time discretization

The motion of the deformed membrane is discretized according to an updated Lagrangian formulation of the principle of virtual work /2/.The external load is applied increment by increment.By application of the principle of virtual work at time $t+\Delta t$ we obtain in terms of plane stress :

$$\int_{\Omega_{t+\Delta t}} (\sigma+\Delta\sigma)^{\alpha\beta} \overset{*}{V}_{\beta/\alpha} (h+\Delta h) \, ds = \int_{\Omega_{t+\Delta t}} T(t+\Delta t).\overset{*}{V} ds - \int_{\partial\Omega_{t+\Delta t}} F(t+\Delta t).\overset{*}{V} dl \qquad (5)$$

for any arbitrary velocity field $\overset{*}{V}$,

(/α denotes covariant derivation with respect to the material coordinate θ^α),

σ Cauchy stress tensor , T external applied surface forces and F external applied forces.This equilibrium equation has only one unknown quantity which is the displacement field $\mathbf{u}$ of sheet material points providing that relations between $\mathbf{u}$ and $\Delta\sigma$,

Δh, **T** and **F** are established.So we have to present the incremental constitutive law which permits to connect $\Delta\sigma$ to **u**.

2.2.2 Incremental flow rule

The basic idea is a semi-implicit integration scheme of the plastic deformation rate /1/.
In the case of elasto-viscoplasticity, and under the assumption of small elastic strains we have /2/ :

$$\begin{cases} \Delta L = \Delta L^e + \Delta L^p \\\\ \Delta\sigma = E\ \Delta L^e \\\\ \Delta L^p = \Delta t\ [\ (1-\eta)\ D^p(\sigma) + \eta\ D^p(\ \sigma+\Delta\sigma)\ ] \end{cases} \qquad (6)$$

where ΔL^e and ΔL^p are respectively the elastic and the plastic part of the incremental strain tensor , E is the elasticity tensor , D^p is the viscoplastic strain rate tensor which derives from a viscoplastic potential Q :

$$D^p = \frac{\partial Q}{\partial \sigma} \qquad (7)$$

η is an arbitrary fixed parameter in [0,1] which makes the integration scheme more or less implicit .The usual value fo η is 0.5 .
For application to hot forming processes such as superplastic forming, the elastic effects can be neglected and the set of equations (6) can be restricted to /4/ :

$$\Delta L = \Delta t\ [\ (1-\eta)\ D^p(\sigma) + \eta\ D^p(\ \sigma+\Delta\sigma)\ ] \qquad (8)$$

For the free elastic-plastic loading we get the following set of equations /2/ .

$$\begin{cases} \Delta L = \Delta L^e + \Delta L^p \\\\ \Delta\sigma = E\ \Delta L^e \\\\ \Delta L^p = \Delta l^p[\ (1-\eta)\ \dfrac{\partial f(\sigma,\bar\varepsilon)}{\partial \sigma} + \eta\ \dfrac{\partial f(\sigma+\Delta\sigma,\bar\varepsilon+\Delta\bar\varepsilon)}{\partial \sigma}\ ] \\\\ \Delta\bar\varepsilon = \sqrt{\dfrac{2}{3}\Delta L^p_{\alpha\beta}\ \Delta L^{p\alpha\beta}} \\\\ f(\sigma+\Delta\sigma,\bar\varepsilon+\Delta\bar\varepsilon) = 0 \end{cases} \qquad (9)$$

f is the yield criterion , $\bar\varepsilon$ the equivalent plastic strain .

When elasticity is taken into account , the local set of equations (6) or (9) are solved by the Newton Raphson method.For pure viscoplasticity,the new stress tensor $\sigma + \Delta\sigma$ is given directly by equation (8) .

2.2.3 Spatial discretization : Finite Element Method

The membrane surface Ω is meshed with finite isoparametric membrane elements with three degrees of freedom at each node which are the cartesian components of the displacement /3/ . The equilibrium equation at time $t + \Delta t$ is the space discretized virtual work equation:

$$R_{Nk}(\mathbf{u}) =$$

$$\int_{\Omega_{t+\Delta t}} (\sigma + \Delta\sigma)^{\alpha\beta} B_{\alpha\beta Nk} \, (h + \Delta h) \, ds \; - \int_{\Omega_{t+\Delta t}} T_k(t+\Delta t).\psi_N ds \; - \int_{\partial\Omega_{t+\Delta t}} F_k(t+\Delta t).\psi_N dl = 0$$

(10)

for any degree of freedom (N,k), ψ_N is the value of interpolation function of node N,coefficients B are given by the relation between the covariant derivatives and the nodal cartesian components of the velocity :

$$V_{\alpha/\beta} = B_{\alpha\beta Nk} V_{Nk} \quad \text{with} \quad B_{\alpha\beta Nk} = \frac{\partial\psi_N}{\partial\theta^\alpha} I_k \cdot g_\alpha$$

where I_k are the cartesian basis vectors .
The global set of nonlinear equations (10) is solved by a Newton Raphson Method in which the tangent stiffness matrix is calculated as exactly as possible /2/ .

3. Numerical results

3.1 THIN SHEET SUPERPLASTIC FORMING

The basic forming operation is sheet bulging by a one-sided gas pressure in a shaped mould .Although the technology seems conceptually simple , the influence of the process parameters such as pressure ,die geometry ,lubrication is very important and not well established.The problem consists ,in this case,in computing the final thickness as well as in determining the pressure cycle.The optimal pressure curve must satisfy two main objectives :
- the equivalent plastic strain rate $\dot{\bar{\varepsilon}}$ must remain in the superplastic range .
- the forming time must be minimum for quality and productivity reasons .
That is why the automatic pressure computation that is perfomed at each time increment regulates the maximum value of $\dot{\bar{\varepsilon}}$ around the prescribed upper bound of the superplastic range .

3.1.1 2-D case

The figure 1 shows the optimal pressure cycles for plane strain bulging of Ti-6Al-4V sheets ($\dot{\bar{\varepsilon}}_{max}$ value = 3.10^{-4} s^{-1}).Three mould geometries have been studied : 180 , 70 , 30 mm width and 30 mm height with a Coulombic friction $\mu = 0.2$. In the middle part of the box the sheet is in plane deformation , and the process is computed with a 2-D model /8/.The curves show the important influence of the mould geometry on the pressure cycles.

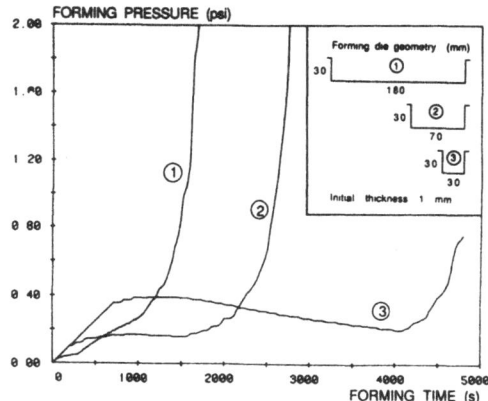

figure 1 : Optimum pressure cycles in plane strain for three mould geometries .

In figure 2 are plotted the numerical and experimental thickness profiles for the 70 mm wide mould .We observe a good agreement with the realistic Coulomb friction coefficient $\mu = 0.2$.

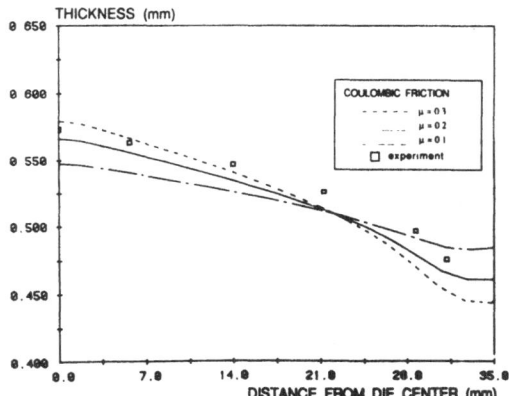

figure 2 : Thickness profile at the end of the process .

3.1.2 Three dimensional forming operation

We have chosen a part of a rather simple shape with large 3-dimensional plastic deformation : it is a square box which is 18 inches wide and 6 inches deep . All tool edges are rounded with a radius of 1 inch .The initial thickness of the 7475 aluminium alloy sheet is 0.06 inch and the Coulomb friction coefficient is $\mu=0.3$. The deformed mesh ,at the end of the forming (5945 s) , is pictured in figure 3.The minimum thickness in the corner is 0.023 inch.

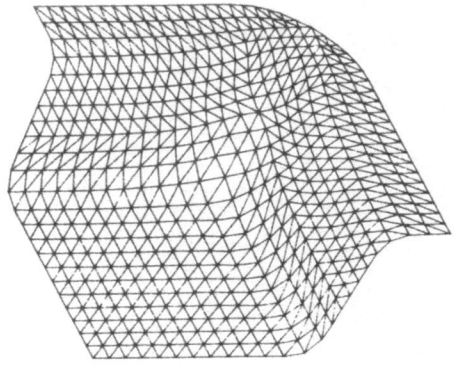

figure 3 : Deformed mesh . final sheet (1/4) .

The figure 4 gives the optimal pressure cycle and figure 5 the regulated $\dot{\overline{\epsilon}}_{max}$ value around $2.10^{-4}s^{-1}$.

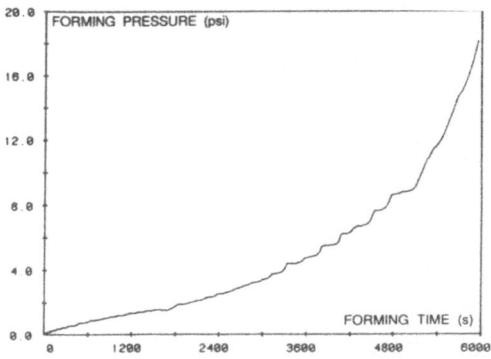

figure 4 : Optimum pressure cycle .

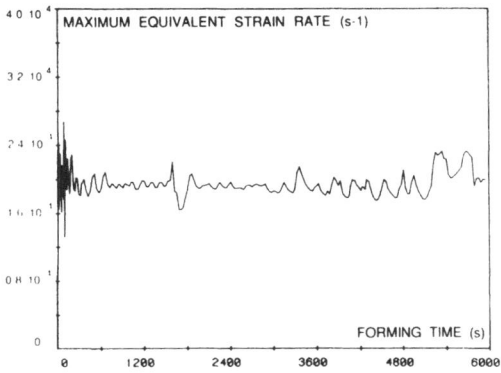

figure 5 : Regulated maximum equivalent strain rate .

3.2 PUNCH STRETCHING

The second chosen test is an hemispherical punch stretching.The sheet rheology is represented by an elastic-plastic constitutive law. The experimental test have been made for an HSLA Nb-alloyed steel (figure 6) /5/.
The material parameters are :

Young's Modulus	E = 208 000 MPa
Poisson's coefficient	$\nu = 0.3$
Anisotropy coefficients:	$R_0 = 0.86$; $R_{90} = 1.15$; $R_{45} = 1.27$

Figure 7 gives the test geometry . The thickness sheet is 0.7 mm.An annular blank-holder with radius 130 mm prevents all sheet displacement beyond this position .The strain components were calculated at 15 points initially regularly spaced along a meridian.For the numerical simulation both die and punch surfaces were discretized in triangular elements. The sheet material is considered an isotropic elastic-plastic work-hardening material.Because of axial symmetry, only a narrow sector $\pi/16$ is studied : it is meshed with 29 triangular elements (figure 7) and a no displacement condition is assumed at the boundary (r=130 mm).

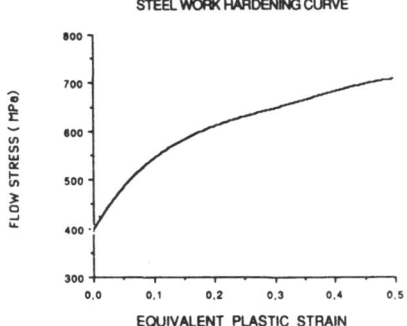

figure 6: Work-hardening curve .

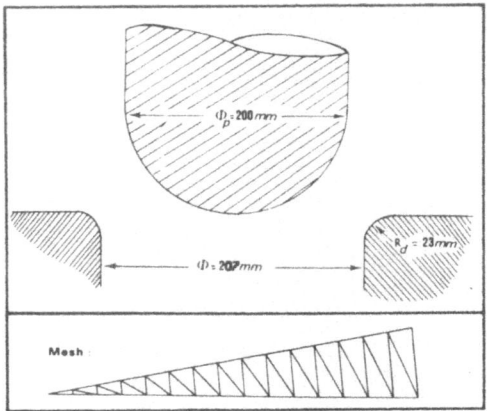

figure 7: Geometrical data .

Concerning friction , previous experimental identification of Coulomb's coefficient has given a quasi constant value $\mu = 0.15$ during the process.But in the numerical model friction is simulated through a plastic friction law(Tresca law) :

$$T_f = \bar{m}\, \frac{\sigma_0(\mathcal{E})}{\sqrt{3}}\, \frac{u_r}{||\, u_r\, ||}$$

where the frictional stress T_f is proportional to the shear yield stress $\dfrac{\sigma_0}{\sqrt{3}}$ with friction coefficient $\bar{m}$, and u_r is the relative displacement vector in contacting element.A short approximate calculation of the friction stresses in the cases of Coulomb and Tresca law gives an equivalent value of $\bar{m}$ about 0.005.Three $\bar{m}$ values have been tested : 0.003 ; 0.005 ; 0.007 .For these cases $\bar{m}$ is the same on the die and on the punch . The results are gathered on figure 8 . it appears clearly that the influence of $\bar{m}$ coefficient is important on the area contacting the punch and it is in a good agreement with the experimental data (the 0.007 value gives the best results) .

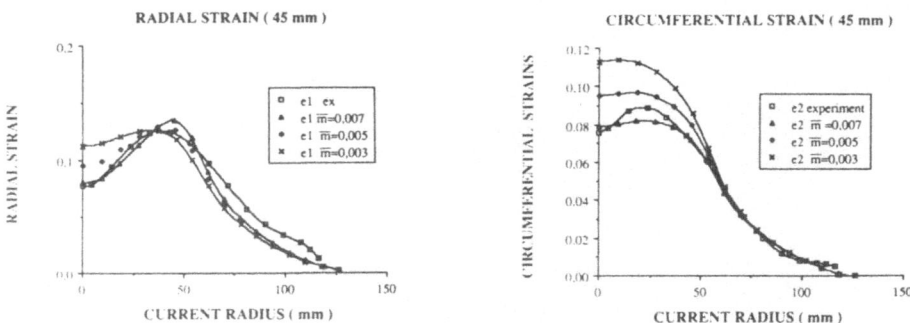

figure 8 a) Punch stretching with hemispherical punch; Strain distribution .

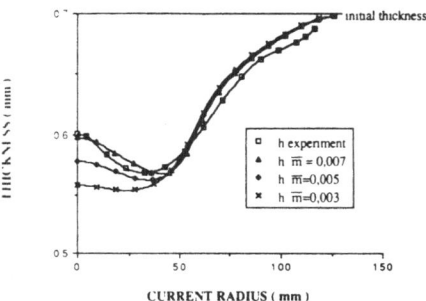

figure 8 : b) Punch stretching with hemispherical punch .Thickness distribution .

4.Conclusion

We have presented in this paper a membrane mechanical approach for solving thin sheet forming.The finite element method is used with an incremental updated Lagrangian procedure.

We have applied this formulation first to a simulation of superplastic forming with viscoplastic flow rule . The numerical results of 2-D approximation are in a good agreement with experimental data in terms of thickness distribution .The optimal pressure cycle is determined by the computer code . The application of this calculation is done in 2-D and in 3-D cases .

In a second time , elastic-plastic deformation is studied during punch stretching process with an hemispherical punch .The finite element model takes into account the Von-Mises flow rule and work-hardening . Friction effects are introduced in the numerical model.The present calculations are in a good agreement with experimental data with a suitable choice of the friction coefficients .

196

References

1/ BRAUDEL,H.J. , ABOUAF,M. , CHENOT,J.L. : An implicit and Incremental Formulation for the Solution of Elastoplastic problems by the F.E.M. .Comp. and Struct., Vol 22, p.801 , (1986).

2/ MASSONI,E. , BELLET,M. , ABOUAF,M. , CHENOT ,J.L. : Large Displacements Numerical Calculation of 3-Dimensional Elasto-plastic and Elasto-viscoplastic membrane by the F.E.M., Num. Meth. for Non lin. Prob. publ. by Pineridge Press, Vol 5, p.480 , (1986).

3/ MASSONI,E. , BELLET,M. , CHENOT,J.L. : Elastic-Plastic Analysis of Thin Plates by the Finite Element Method in a Deep Drawing Process , Int.Conf.on Computational Plasticity , Barcelone ,Pineridge Press , II , p. 901 , (1987).

4/ BELLET,M. , MASSONI,E. , CHENOT,J.L. : A Viscoplastic membrane formulation for the 3-Dimensional Analysis of Thin Sheet Metal Forming. Int.Conf.on Computational Plasticity , Barcelone ,Pineridge Press , II , p. 917 , (1987).

5/ MASSONI,E. , BELLET,M. , CHENOT,J.L. , DETRAUX,J.M. , de BAYNAST,C. : A Finite Element Modelling for Deep Drawing of Thin Sheet in Automotive Industry. Advanced Technology of Plasticity . Edited by K.lange, Springer - Verlag, Berlin , II , p. 719 ,(1987).

6/ MATTIASSON,K. , SARAN,M. , MELANDER,A. , SCHEDIN,C. , GUSTAFSSON,C. : Finite Element Simulation of Deep Drawing of low and high strength Steel.Advanced Technology of Plasticity . Edited by K.Lange, Springer-Verlag, Berlin , 1, p.557, (1987).

7/ NAKAMACHI,E. : Finite Element Modeling of the Punch Press Forming of Thin Elastic-Plastic Plates, Numiform 86 . Edited by K.Mattiasson,A. Samuelsson, R.D.Wood and O.C. Zienkiewicz, Balkema, Rotterdam , p.333, (1986).

8/ BELLET M. : Modélisation numérique du formage superplastique de tôles Thesis.Ecole Nationale Supèrieure des Mines de Paris.Mars (1988).

9/ MASSONI.E : Modélisation numérique par éléments finis de l'emboutissage. Thesis.Ecole Nationale Supèrieure des Mines de Paris.Mars (1987).

PART 4

FORGING AND DRAWING

COMPARISON OF SIMULATION TECHNIQUES
AND INDUSTRIAL DATA IN FORGING APPLICATIONS

G. SURDON, M. BAROUX
LUCHAIRE, Division Automobile
02650 CREZANCY FRANCE

Abstract

This paper presents applications of simulation techniques in an industrial environment. The experimental simulation by model material and numerical simulation by a FEM code have been used. A comparison with real data shows how interesting it is to use these techniques in industry. It appears that the main difficulty concerns the knowledge of boundary conditions, such as friction, during metal forming.

1. Introduction

Forging is one of the metal forming processes most based on prior art, experience and empirical rules developed by engineers through many years of practice. The forging designer mainly seeks to reduce costs by minimizing the number of processing steps between initial and final shapes. By using prototypes, he can check the forging sequence, that is:
- the starting slug size and shape
- the intermediate shapes
- the mechanical and metallurgical state of the material
- the absence or the risk of defects
- the resistance of the forming tools.

However, optimizing immediately the different processing steps is difficult through this technique, as changes in the shape of the material have non linear effects on such variables as stress, strain and temperature which in turn affect material properties, and depend on the contact conditions with the dies. Hence, besides this conventional "trial and error" technique, numerical and model material simulation appear attractive methods to analyse the forging process. The major purposes are to reduce development time, optimize the conditions of industrial production (deformation steps, stresses on tools, load) and predict the influence of processing parameters (starting and intermediate shapes, friction conditions, flow stress and temperature).

The present paper describes industrial applications of the simulation techniques and compares them with industrial data.

J. L. Chenot and E. Oñate (eds.), Modelling of Metal Forming Processes, 199–205.
© 1988 by Kluwer Academic Publishers.

2. Experimental simulation by model material

The basic idea is that it is easier to analyse the plastic flow of a soft model material such as plasticine or wax than of a hot metal such as steel. Three rules are required by the theory of similarity /1,2/:
- workpiece and dies must be designed to scale
- the mechanical behaviour of the real and model material must be similar. For hot metal forming, the power law viscoplastic behaviour is very often used; in this case, the strain hardening coefficient n and the strain rate sensitivity m must have the same value for hot steel and model material. Flow stress is written as:

$$\sigma_0 = A \ \bar{\varepsilon}^n \ \dot{\bar{\varepsilon}}^m$$

where A is a stress factor, $\bar{\varepsilon}$ is the generalised strain, $\dot{\bar{\varepsilon}}$ is the generalised strain rate. A, n, m depend on temperature.
- the friction coefficient $\bar{m}$ at workpiece-die interface must have the same value for the forged metal and the model material. The friction stress will be written as:

$$\tau = \bar{m} \ \sigma_0 / \sqrt{3}$$

This condition is one of the similarity rules most difficult to respect, which is all the more damageable as friction has a very strong influence in forging.
To simulate the hot forging of steel, we have chosen wax /3/ as a model material in the following experiments. The manufacturers (ICAM, CETIM, University of VALENCIENNES) obtain values of n in the range -0.4 / 0.1 and of m in the range 0.15 to 0.4, depending on the chemical composition. The rheology of the wax used in our experiments is shown in figure 1. Temperature gradients are supposed sufficiently small to be neglected, although OUDIN et al. /4/ present a formulation with thermal phenomena.

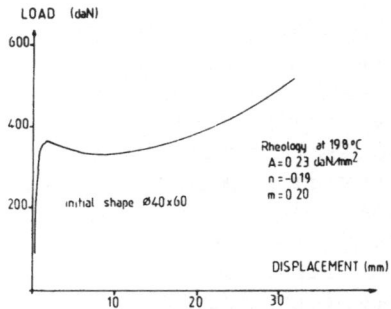

Figure 1 : wax rheology as obtained by an upsetting test

3. Numerical simulation of hot forging

Finite Element Method based codes for investigating the forging process are being developed in Universities and research centres all around the world / see for example 5-

7/. Most of them include in their mathematical formulation the flow stress as a function of strain, strain rates and temperature. Usually, these codes predict load-stroke curves, deformed geometries, material velocities, strain fields, strain rate fields and the stresses on the tools at each incremental displacement of the dies. The displayed results constitute a useful tool for the designer to enforce the desired quality standard.

The FORGE 2 code developed by CEMEF /6/ is now being used by LUCHAIRE at CREZANCY Works. This code is based on 2D axisymmetric or plane strain finite elements; the material is assumed to be incompressible viscoplastic, with an associated friction law. Isothermal deformation is assumed. The workpiece is discretized in linear or quadratic quadrilateral elements, and the dies are modelled as rigid moving bodies. There is presently no coupling of thermal effects (heat transfer to the dies) and material rheology. The large strains necessitate remeshing during forging simulation.

4. Industrial applications of simulation techniques

The FORGE 2 code and model material simulation were applied to the three following forging sequences. Each example illustrates different aspects of the simulation techniques, such as the prediction of load-stroke curves, processing defects, and the final microstructural state of the material.

4.1 HOT FORGING OF A HOLLOW PART

The first application presents the forging of a hollow part out of low alloy carbon steel. Three forming steps are necessary to obtain this product on a hydraulic automatic press:
- prepiercing of the cylindrical slug
- then a piercing of the first preform
- final drawing

Only the first two steps were simulated. The furnace temperature was 1200°C and the temperature field was supposed uniform in the slug and in the different preforms. A graphitized oil was used as a lubricant.The load-stroke curves were measured by means of pressure transducers and recorded on a multi-track recorder. Table 1 gives the different conditions and numerical data on the simulations.

TABLE 1 : Simulation conditions and numerical data for the forging of a hollow part

	Temperature (°C)	stress factor A (daN/cm2)	m	lubricant	scale
wax simulation	20±1	0.23	0.2	white lubricant and talc	1/2.5
FORGE2	1200±20	4.0	0.2	$\overline{m}$=0.3	1/1

Figure 2 compares the load stroke curves obtained by measurement under industrial conditions, numerical and model material simulation.

For the first forming step, the simulation curves are in good agreement with the experimental one. However, for the second forming step, a large discrepancy appears between the three curves, with simulations underestimating the load by 20% to 50%. The reasons for such a poor simulation sem to be a poor description of process conditions, namely:

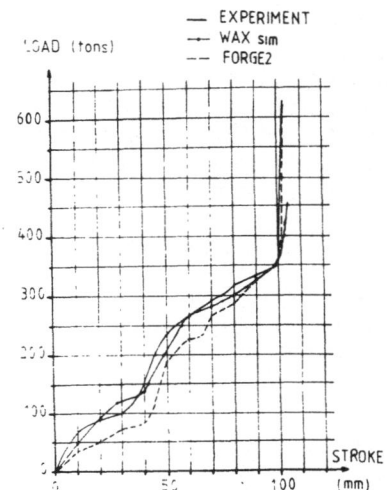

Figure 2 Load-stroke curves for the first forging step.

 - the effect of temperature gradients in the preform after the first forming step
 - the effect of frictional conditions.

During the process, the slug and later on the preform are allowed to cool for varying times inside and out of the dies, which are cooled by circulating water. For the first step, we can neglect radiative cooling of the slug during the transfer cycle; furthermore, the forging time is small enough to suppose that thermal transfer has no significant effect on the temperature distribution. These remarks are no more valid for the second forging step: the transfer and forging times are greater, and moreover, the piercing process involves a variable and high friction between workpiece and dies. Hence, knowing accurately the value and the evolution of friction is a critical point for simulation techniques.

4.2 SIMULATION OF MATERIAL PROCESSING DEFECTS

This second application concerns the last deformation step of a part with hollow shaft and flange. A mechanical multi-step press is used for this test.

The main purpose of this example is to show the prediction of material processing defects by simulation. During the last deformation step, a fold appears on the inner surface of the flange as shown in figure 3. Simulation techniques displayed the same phenomenon, and geometry was changed so as to avoid the defect. The following results are only qualitative.

For industrial tests, graphitized water is used as a lubricant; under these conditions, the friction coefficient has been experimentally estimated to be $\bar{m}=0.3$. A first calculation with $\bar{m}=0.3$ does give a fold defect, but its amplitude is too weak. So computation was restarted with $\bar{m}=0.1$ to amplify the fold. Figure 4 and 5 show that in this case, a satisfying agreement was found between simulation and industrial results.

Figure 3: the fold defect on the inner surface of the flange (experiment on metal)

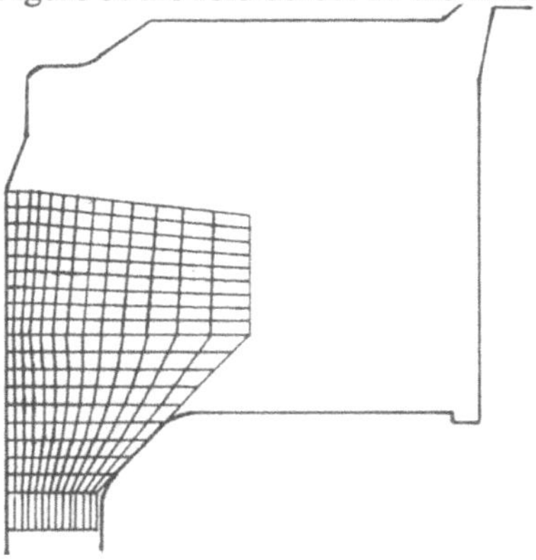

Figure 4A: initial FEM mesh

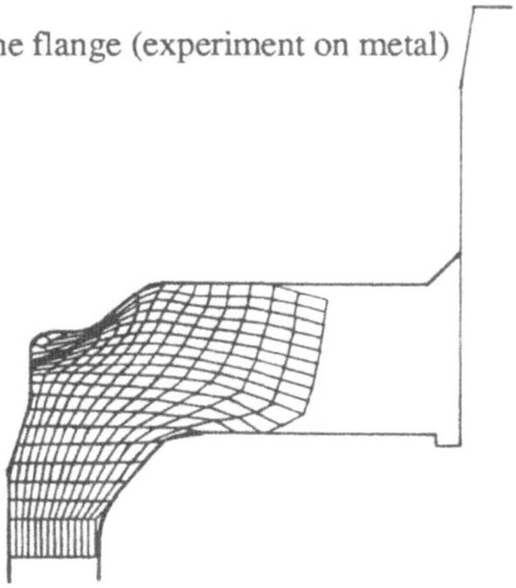

Figure 4B: deformed FEM mesh: a fold defect appears on the inner surface

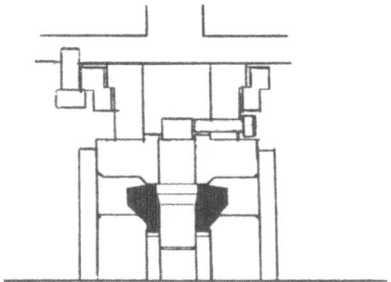

Figure 5A: dies and wax preform

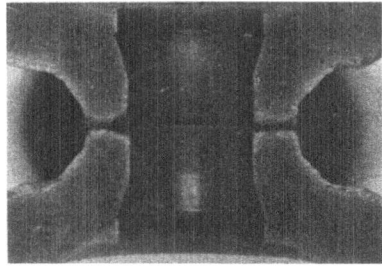

Figure 5B: after deformation, the defect appears

4.3 PREDICTION OF MICROSTRUCTURAL STATE AFTER FORGING

The FEM is able to simulate load-stroke curves and material flow. From these results, the distribution of such microstructural parameters as grain size can be predicted. To illustrate this point of view, this last application presents the forging of an automotive transmission part out of carbon steel.

After hot forging, the microstructure was observed through optical microscopy. A comparison with strain maps computed by the FEM code (see figure 6) shows a correct correlation between the measured average grain size I and the computed generalised strain $\bar{\varepsilon}$:

- for $0.8 < \bar{\varepsilon} < 1.5$, I=4.2
- for $1.5 < \bar{\varepsilon} < 2.0$, I=5
- for $2.0 < \bar{\varepsilon} < 3.9$, I=6

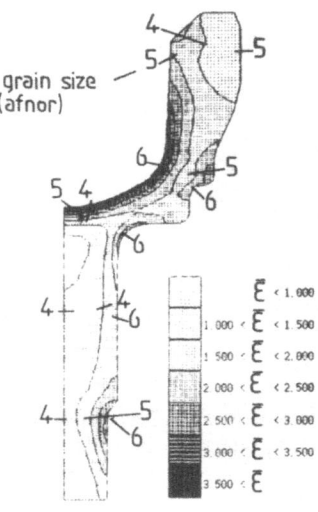

Figure 6 : comparison between the FEM-computed strain distribution and experimental grain size

5. Conclusions

The most significant aim of the simulation methods is to assist the forming enginneer in defining the different steps needed to transform the initial simple geometry into the final goemetry without material defects. However, metal flow stress, die-workpiece friction and heat transfer phenomena are at present difficult to incorporate accurately, which sometimes perturb the results of simulation techniques.

References

/1/ J.DANKERT, T. WANHEIM: model material technique applied in the analysis of the forging of a specimen of complicated shape. Scand. J. Met. 6 (1977)186-190

/2/ A. LE FLOC'H: Plasticité et mise en forme des métaux . Séminaire CEMEF-ENSMP (1986) T. 1

/3/ J. OUDIN, Y. RAVALARD, S. ROMMENS: on the contribution of waxes to the simulation of metal forming processes . Proc. NAMRC VIII (1980)166-170

/4/ J. OUDIN, Y. RAVALARD, S. ROMMENS: on the simulation of hot forming of metals . Proc. NAMRC VIII (1980) 464-466

/5/ T. ALTAN : Process simulation of hot die forging processes. Proc. Conf Advanced Technology of Plasticity, p 1021-1034 (STUTTGART,1987) K. LANGE ed. Publ. SPRINGER Verlag.

/6/ Y. GERMAIN, J.L. CHENOT, P.E. MOSSER: Finite element analysis of shaped lead-tin disk forging. Proc. NUMIFORM 86 (Gothenburg 1986); K.MATTIASSON et al. eds. Publ. BALKEMA

/7/ G. SURDON, J.L. CHENOT: Finite element calculation of three dimensional hot forging. Proc. NUMIFORM 86 (Gothenburg 1986); K.MATTIASSON et al. eds. Publ. BALKEMA

FINITE ELEMENT CALCULATION OF HOT FORGING
WITH CONTINUOUS REMESHING

J .P. CESCUTTI, E. WEY and J.L. CHENOT
CENTRE DE MISE EN FORME DES MATERIAUX
ECOLE NATIONALE DES MINES DE PARIS
SOPHIA ANTIPOLIS, 06560 VALBONNE, FRANCE
P.E. MOSSER, SNECMA,
291 av. d'Argenteuil, 92234 GENNEVILLIERS, FRANCE

One of the most crucial problems for the finite element modelling of
non-stationary large plastic deformations in metal forming is the
dramatic evolution of the mesh. In many industrial configurations the
elements are so distorted after few incremental steps that accurate
calculation is no longer possible, without remeshing. In this work, it
is shown that a continuous remeshing procedure with adaptivity,
compatible with the classical flow formulation approach, allows to
perform very large deformations without any intervention of the
operator. A practical example is analyzed for the forging of a gear
tooth which demonstrates that the total process may be calculated
completely automatically, saving more than 10 remeshing steps.

1. Introduction

Since the early work on visco-plastic finite element computation (/1/,
/2/, /3/), modelling of the hot forging process has received an increasing interest in several laboratories.
    Most of these studies concern plane strain problems or axisymme-
trical configurations, and recently simple three dimensional calculations are reported /4/, /5/ and relatively complex industrial shapes
have been presented /6/, /7/. But one of the most limiting problem,
apart from computing time, seems to be the evolution of the mesh if an
updated lagrangian description is used. Even for 2D cases, the forging
process may involve such a localized deformation, associated with high
material shear, that remeshing is absolutely necessary to prevent
degeneracy of the elements. For complex configurations several
remeshing steps must be performed, which are a tedious task if performed manually.
    On the other hand a lot of work has been devoted to the construc-
tion of meshes with optimum geometric properties, or with some degree
of adaptivity to the solution (see /8/, /9/, /10/, /11/, /12/). For
processes involving very large deformations the mesh should be updated
periodically using an optimality criterion in some sense but this
technique involves interpolation of the solution from the old distorted

207

J. L. Chenot and E. Oñate (eds.), Modelling of Metal Forming Processes, 207–216.
© 1988 by Kluwer Academic Publishers.

mesh to the new one. This procedure is time consuming and produces additional errors which may result in discontinuity on the force in the forging process. Recently a continuous remeshing technique has been suggested /13/, which allows to keep a smooth and adaptive mesh during all the process. This method has been illustrated in the simple case of upsetting of a rectangular parallelepipedic block in /14/.

The objective of this work is to present a more complex industrial example where very large deformations and distorsions are involved.

2. Theoretical formulation

The material is assumed to be homogeneous, incompressible and to obey the visco-plastic Norton Hoff law :

$$\mathbf{s} = 2 \ K \ (\sqrt{3} \ \dot{\bar{\varepsilon}})^{1-m} \ \dot{\varepsilon} \tag{1}$$

$$tr(\dot{\varepsilon}) = 0 \tag{2}$$

where $\mathbf{s}$ is the stress deviator

$\dot{\varepsilon}$ is the strain rate tensor

$\dot{\bar{\varepsilon}} = (2/3 \ \dot{\varepsilon}_{ij}{}^2)^{1/2}$

K is the consistancy which depends on $\bar{\varepsilon}$ for work hardening

m is rate sensitivity

$\bar{\varepsilon} = \int \dot{\bar{\varepsilon}} \ dt$ is the cumulated strain.

The viscoplastic friction law we have used is quite similar and may be written as :

$$\boldsymbol{\tau} = - \ \alpha \ K \ \boldsymbol{\Delta v} \ | \ \boldsymbol{\Delta v} \ |^{p-1} \tag{3}$$

$\boldsymbol{\tau}$ is the shear stress tangential to the tool surface

$\boldsymbol{\Delta v}$ is the velocity difference between the tool and the part (which must be tangential to the tool surface)

α and p are coefficients which may be identified by laboratory tests (such as ring compression test)

With a penalty approach the problem is equivalent to finding the minimum of the functional with respect to the admissible velocity field v :

$$\phi \ (\ v \ ) = \int_{\Omega} \frac{K}{m+1}(\sqrt{3} \ \dot{\bar{\varepsilon}})^{m+1} \ d\omega + \frac{1}{2} \ \rho \int_{\Omega} \ (div \ v \ )^2 \ d\omega$$

$$+ \int_{\partial\Omega_f} \frac{\alpha \ K}{p+1} \ | \ \boldsymbol{\Delta v} \ |^{p+1} \ ds \tag{4}$$

with $\boldsymbol{\Delta v.n} = 0$ on $\partial\Omega_f$ with normal $\mathbf{n}$.

Here Ω represents the interior of the part being processed

$\partial\Omega_f$ is the surface of the part which is in contact with the tools.

3. Space discretization and time integration

The unknown velocity field v is discretized in the classical way using shape functions N_i.

$$\mathbf{v} = \sum_{i=1}^{n} \mathbf{V}_i \, N_i \tag{5}$$

where n is the number of the nodes

$\mathbf{V_i}$ is the velocity vector at node i

If we suppose that the configuration $\Omega(t)$ is known at time t, the approximate velocity field is calculated by minimizing the discrete form of the functional (4), with respect to the nodal velocity components. On the tool surface the normal nodal velocity is imposed as equal to that of the tool.

For the quadrilateral 4 nodes elements, in two dimensional or axisymmetrical problems, the first contribution to the functional problems is calculated with 4 Gauss integration points by element. The penalty term is subintegrated with 1 Gauss point, and the boundary integral with 2 Gauss points. The non linear equations obtained by equating to zero the gradient of the functional (4) are solved by the Newton Raphson method with optimal search.

Now the time integration is performed with the simple Euler one step explicit scheme as a first approach. If the configuration $\Omega(t)$ and the velocity field v (t) are known at time t, the next configuration at time $t+\Delta t$ is approximated by the symbolic relationship :

$$\Omega \ (t+\Delta t) = \Omega \ (t) + \mathbf{v} \ (t).\Delta t$$

In fact each node with coordinate X (t) is moved with a displacement proportional to the material velocity according to :

$$\mathbf{X} \ (t+\Delta t) = \mathbf{X} \ (t) + \mathbf{V} \ (t).\Delta t \tag{6}$$

As for contact problems, different algorithms have been presented in /15/ ; here we have chosen the simplest one which is able to predict the main physical problems for a node i :

- onset of contact :
   if X_i (t) lies on the free surface and
   $\mathbf{X_i}$ $(t+\Delta t)$ is inside the tool,
   Δt is modified so that $\mathbf{X_i}$ $(t+\Delta t')$ is on the tool surface.
- sliding contact :
   if X_i (t) is on the tool surface and the normal stress is compressive, the velocity is constrained so that $\mathbf{X_i}$ $(t+\Delta t)$ follows the tool surface.
- loss of contact :
   if X_i (t) is on the tool surface and the normal stress is non negative, then the constraint on the velocity is released.

4. Remeshing

4.1. STATIC REMESHING

For quadrilateral 4 nodes elements, a function of the node coordinates has been introduced (see fig. 1).

$$\Pi_e = \sum_e P_e \left[\left(X_B^e - X_A^e + X_D^e - X_C^e \right)^2 + c \left(\left(X_C^e - X_A^e \right)^2 + \left(X_A^e - X_B^e \right)^2 \right) \right] \qquad (7)$$

where
- the index e refers to the Ω_e element number e
- P_e in an adaptative weight
- the first contribution measures how much the element differs from a parallelogram,
- the second term is a function of both the element size and the orthogonality of the diagonals
- c is a constant factor which may be chosen to adjust the adaptivity with respect to geometrical regularity.

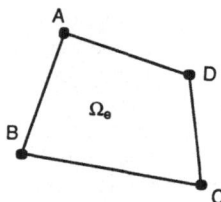

Fig. 1
Element Ω_e

The weights P_e depend on the mean equivalent strain rate $\dot{\varepsilon}_e$ in the element :

$$P_e = \frac{1 + a\,\dot{\varepsilon}_e / \dot{\varepsilon}_m}{1 + a}$$

with $\dot{\varepsilon}_e = \left(\int_{\Omega_e} \dot{\varepsilon}\, d\omega \right) / \int_{\Omega_e} d\omega$

and $\dot{\varepsilon}_m$ is the average of $\dot{\varepsilon}_e$ over the elements, and a chosen between 0 and 5.

The static remeshing procedure is defined by moving the nodal positions in such a way as to minimize the functional Π with respect to the nodal coordinates, i.e. :

$$\frac{\partial \Pi}{\partial X} = 0 \qquad (8)$$

with the constraint that the boundary remains unchanged.

More general functionals have been discussed in /13/ wich may be easily adapted to every kind of element. But equation (7) corresponds to a quadratic functional, the minimization of which is economical in terms of computer time.

4.2. DYNAMIC REMESHING

An abitrary Lagrange-Euler (ALE) description has been reported in /16/ where the mesh velocity V^M is a "numerical velocity", which is different from the kinematic velocity v (Lagrange) or from zero (Euler).

Here we introduce an Optimum Lagrange Euler (OLE) formulation by differentiating (8) with respect to time. We obtain the following equation :

$$\frac{\partial^2 \Pi}{\partial \, x^2} \cdot \mathbf{v}^M = 0 \qquad (9)$$

which expresses that the velocity $\mathbf{v}^M$ preserves the optimality of the mesh (in the sense of the functional Π) during deformation.

However the velocities on the boundary must be equal, and among different hypotheses we have chosen the following :

- $\mathbf{v}_i^M = \mathbf{V}_i$ for each node lying on the free surface

- $\mathbf{v}_i^M \cdot \mathbf{n} = \mathbf{V}_i \cdot \mathbf{n}$ for any node on the tool surface with normal $\mathbf{n}$

At each increment the linear system (9) is solved by few iterations of the conjugate gradient algorithm. The nodal coordinates are updated by changing equation (6) into :

$$\mathbf{X}\,(t+\Delta t) = \mathbf{X}\,(t) + \mathbf{v}^M(t)\Delta t \qquad (10)$$

The cumulated strain at node i is obtained by taking into account the convexion term :

$$\bar{\varepsilon}_i(t+\Delta t) = \bar{\varepsilon}_i(t) + \dot{\bar{\varepsilon}}_i(t)\,\Delta t + \mathbf{grad}\,\bar{\varepsilon}_i(t) \cdot (\,\mathbf{V}_i - \mathbf{v}_i^M)\,\Delta t \qquad (11)$$

5. Application to the forging of a gear tooth

The process has been simplified, and only the last step of the deformation is modelled. A plane strain approximation is presented as a first approach, the diameter of the gear is supposed to be large enough so as to neglect the curvature, and two symmetry axes are introduced (see fig. 2). In addition only the volume which is in the vicinity of the tooth has been considered, so that an upper flat tool is necessary for pressing the material into the die.

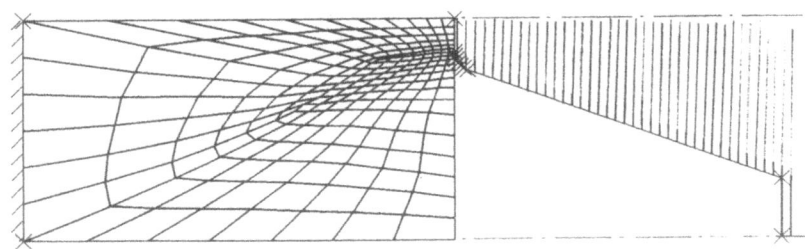

Fig. 2 : Approximate description of the process and initial mesh

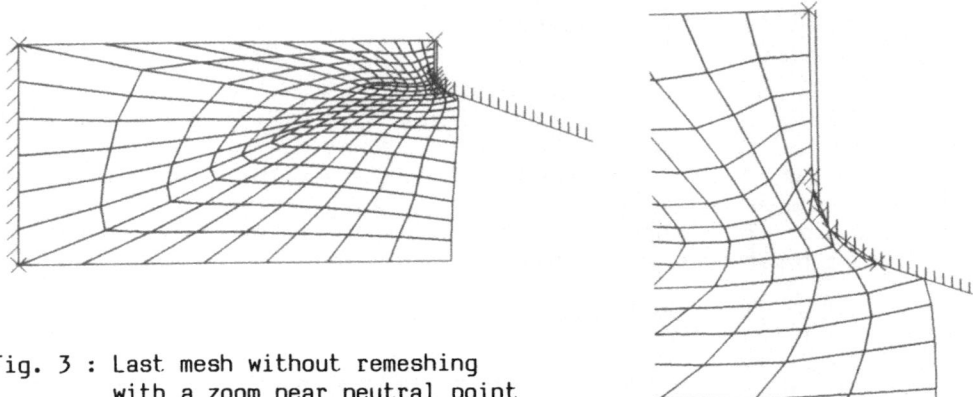

Fig. 3 : Last mesh without remeshing
with a zoom near neutral point

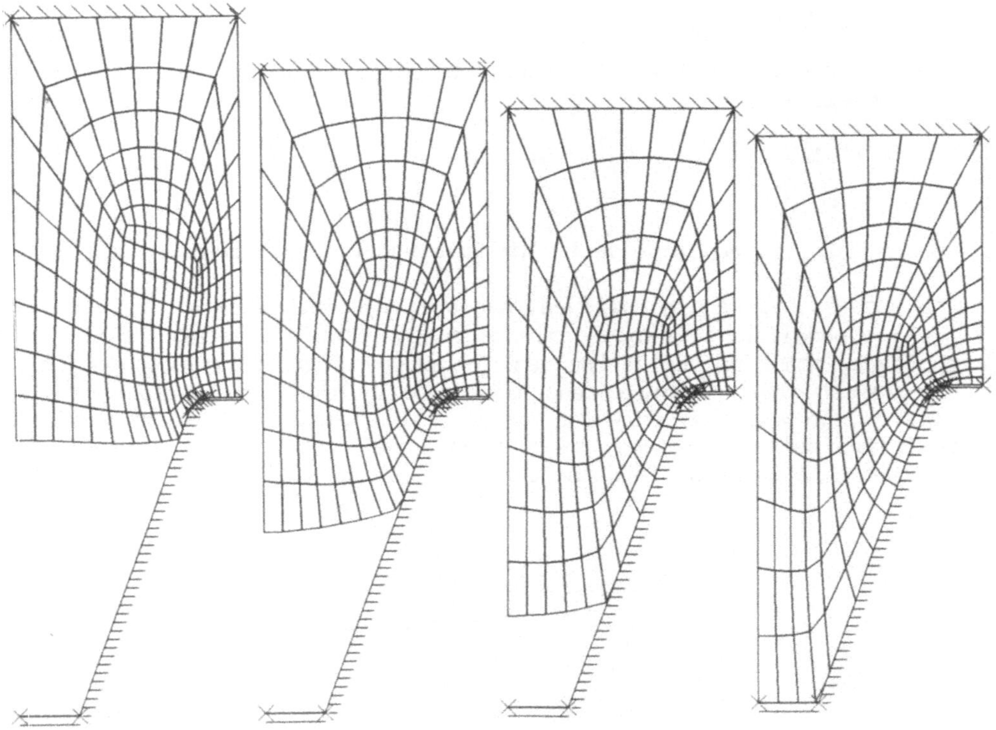

Fig. 4 : Simulation with remeshing

In order to simulate cold forging the following parameters have
been chosen :

$m = 0.02$

$K = K_0 \ (0.4 + \bar{\varepsilon})^{0.2}$

$\alpha = 1.2$

The initial mesh has been constructed using preliminary
calculations which showed that there is a neutral point on the left
upper part of the die (fig. 2).

Keeping this result in mind, an appropriate initial topology of
the mesh has been chosen and the smoothing technique has produced the
mesh shown on fig. 2.

The classical formulation with a kinematic displacement of the
nodes gives a distorted mesh after 28 increments near the neutral point
as shown on figure 3.

After a manual or automatic remeshing the next stage will produce
only additional increments. For the whole process, 10 to 15 remeshing
steps would be necessary.

With the continuous remeshing method as described in section 4.2,
the complete filling of the die for the forming of the gear tooth is
possible without any modification of the mesh structure of the mesh if
we use no adaptative term (a = 0). Figure 4 gives intermediate and last
configurations in this case.

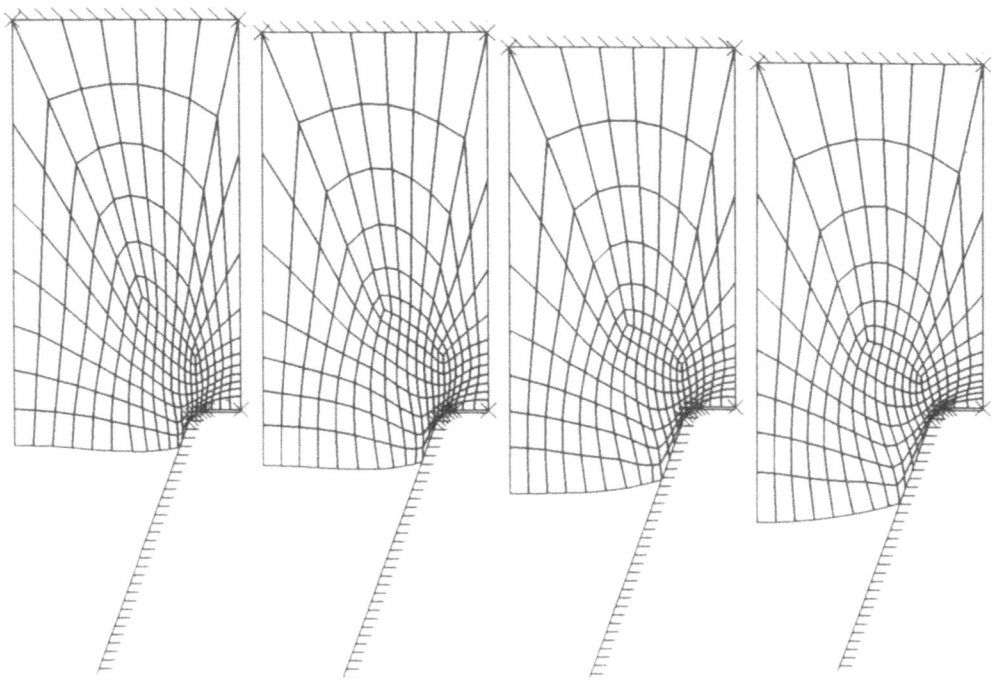

Fig. 5 : Simulation with adaptative remeshing

214

In the case we use an adaptative factor of two (a = 2) only half of the process can be simulated with the same mesh topology. Figure 5 gives intermediate configurations. Figure 6 represents the same configuration as Figure 3 but with adaptative remeshing. We observe that adaptative term allows the elements in the vicinity of the neutral point and the shear line to remain small while the geometric part of the functional prevents high distorsion.

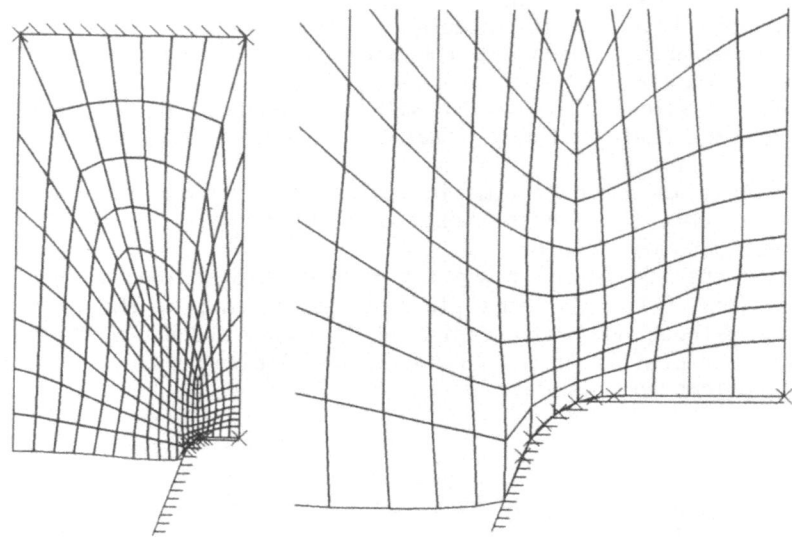

Fig. 6 : Same configuration as on fig. 3 but with adaptative remeshing

And finally, figure 7 gives a map of the iso-cumulated strains, showing that locally this process involves deformation as high as 1000 %.

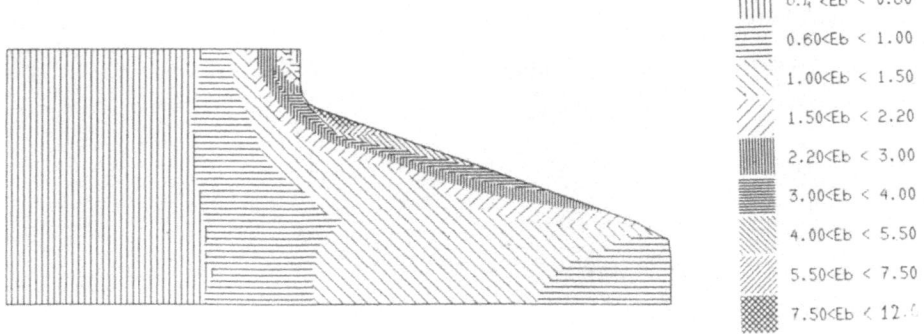

Figure 7 : Map of the iso $\bar{\varepsilon}$

6. Conclusion

A simple formulation of a continuous remeshing procedure has been
presented for 4 nodes quadrilateral elements, which may be easily
extended to 8 nodes hexahedral elements. This procedure appears to be
easy to implement in an industrial code and is economical with respect
to computer time. A very severe test on the forming of a gear tooth has
demonstrated the efficiency of our method. In the future more general
formulations will be analyzed for the extension to any type of element,
and the adaptivity term will be a function of the error estimate.

7. References

/1/ G.C.CORNFIELD and R.H. JOHNSON, Theoretical prediction of plastic
flow in hot rolling including the effect of various temperature
distribution, J. Iron Steel Inst., **211** , 567-573 (1973)

/2/ C.H. LEE ans S. KOBAYASHI, New solutions to rigid-plastic deformation problems using a matrix method, Trans. ASME, J. Eng.
Ind., **95** , 865-873 (1973)

/3/ O.C. ZIENKIEWICZ and P. N. GODBOLE, Flow of plastic and viscoplastic solids with special reference to extrusion and forming
processes, Int. J. Numer. Methods Eng., **8** , 3-16 (1974)

/4/ J.X. SUN and S. KOBAYASHI, Analysis of block compression with
simplified three-dimension element, Adv. Technology Plasticity,
2 , 1027-1034 (1984)

/5/ G. SURDON and J.L. CHENOT, Finite element calculation of three-
dimensional hot forging, Int. J. Numer. Methods Eng., **24** ,
2107-2117 (1987)

/6/ G. SURDON and J.L. CHENOT, Finite element calculation of three
dimensional hot forging, Proc. Int. Conf. on Numerical Methods
in industrial forming proceses, Gothenburg (1986)

/7/ S. KOBAYASHI, Process desing in metal forming by the finite
element method, Proc. Int. Conf. on Technology of plasticity,
Stuttgart (1987)

/8/ J.U. BRACKBILL, J.S. SALTZMAN, Adaptive Zoning for Singular
Problems in two Dimensions", J. Comput. Physics, **46** , 342-368
(1982)

/9/ A.R. DIAZ, N. KIKUCNI, J.E. TAYLOR, A method of Grid Optimization
for Finite Element Methods, Comp. Meth. in Appl. Mech. Eng., **41** ,
29-45 (1983)

/10/ N. KIKUCHI, Adaptive Grid-Design Methods for Finite Element
Analysis, Comp. Meth. in Appl. Mech. Eng., **55** , 129-160 (1986)

/11/ L. DEMKOWICZ, J.T. ODEN, On a Moving Mesh Strategy based on an
Interpolation Error Estimate Technique", Int. J. Eng. Sci., Vol
24 n° 1, 55-68 (1986)

/12/ O.C. ZIENKIEWICZ, Y.C. LIU and G.C. HUANG, Error estimation and
adaptivity in flow formulation for forming problems, Int. j.
numer. methofs, eng., **25** , 23-42 (1988)

216

/13/ J.P. CESCUTTI and J.L. CHENOT, A geometrical continuous remeshing
     procedure for application to finite element calculated of
     non-steady state forming processes, Proc. Int. Conf. on Numerical
     Methods in Engineering. Theory and Applications, Swansea (1987)
/14/ J.P. CESCUTTI, N. SOYRIS, G. SURDON and J.L. CHENOT, Thermo-mecha-
     nical finite element calculation of three-dimensional hot forging
     with remeshing, Proc. Int. Conf. on Technology of plasticity,
     Stuttgart (1987)
/15/ J.L. CHENOT, Finite element calculation of unilateral contact with
     friction in non steady-state processes, Proc. Int. Conf. on
     Numerical Methods in Engineering. Theory and Applications, Swansea
     (1987)
/16/ J. DONEA, P. FASOLI- STELLA, , S. GIULIANI, J.P. HALLEUX and A.V.
     JONES, An Arbitrary Lagrangian Eulerian finite element method for
     transient dynamic fluid structure interaction problems, SMIRT-5
     Conference, Berlin (1979)

FINITE ELEMENT APPLICATIONS
IN FORMING BILLET AND P/M PREFORMS

Yong-Taek Im
The Ohio State University
Columbus, Ohio 43210

and

Shiro Kobayashi
University of California
Berkeley, California 94210

ABSTRACT

The finite element simulation of various metal forming processes assists in producing defect-free final products from billet and sintered powdered-metal(P/M) preforms. In this paper a rigid-thermoviscoplastic finite element program, including the effect of compressibility of the material and temperature on deformation was applied to forming of sintered P/M preforms in ring compression, die pressing, closed-die forging at room temperature, and non-isothermal plane-strain compression. Possible fracture sites in such sintered P/M forming processes were predicted from the numerical simulations and compared with the experimental observations available in the literature. Also, the computed macroscopic densification and forging pressure required were compared with the available data. Finally, the effect of temperature on metal flow has been investigated in conventional hot forging.

INTRODUCTION

During the last two decades, the finite element method has been developed and applied with great success to various problems in the metal forming, specifically in forging, extrusion, drawing, rolling, nosing, sheet metal forming at room and elevated temperatures [1-4]. Recently, the application was extended to deformation analysis of sintered powdered-metal preforms [5-7]. The finite element simulation of such metal forming processes assists in understanding metal flow under practical metal forming conditions and producing defect-free final products. In the present paper some highlights of the applications of a rigid-thermoviscoplastic approach in forming billet and P/M preforms are presented. Since detailed information of theoretical formulation is available in the literature, it is omitted here.

In the sintered P/M forming processes, the hydrostatic pressure affects the plastic deformation. As a result of this, there is a volume change during deformation. The apparent volume change was, therefore, predicted in ring

J. L. Chenot and E. Oñate (eds.), Modelling of Metal Forming Processes, 217–225.
© *1988 by Kluwer Academic Publishers.*

compression and compared with the volume of the deformed ring with fully dense material. Also, the mode of densification in ring compression and isostatic die pressing at room temperature was investigated. In such processes, the preforms are more susceptible to fracture than are solid preforms. Thus, flow and fracture are of particular importance in forming these preforms. Using the F.E.M. simulations, possible fracture sites were predicted by finding the element where the porosity of that element is larger than the limiting value, where the preforms lose their mechanical strength. The prediction compared well with experimental observations in upsetting and closed-die forging of a pulley blank at room temperature [8-10].

At elevated temperatures, metal flow is sensitive to temperature and strain-rate. In order to investigate this, a non-isothermal plane-strain compression of sintered P/M preforms was simulated. The computed results were compared to the experimental data from reference [11] in terms of macroscopic densification, forging pressure, and relative density distribution. The predicted relative density distribution showed excellent agreement with the experimental hardness distribution in the same reference [11]. Finally, a non-isothermal spike forging had been simulated to investigate the effect of heat loss during transfer of the billet from furnace to the dies and dwell on die filling.

These studies clearly demonstrate that the F.E.M. analysis can be used reasonably well to predict the effect of various process parameters on metal flow involved in billet and sintered P/M forming processes without expensive experimentation and die try-outs.

NUMERICAL APPLICATIONS

RING COMPRESSION OF P/M BILLET

The compression of the sintered P/M aluminum rings was simulated. The ring geometry was 6:3:2 (outer diameter: inner diameter: height) and the initial height was 1 cm. The initial relative density was assumed to be uniform and 0.8. The friction condition was approximated by the constant shear model and the shear friction factor used in simulations was 1.0. The grid distortions during compression of fully dense and porous rings are compared at 40 % reduction in Fig. 1. From this figure, it can be easily seen that densification occurs during compression of the porous ring and thus the apparent volume of the porous ring decreased during deformation.

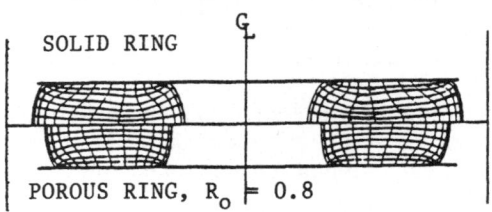

Fig. 1 Comparison of grid distortions for solid and porous materials at 40% reduction in height for m = 1.0.

The relative density distributions are plotted in Fig. 2 at 20, 40, and 60% reductions. At an early stage of compression like 20% reduction, the figure shows that densification occurs over the whole ring and was effective near the edge of the die-workpiece interface. Less densification takes place in the central area under the contacting die area due to the friction effect and near the free surface along the equatorial surface due to the tensile hydrostatic stress. As deformation continues, densification increases, except near the outer free surface of the equatorial plane. At 60% reduction, the relative density reaches almost full density in the most region of the deformed ring. In the element near the free surface, however, the relative density changes from the initial value of 0.80 to 0.83, and 0.78 at reductions of 20-40% and 60%, respectively. This indicates possible fracture at this area during deformation.

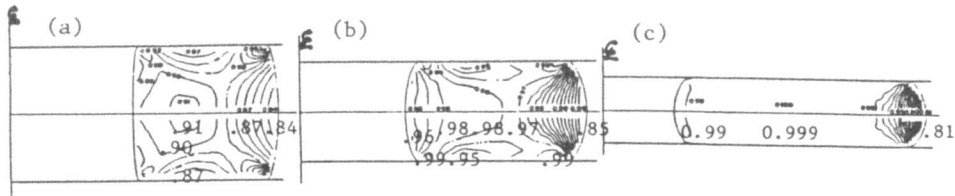

Fig. 2. Relative density distributions at (a) 20%, (b) 40%, and (c) 60% reductions in height for a porous material with m = 1.0.

ISOSTATIC DIE PRESSING OF P/M BILLET

In the isostatic die pressing, a sintered P/M cylindrical preform was compressed and densified in a cylindrical container. The dimension of the container was 2.0 in. x 6.0 in. (diameter x height). The material behavior of the base metal was assumed to be strain-rate sensitive and represented by $\sigma = \bar{\varepsilon}^{0.1}$. The shear friction factor used in simulation was 0.1. The ram velocity was assumed to be 3.0 in/sec. The initial relative density was assumed to be uniform and 0.8.

Fig.3 shows the relative density distribution at several stages of deformation(the upper half of the central cross-section of the preform is shown). It is observed that the densification is maximum near the outside corner contacting with the dies and minimum near the outside radius at the central part. As deformation increases the relative density becomes more uniform in the figure. A similar result was obtained numerically and experimentally in other studies [5,12].

CLOSED-DIE FORGING OF P/M PREFORMS

A closed-die forging of a pulley blank was simulated with two preforms as shown in Fig. 4. The geometry was selected from the work by Downey et. al. [10]. Since the geometry was axi-symmetric, one half of the center cross-section of the preforms and the dies was used for computation. The preforms were made of the aluminum powder and the initial relative density was

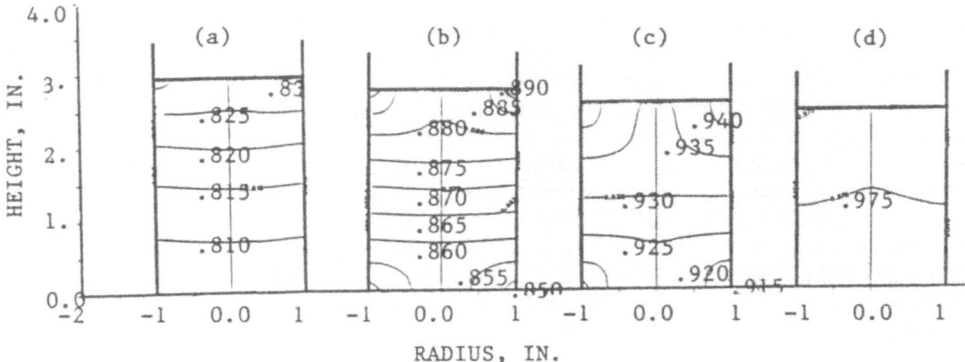

Fig. 3. Change of relative density distribution in isostatic die pressing at (a) 2%, (b) 8%, (c) 14% and (d) 18% reductions (figure represents the upper half of the central cross-section of the cylindrical P/M billet): initial relative density = 0.8, shear friction factor = 0.1.

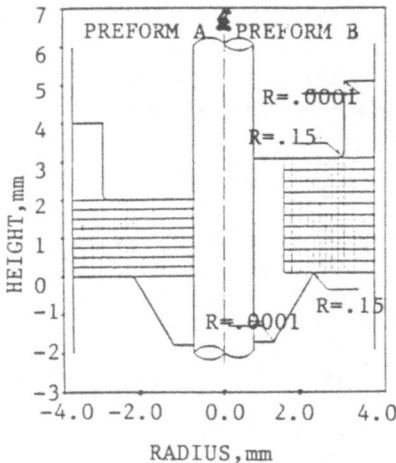

Fig. 4. Preform shapes and mesh systems in axisymmetric closed-die forging of a pulley blank.

assumed to be 0.780 for both preforms. The shear friction factor was assumed to be constant throughout the deformation as 0.1.

In Fig. 5(a), the distributions of relative density and hydrostatic stress at the 38% reduction in height are given for Preform A. The patterns of these two are similar to each other. According to the computed results, the relative densities of elements around free surfaces in the hub section decreased up to 30% reduction due to the tensile stress states at those area. However, a transition in the velocity field around the rim section occurred from upsetting to extrusion mode around 30% reduction in height. As a result of this change, the relative density of the element near the side surface of the upper die was below the limiting value at 38% reduction, where the preform loses its mechanical strength. The relative density decreased further at the same element, indicating that crack probably starts from that element.

For Preform B the complete filling was obtained as shown in Fig. 5(b). This figure shows that the relative density in most of the forged hub section at the final stage reached full density, 0.999 while it was around 0.84 near the tips of the rim section. This indicates that the mechanical strength near those area will be weak compared to the rest of the forged part.

The computed results agreed well with the experimental observations [10] regarding the proper preform shape which produced defect-free pulley blank. This agreement verified the prediction of the probable locations of fracture during forging of the sintered P/M preforms using the F.E.M simulations.

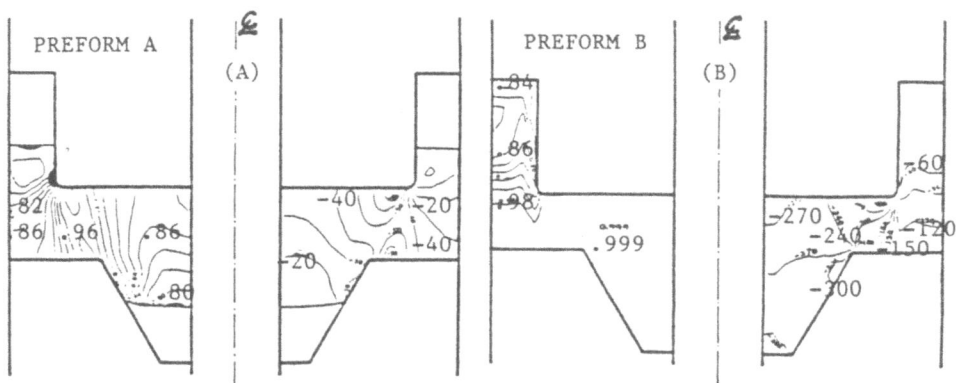

Fig. 5   Relative density (left) and hydrostatic stress (right) distributions at the final stage of forging (a) preform A (darkened places reflect the possible fracture sites) and (b) preform B.

NON-ISOTHERMAL PLANE-STRAIN COMPRESSION OF P/M BILLET

The upsetting of long bar between the flat dies at elevated temperatures was simulated. Because of the longitudinal constraint, the plane-strain condition was satisfied in the central cross-section. The long bar was made of sintered iron powder and the dimension was $10 \times 10 \times 100$ mm^3. Two cases of initial relative densities, 0.743 and 0.802, were chosen from the experimental studies, given in reference [11]. The two shear friction factors were taken as 1.0 and 0.5 for both cases. The temperatures of the billet and the dies were initially assumed to be 1433 and 293 $^\circ$K, respectively. The material property of the base metal was assumed to be rigid-plastic and the yield stress was taken as 198 MPa from a handbook. The thermal properties of the base metal were taken from a handbook and were assumed to be independent on temperature because of lack of information although program can handle the temper-

222

ature sensitive material properties. Due to the geometrical symmetry and the plane-strain condition, only a quarter section of the central cross-section of the bar was used for computation.

The overall change of the average relative density is compared between the experiment and the prediction in Fig. 6. From this figure it can be seen that the predictions are excellent for both cases, implying that friction was insignificant in densification. It is also noted that the densification takes place mainly at an early stage of upsetting. As deformation continues, the densification becomes saturated and lateral flow increases. Although the preform with loose density densifies more rapidly than that with dense density does, it does not reach the same level of densification at the same deformation level.

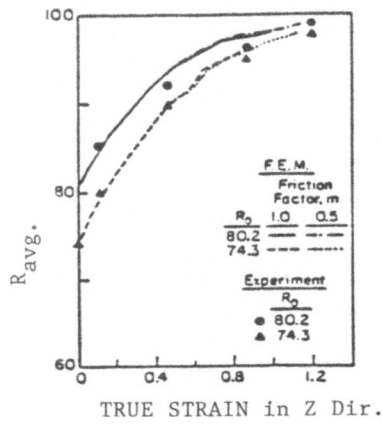

Fig. 6. Changes of average relative density during compression.

In Fig. 7, the local hardness distribution, obtained from the experimental results of reference [11], is reproduced. This information was correlated with the computed relative density distributions as shown in Fig. 8 for both initial preform densities with two different frictions at the 40% reduction in height. By comparing these two figures an excellent agreement between the relative density and the hardness distributions was obtained as expected.

NON-ISOTHERMAL SPIKE FORGING OF CYLINDRICAL BILLET

In this study two simulations, without(Case I) and with(Case II) transfer and dwell periods were conducted to determine the effect of heat loss on metal flow. The geometries of the billet and the dies are given in Fig. 9. The billet material was Ti 6242(β-microstructure) and the billet size 2x2.25 in^2. (diameter x height). The temperatures of the billet and the dies were 1750 and 570 °F, respectively. The ram velocity was 1 in/sec, representing a hydraulic press. The shear friction factor was selected as 0.3 to represent the friction condition. The thermal properties and the flow stress used in computations

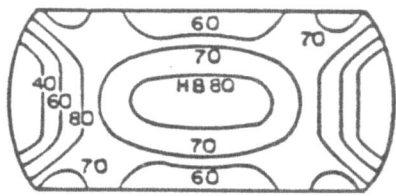

Fig. 7 Experimental local hardness distribution [11] in the center cross-section of a bar.

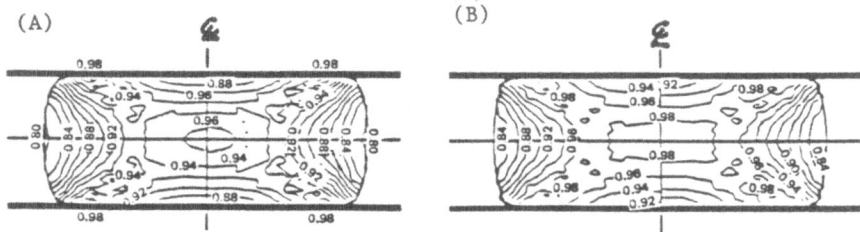

Fig. 8. Relative density distributions at 40% reduction in height with two friction conditions: left (m = 0.5), right (m = 0.1), (a) Case I (R_0 = 0.743) and (b) Case II (R_0 = 0.802).

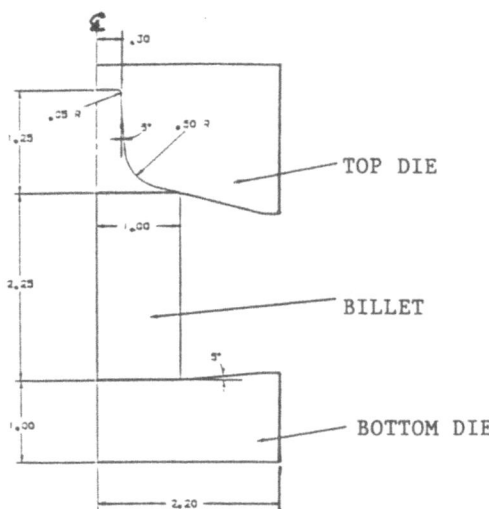

Fig. 9 Schematic of dies and billet in spike forging (dimensions are all in inches).

were taken from the literature and are summarized elsewhere [13].

Grid distortion at the final stage of deformation is given for Case I in Fig. 10(a) and for Case II in Fig. 10(b). It was observed that more bulging was formed in Case II during deformation, *i.e.*, when the effects of heat transfer prior to forging were considered. Such a effect was apparent in further stage of deformation as shown in Fig. 10. The comparison of these two figures shows that complete filling takes place at an earlier stroke position, at 1.766 in. for the Case II than for the Case I, at 1.811 in.. Thus, heat loss has essentially the same effect as increasing friction in the web area of the die cavity and results in reducing flash formation.

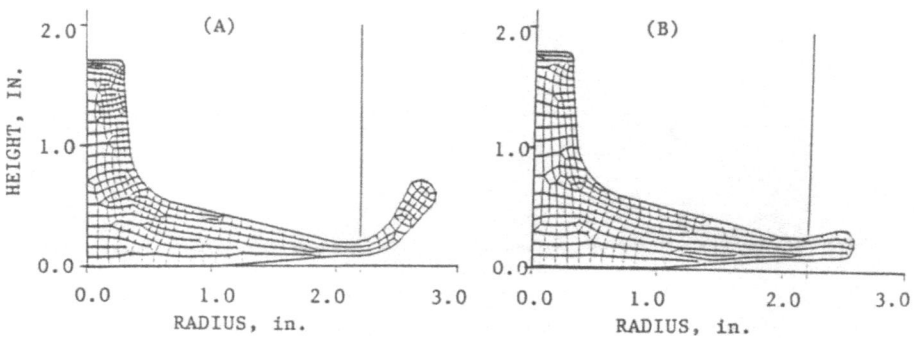

Fig. 10   Grid distortions at complete die filling for (a) Case I (die stroke = 1.811 in.) and (b) Case II (die stroke = 1.766 in.).

CONCLUDING REMARKS

Based on this study, it can be concluded that the Finite Element Technique is an excellent tool for investigating the effect of heat transfer upon metal flow in conventional hot forming and the effect of compressibility on the macroscopic mechanical deformation and possible prediction of fracture sites in sintered P/M forming at room and elevated temperatures. For further elaborated analysis, additional experimental and theoretical studies on the characterization of the mechanical properties and the thermal properties of the workpiece materials are required.

REFERENCES

1.   Kobayashi, S., "A Review on the Finite-Element Method and Metal-Forming Process Modelling,"" *J. Appl. Metalworking*, 2, p.163, 1982.

2.   Pittman, J.F.T., *et. al.* (editors), <u>Numerical Analysis of Forming Processes</u>, A Wiley-Interscience Publication, John Wiley & Sons, 1984.

3.   Lange, K. (editor), <u>Simulation of Metal Forming Processes by the Finite Element Method</u>, Workshop Proceedings, Stuttgart, 1985.

4.   Monograph of the Publications of Centre De Mise En Forme Des Materiaux, Groupe Thermo-Mecanique Plasticite, Sophia-Antipolis-06560 Valbonne, France, Private Communication, 1986.

5.   Oh, S.I. and Gegel, H.L., "ALPIDP - Modeling of P/M Forming by the Finite Element Method," Private Communication, 1986.

6.   Im, Y.T. and Kobayashi, S., "Finite-Element Analysis of Plastic Deformation of Porous Materials," Metal Forming and Impact Mechanics(edited by Reid, S.R.), p.103, Pergamon Press, 1985.

7.   Im, Y.T. and Kobayashi, S., "Coupled Thermo-Viscoplastic Finite Element Analysis of Plane-strain Compression of Porous Materials," *Advanced Manufacturing Processes*, $\underline{1}$(2), p.269, 1986.

8.   Im, Y.T. and Kobayashi, S., "Analysis of Axisymmetric Forging of Porous Materials by the Finite Element Method," *ibid*, $\underline{1}$(3&4), p.473-499, 1986.

9.   Kuhn, H.A. and Lawley, A. (editors), Powder Metallurgy Processing, Academic Press, 1978.

10.   Downey, C.L. and Kuhn, H.A., "Designing P/M Preforms for Forging Axisymmetric Parts," *Int. J. Powder Met. & Tech.*, $\underline{11}$, p.255, 1975.

11.   Fischmeister, H.F., Aren, B., and Eastering, K.E., "Deformation and Densification of Porous Preforms in Hot Forging," *Powder Metallurgy*, $\underline{14}$(27), p.144, 1971.

12.   Schacher, H.D., Kaltmassivumformen von Sintermetall, Ph.D. Dissertation, Institute fur Umformtechnik, Universitat Stuttgart, 1978.

13.   Im, Y.T., *et. al.*, "Applications of FEM to Simulations of Non-isothermal Forging Processes," Technical Report No. ERC/NSM-87-15, ERC for Net Shape Manufacturing, The Ohio State University, Columbus, Ohio, USA, 1987.

THREE DIMENSIONAL FINITE ELEMENT CALCULATION OF THE FORGING OF A CONNECTING ROD

N.SOYRIS*,J.P.CESCUTTI*,T.COUPEZ*,G.BRACHOTTE**,J.L.CHENOT*
* Ecole Nationale Supérieure des Mines de Paris
Sophia-Antipolis 06560 Valbonne FRANCE
** Peugeot S.A. Service Méthodes 25318 Montbéliard cedex FRANCE

Abstract

The three dimensional finite element program FORGE3 is used to simulate the forging of a connecting rod. The material behaviour is assumed to follow the Norton-Hoff law (incompressible and viscoplastic).The thermal effects are introduced by a coupled thermo-viscoplastic analysis with the possibility of taking into account possible phase transformations.

1. Introduction

Computerized analytical design tools are more and more necessary to the manufacturing process engineers. They give them the opportunity of lowering the process development wastes and costs. During the last few years, mathematical models of the forging process have been developed. The finite element method seems really well adapted for this kind of problems.

The previous finite element computer calculations, for plastic and viscoplastic material, considered only the plane or axisymmetric deformations (/1-4/). But for most finished products, that is for complex geometries, the metal flow is three dimensional. Then a finite element analysis has been introduced for cubic blocks or horizontal cylinders compression (/5-7/) and also for the hot forging simulation of an industrial pinion /8/.

Nevertheless, none of these formulations takes into account the temperature influence. In hot forging, that means at high temperature, heat is lost by conduction in the contact with the dies and by radiation and convection through the other boundaries; the plastic work dissipation and the friction on the other hand produce heat. The temperature distribution will be affected, the material properties can be changed. As for the mechanical problem, the first works about coupled thermo-plastic analysis appeared for axisymmetric or plane deformation problems (/9-13/), and then in the three dimensional case, but with simplified assumptions, for the thermal resolution and with simple geometries such as cubic blocks and horizontal cylinders /14/.

The purpose of this paper is to illustrate the introduction of thermal effects in the mechanical simulation with minimal assumptions, and also to show that it is possible to simulate a complex geometry forging.

J. L. Chenot and E. Oñate (eds.), Modelling of Metal Forming Processes, 227–236.
© 1988 by Kluwer Academic Publishers.

2. Mechanical formulation

2.1. BASIC EQUATIONS

Consider a body of volume Ω at time t with the velocity $\vec{V}^d$ prescribed on a portion of the surface S_v and the stress vector $\vec{f}^d$ on S_f.

The material is assumed to be homogeneous, isotropic and incompressible.

In hot forming processes the elastic part of the deformation is small enough compared to the visco-plastic one and can be neglected.

The material obeys the Norton-Hoff law. We have:

the constitutive law $\qquad\qquad \sigma'_{ij} = 2 K \dot{\varepsilon}_{ij} (\sqrt{3} \; \dot{\bar{\varepsilon}})^{m-1} \qquad\qquad$ on Ω

the incompressibility condition $\qquad \mathrm{div} \; \vec{V} = \mathrm{tr} \; \dot{\varepsilon} = 0 \qquad\qquad$ on Ω

where σ_{ij} is the deviatoric stress tensor, $\dot{\varepsilon}_{ij}$ the strain rate tensor, $\dot{\bar{\varepsilon}}$ the equivalent strain rate, m the strain rate sensitivity ($0.1 < m < 0.3$), and K the material consistency.

The friction law between the dies and the workpiece is similar to the constitutive one:

$$\vec{\tau} = - \alpha K \, \| \Delta\vec{V}_t \|^{p-1} \, \Delta\vec{V}_t \qquad\qquad \text{on } S_v$$

where $\vec{\tau}$ is the shear stress vector, $\Delta\vec{V}_t$ the relative tangential velocity between the die surface and the workpiece, $\|\Delta\vec{V}_t\|$ its euclidian norm, α and p are parameters depending upon the type of contact.

The variational method leads to the functional $J(\vec{V})$ which is minimum for the velocity field solution $\vec{V}$:

$$J(\vec{V}) = \int_{\Omega} \frac{K}{m+1} (\sqrt{3} \; \dot{\bar{\varepsilon}})^{m+1} \, d\Omega - \int_{S_f} \vec{f}^d \vec{V} \, dS_f + \int_{S_v} \frac{\alpha K}{p+1} \| \Delta\vec{V}_t \|^{p+1} \, dS_v + \frac{1}{2}\rho \int_{\Omega} K(\mathrm{div}\vec{V})^2 \, d\Omega$$

where ρ is a large positive constant ($\approx 10^7$).

2.2. FINITE ELEMENT FORMULATION

The functional $J(\vec{V})$ is calculated with $\vec{V}$ discretized using linear isoparametric eight nodes hexaedral elements.

The velocity expression is $\vec{V} = \sum_{i=1}^{8} \Psi^i \vec{V}^i$ where Ψ^i is the shape function related to the i-nodal point, $\vec{V}^i$ the velocity at the ith nodal point.

The minimization of the discretized functional gives a set of non linear algebraic equations which is solved by the Newton-Raphson method.

Unsteady forging process is analyzed by using small pseudo-steady deformation steps. Ω and $\vec{V}$ are known and fixed at time t. The configuration at time $t + \Delta t$ is calculated with an explicit eulerian integration: $\Omega(t + \Delta t) = \Omega(t) + \vec{V}(t) \cdot \Delta t$

3. Coupled Thermo-viscoplastic analysis

3.1. TEMPERATURE INFLUENCE ON THE MECHANICAL FORMULATION

The analysis, §2, has been done with the assumption that the introduced parameters were constant. In fact, they are all strain, strain rate and temperature dependent.

It has been supposed that the consistency is thermo-dependent by the following law: $K = K_0 \exp(\beta/T)$ and the other parameters constant.

The workpiece dilatation has also been neglected.

3.2. HEAT CONDUCTION EQUATION

In a Lagrangian formulation and using an isotropic Fourier law for the heat flux, the differential heat conduction equation is:

$$\rho c \frac{\partial T}{\partial t} = \text{div} (k \overrightarrow{\text{grad}}T) + \dot{W} \qquad \text{on } \Omega \qquad (1)$$

where k is the thermal conductivity, c the specific heat, ρ the density.
The power $\dot{W}$ dissipated during the plastic deformation has the following expression:

$$\dot{W} = f \sum_{i,j} \sigma'_{ij} \dot{\varepsilon}_{ij} = f K (\sqrt{3} \dot{\varepsilon})^{m+1}$$ where the constant f represents the part of the plastic

deformation energy which really becomes heat. We considered $f \approx 1$.

3.3. BOUNDARY CONDITIONS

The initial temperature distribution can be uniform or resulting from a previous cooling finite element calculation.

On the contact surface Σ_c, there is conduction with the dies and also friction heat flux. Then: $k \overrightarrow{\text{grad}} T . \overrightarrow{n}_c = - h_{cd} (T - T_d) + \Phi_{fr}$ on Σ_c (2)
where $\overrightarrow{n}_c$ is the unit normal vector to the contact surface, T_d the die temperature which has been supposed constant, and h_{cd} the heat transfer coefficient. Φ_{fr} the friction heat flux:

$$\Phi_{fr} = \frac{b}{b+b_d} \alpha K \|\Delta \overrightarrow{V}_t\|^{p+1}$$ with $b = \sqrt{k\rho c}$ for the workpiece and b_d for the die, is

assumed constant during a time step and up-dated at each new one.

On the free surface Σ_f, radiation and convection are dominant:

$$k \overrightarrow{\text{grad}} T . \overrightarrow{n}_f = - h_f (T - T_a) + \Phi \qquad \text{on } \Sigma_f \qquad (3)$$

where $\overrightarrow{n}_f$ is the unit normal vector to the free surface, T_a the air temperature, Φ a constant flux which can possibly be prescribed and h_f has the following expression: $h_f = h_c + h_r$ with h_c the heat transfer coefficient, $h_r = \varepsilon\sigma (T^2 + T_a^2)(T + T_a)$ where ε is the material emissivity and σ the Stefan constant.

3.4. NUMERICAL RESOLUTION

The numerical discretization of the equation (1) and its associated boundary conditions (2) and (3) is done using a weighted residual method and the Galerkin's method (/15/ to /17/).As for the velocity, the unknown function T can be approximated throughout Ω at any time t, by the relationship: $T = \sum_{i=1}^{8} \Psi^i T^i$ .The equation (1) then leads to the differential equations set which can be written in a matrix form as: $C\dfrac{dT}{dt} + K\,T + Q = 0$ (4), where

T is the T^i nodal temperature matrix.The set of equations (4) is non-linear since the material propreties $(\rho c, k, h_f)$ and therefore the matrices C, K and Q are thermo-dependent.

3.5. TIME-STEPPING SCHEME

To solve the set of non-linear differential equations (4), two-step schemes that are consistent to the second order are used /18/. As the time intervals Δt may slightly vary during the forging simulation, the temperature T and its derivative $\dfrac{dT}{dt}$ are written as

$$T = \alpha\, T_{t-\Delta t_1} + (\tfrac{3}{2} - 2\alpha - \gamma)\, T_t + (\alpha - \tfrac{1}{2} + \gamma)\, T_{t+\Delta t_2}$$

$$\frac{dT}{dt} = (1-\gamma)\frac{T_t - T_{t-\Delta t_1}}{\Delta t_1} + \gamma\frac{T_{t+\Delta t_2} - T_t}{\Delta t_2}$$

If we choose for C, K and Q the same kind of expressions as we use for T,to solve (4), it becomes necessary to introduce an iterative procedure. In order to avoid iterations, and not to increase the CPU time too much, a linearized technique has been tested Each non linear parameter A has been written as: $A^* = (\tfrac{1}{2} - \gamma)\, A_{t-\Delta t_1} + (\tfrac{1}{2} + \gamma)\, A_t$

Depending upon α and γ choice, different kinds of schemes may be used: Lees scheme $(\alpha = \tfrac{1}{3}, \gamma = \tfrac{1}{2})$, Dupont scheme $(\alpha = \tfrac{1}{4}, \gamma = 1)$ or implicit scheme $(\alpha = 0, \gamma = \tfrac{3}{2})$.

For some materials the heat capacity as well as the thermal conductivity may change very sharply within a narrow range of temperatures. The integral averaging technique /19/ allows to take into account a large difference of temperature inside one element.The average value $\bar{u}$ of the function u(T) in the interval T_1, T_2 may be expressed by $\bar{u} = \dfrac{\displaystyle\int_{T_1}^{T_2} u\, dT}{T_2 - T_1}$ .The enthalpy : $H_T = \displaystyle\int_{T_0}^{T} \rho c\, dT$ /16/ and the following function has been chosen: $G_T = \displaystyle\int_{T_0}^{T} \frac{1}{k}\, dT$ /19/.

3.6. RESOLUTION ALGORITHM

In fact the mechanical problem and the thermal one are solved independantly at each time step. At time t: the temperatures T_t et $T_{t-\Delta t_1}$, the consistency K_t and the domain Ω_t are known; then the viscoplastic resolution is computed: the velocity $\vec{V}_t$, the friction flux Φ_{fr}, the plastic work rate $\dot{W}$ and the domain $\Omega_{t+\Delta t_2}$ are obtained; at last the thermal calculation is computed: the temperature $T_{t+\Delta t_2}$ and the consistency $K_{t+\Delta t_2}$ are found.

4. Forging of a connecting rod

4.1 THE MESHES

Because of the symmetry only half of the workpiece is studied.The figure 1 shows, in position at the beginning of the forging process:
  the material mesh with 624 eight nodes elements and 960 nodes
  the upper die mesh with 2070 triangle elements and 1112 nodes
  the lower die mesh with 2053 triangle elements and 1101 nodes.

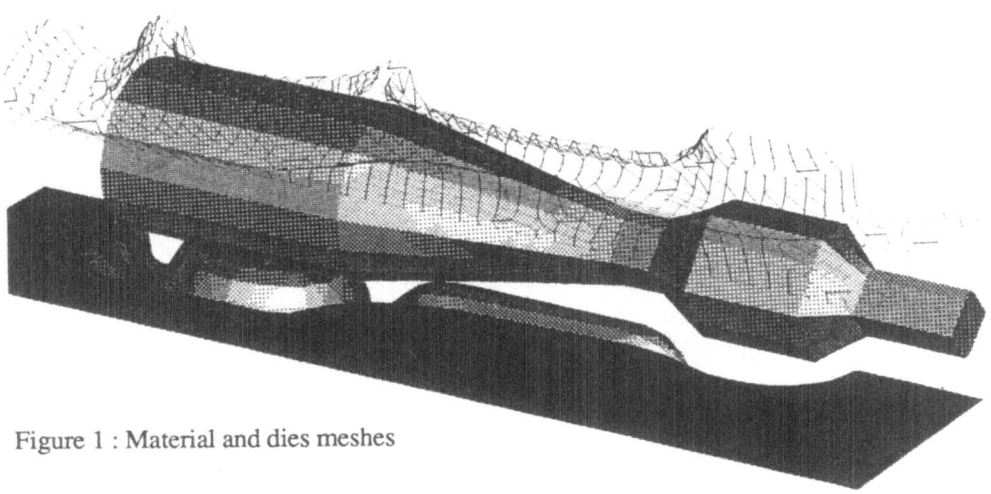

Figure 1 : Material and dies meshes

4.2. THE PARAMETERS VALUES

The mechanical parameters are:
  $m = p = 0.15$ $\qquad\qquad$ $\alpha = 0.3$
  $V_{upper\ die} = -50$ mm/s $\qquad$ $V_{lower\ die} = +50$ mm/s
The time intervals are calculated so there is 1% of deformation at each time step.
The thermal parameters are:
  $K_0 = 1.55\ 10^6$ Pa/s^m $\qquad\qquad$ $\beta = 4.2\ 10^3$ °K^{-1}
  $\rho = 7.8\ 10^3$ kg/m^3 $\qquad\qquad$ $c = 7.0\ 10^2$ J/kg.°K $\qquad$ $k = 2.3\ 10^1$ W/m.°K

$$b = 1.12 \ 10^4 \ \text{I.S.} \qquad b_d = 1.24 \ 10^4 \ \text{I.S.}$$
$$h_{cd} = 6 \ 10^3 \ \text{W/m}^2.°\text{K} \qquad T_d = 120°\text{C}$$
$$h_c = 6 \ \text{W/m}^2.°\text{K} \qquad \varepsilon = 0.7$$

The initial temperature is uniform: $\quad T = 1200°\text{C}$.

The Dupont scheme has been used: $\quad \alpha = 0.25 \qquad\qquad\qquad \gamma = 1$

4.3. RESULTS

At the beginning of the forging process simulation, the time step is $\Delta t = 3.652 \ 10^{-3}$s and the height between the two dies is $h = 36.5$ mm. The program has stopped at the step n°136 (that is for 72% of reduction) because of some elements degeneracy. At this stage $\Delta t = 9.4 \ 10^{-4}$ s and h = 9.3 mm.

The figure 2 shows the deformed material for the steps n° 50, 100 and 130.

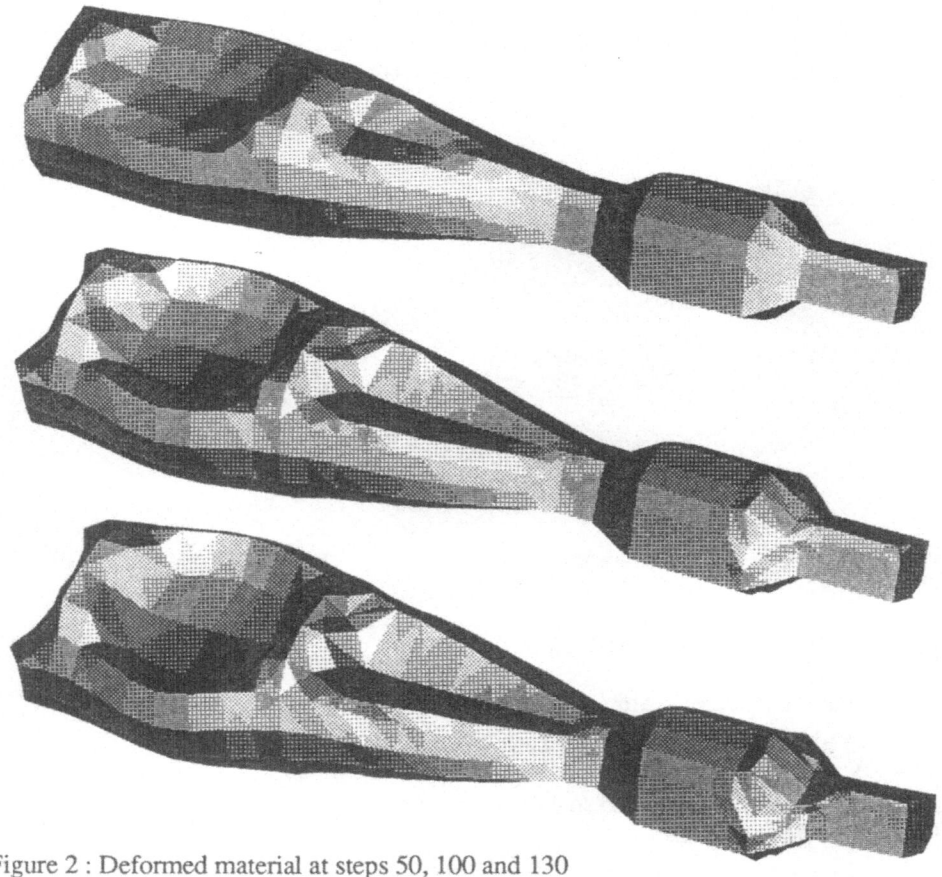

Figure 2 : Deformed material at steps 50, 100 and 130

The figure 3 shows the deformed material mesh and the evolution of the strain rate at the steps n°50, 70 and 130. The mesh is about to degenerate at step 130.

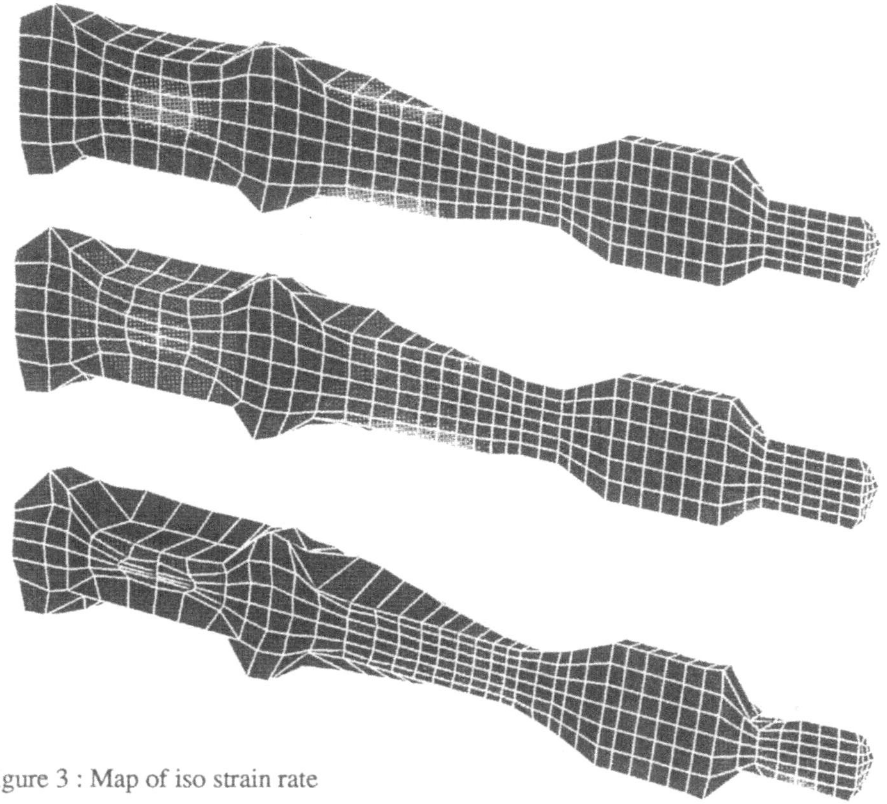

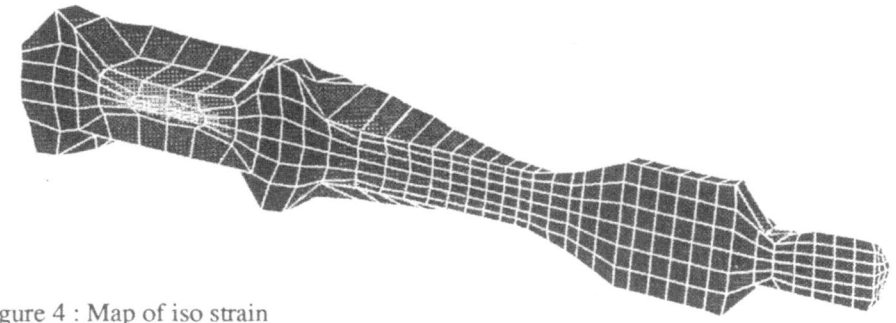

Figure 3 : Map of iso strain rate

At the end of the forging (step n°130), the strain iso-values in the symmetric plane (figure 4) are high in the head of connecting rod and negligible elsewhere.

Figure 4 : Map of iso strain

234

At step 130, the temperature iso-values in the symmetric plane (figure 5) give an almost uniform temperature in the volume, the coolest parts are on the contact surfaces.

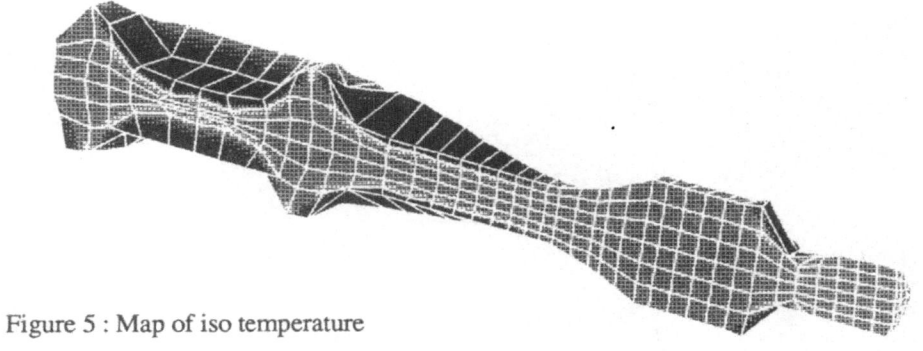

Figure 5 : Map of iso temperature

The figure 6 shows the evolution of the strain iso-values during the forging process in a plane located as indicated (step n° 70, 100 and 130).

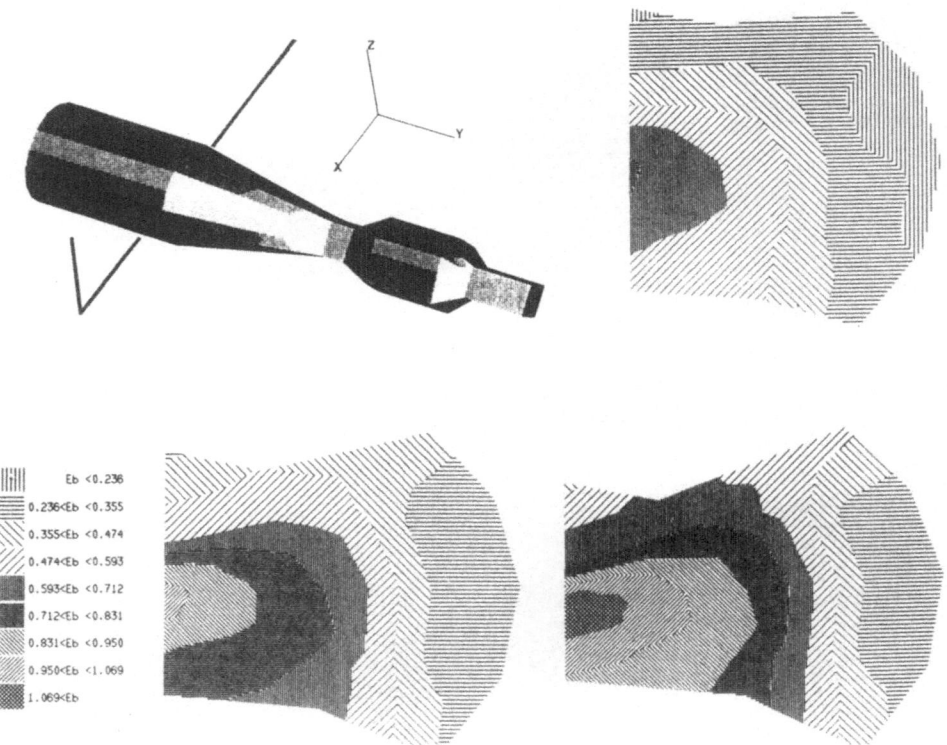

Figure 6 : Sections : Map of iso stain

The figure 7 shows in the same plane the temperature evolution during the forging process.

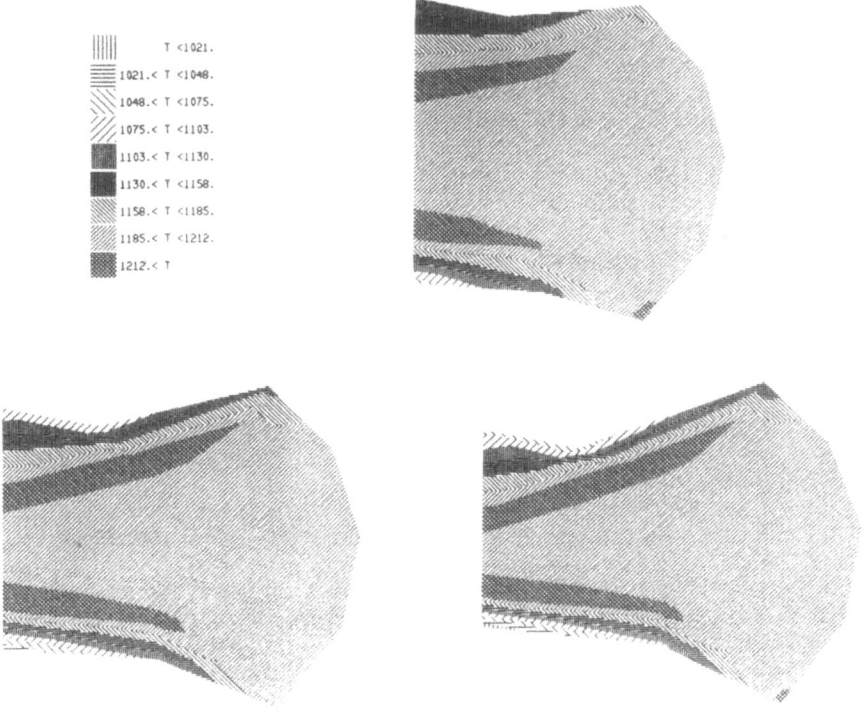

Figure 7 : Sections : Map of iso temperature

4.4. CONCLUSION

The program FORGE3 is able now to treat industrial cases.The mesh ,in this example, was too coarse, the results would really be better with a finer one. The introduction of the thermal effects gives interesting indications on the temperature distribution, nevertheless for steel fast forging it appears that the convection and the radiation on the free surfaces do not have too much influence, it would absolutely be different for low speed aluminium forging. It seems necessary now to test this code in more and more difficult and different industrial cases. The introduction of the remeshing procedure would allow to go on until the end of the forging process.

Acknowledgements

We would like to thank the companies Peugeot, Pechiney, Forges de Courcelles, Teksid and the French Ministère de la Recherche et de la Technologie for their financial support.

236

References

/1/ G.C.Cornfield and R.H.Johnson: "Theoretical prediction of plastic flow in hot rolling including the effect of various temperature distribution" J. Iron Steel Inst 211, pp 567-573; 1973

/2/ C.H.Lee and S.Kobayashi: "New solutions to rigid-plastic deformation problems using a matrix method" Trans ASMF, J. Eng. Ind. 95, pp 865-873; 1973

/3/ O.C.Zienkiewicz and P.N.Godbole: "Flow of plastic and visco-plastic solids with special reference to extrusion and forming processes" Int. J. Num. Meth. Eng. 8, pp 3-16; 1974

/4/ Y.Germain,J.L.Chenot,P.E.Mosser: "Finite element analysis of shaped lead-tin disk" Num. Meth. in Ind. Form. Proc. pp 271-276; 1986

/5/ K.Mori,K.Osakada,K.Nakadoi and M.Fukuda: "Simulation of three-dimensional deformation in metal forming by the rigid-plastic finite element method" Adv. Tech. of Plasticity vol. 2, pp 1009-1014; 1984

/6/ J.J.Park and S.Kobayashi: "Three-dimensional finite element analysis of block compression" Int. J. Mech. Sci. vol.26, n °3, pp 165-176; 1984

/7/ G.Surdon,J.L.Chenot: "Finite element calculation of three-dimensional hot forging" Int. J. for Num. Meth. in Eng. vol 24, pp 2107-2117; 1987

/8/ G.Surdon,J.L.Chenot: "Finite element calculation of three-dimensional hot forging" Num. Meth. in Ind. Form. Proc. pp 287-292; 1986

/9/ N.Rebelo,S.Kobayashi: "A coupled analysis of viscoplastic deformation and heat transfer.Theorical considerations, applications" Int. J. Mech. Sci. vol.22, pp 699-718; 1980

/10/ J.H.Argyris and J.S.Doltsinis: "On the natural formulation and analysis of large deformation coupled thermomechanical problems" Comp. Meth. in Appl. Mec. and Eng. 25, pp 195-253; 1981

/11/ S.Kobayashi: "Thermoviscoplastic analysis of metal forming problems by the finite element method" Numerical Analysis of Forming Processes John Wiley &Son, pp 45-69; 1984

/12/ C.Theodosiu,E.Soos and I.Rosu "A finite element model of the hot working of axially symmetric products. 2 Determination of the velocity and temperature fields during hot extrusion" Rev. Roum.. Sci. Techn. Mec. appl. 29; 1984

/13/ R.Stafford: "Simulation and design of high precision unit processes via numerical methods" Structural Dynamics Research Corporation Milford Ohio; 1987

/14/ J.P.Cescutti,N.Soyris,G.Surdon,J.L.Chenot: "Thermo-mechanical finite element calculation of three-dimensional hot forging with remeshing" Adv. Tech. of Plasticity vol. 2, pp 1051-1058; 1987

/15/ O.C.Zienkiewicz "Finite element methods in thermal problems" Numerical Methods in Heat Transfer John Wiley & Sons; 1981

/16/ G.Comini,S.Del Guidice, R.W.Lewis,O.C.Zienkiewicz: "Finite element solution of non linear heat conduction problems with special reference to phase change" Int. J. for Num. Meth. in Eng. vol 8, pp 613-624; 1974

/17/ J.L.Marcelin, M.Abouaf ,J.L.Chenot: "Analysis of residual stresses in hot-rolled complex beams" Comp. Meth. in applied Mech. and Eng. 56 pp 1-16; 1986

/18/ M.A.Hogge: "A comparison of two and three level integration schemes for non-linear heat conduction" Numerical Methods in Heat Transfer John Wiley & Sons; 1981

/19/ E.C.Lemon: "Multidimensional integral phase change approximations for finite conduction codes" Numerical Methods in Heat Transfer John Wiley & Sons; 1981

NUMERICAL IDENTIFICATION OF FORGING PARAMETERS

P.GROCHE and U.WEISS
Institut für Umformtechnik und Umformmaschinen
Prof. Dr.-Ing. E.Doege
University of Hannover, West-Germany

ABSTRACT

Finding correct values for material- and process parameters is a frequently critical step in the creation of models for numerical simulations. This work involves the identification of die forging parameters by means of experimental investigations and FEM simulations. An algorithm based on sensitivity analyses and the consideration of plausibility regions allows the determination of time and space-dependent parameter sets. In addition to the creation of a process-parameter-database, the obtained results can be effectively used for process control and verification of micromechanical models.

1. INTRODUCTION

One increasingly endeavours to use simulations for optimizing metal forming processes under technological and economical criteria. For practical applications, simulations have to describe experimental observations quantitatively with sufficient accuracy. A crucial criterion for the design of die forging processes is the minimization of deviations between nominal and actual dimensions. Consequently, planning die forging processes with numerical simulations presupposes, that values of process parameters are so exactly known in dependency of time and space, that the inaccuracy of the numerical solution resulting from the values' inaccuracies is negligible compared to the allowed dimensional tolerance. Many of the influential parameters, such as friction and heat transfer coefficient are not measurable as such a measurement would considerably falsify the measuring result. The following two sections will deal with the basis and method for the identification of these parameters. Finally, possible applications of identified

237

J. L. Chenot and E. Oñate (eds.), Modelling of Metal Forming Processes, 237–244.
© *1988 by Kluwer Academic Publishers.*

parameter sets to process control and verification of micromechanical models will be discussed.

2. BASIS FOR THE IDENTIFICATION

Sets of precisely measured quantities and numerical simulations of die forging processes served as basis for the identification. For our experimental and numerical studies we divided the die forging processes into five different phases:
1. steady state conditions inside preform and die
2. feeding the preform into the die
3. preform lying in the bottom die
4. forming
5. cooling the forged piece down to room-temperature.

In the following, we will consider die forging processes of two different geometries. Their preforms and final shapes are shown in Fig.1. Details of the experimental and numerical investigations are given in /1/.

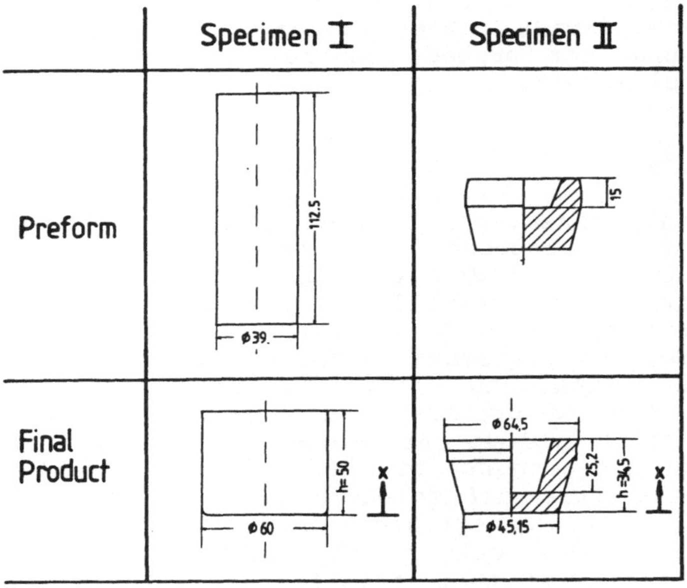

Fig.1: Specimens

2.1 Experimental Investigations

In our experimental investigations we measured quantities with the utmost precision and frequency. Local quantities

were especially valuable for the numerical identification.
Forming forces were recorded continuously. Die, specimen
and oven temperature previous to forming, feeding and forming times, as well as the dimensions of specimens and dies
at the beginning and end of the process were surveyed.

2.2 Numerical Investigations

In order to keep the numerically caused inaccuracies sufficiently small, the following had to be assumed:
- The numerical analyses were carried out by means of the
  Finite Element Method (FEM). Meshes for the two
  specimens and corresponding tools are shown in figure 2.
- Initial geometries and temperatures coincided with the
  experimental ones.
- Material data were supplied as continuous functions of
  temperatures.
- Tool deformations were treated with a linear elastic material law.
- Simulations of forging processes with fins can be
  succesfully carried out with a rigid-viscoplastic
  material law. But because of the vanishing deformation
  velocities in the last, crucial step of the forming
  operation, finless forging processes must be analysed
  with an elasto-plastic material law.

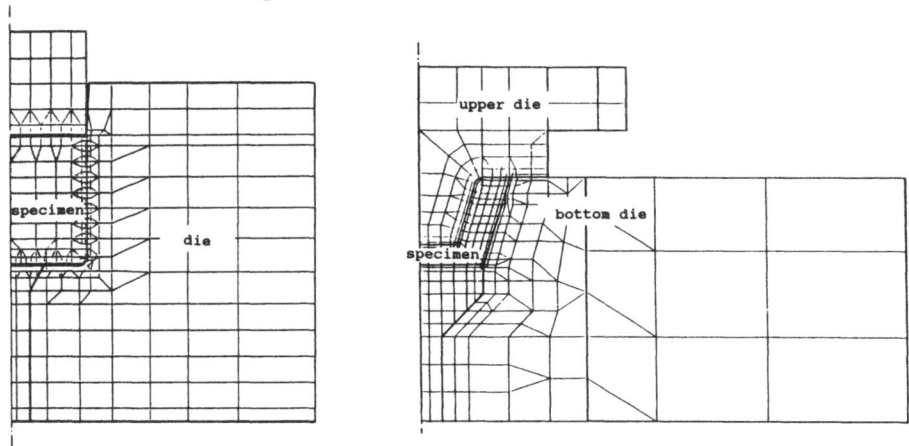

Fig.2: FEM - Meshes

3. PROCESS OF NUMERICAL IDENTIFICATION

Values of many material- and die forging process-parameters
are not known accurately enough. Thus, they must be identified. Therefore we make use of the following two crucial
presumptions:

240

a) Models upon which the numerical simulations are founded
   contain a sufficient number of parameters for fitting
   the experimental values.
b) Experimentally obtained values are correct.

The identification will be more succesful, the more local
and time - defined experimentally gained values are known.

3.1 Principle

In order to determine values and influences of different
parameters, we treat real and mathematical processes together with an algorithm for the adjustment of the model values in the mathematical model as a cybernetic system
(Fig.3). Each set of input data for real processes and numerical simulations is identical. We obtain our results under the general assumption that, should the required values
of the parameters in the mathematical model be correct,
measured and calculated output data would have to be equal.
Thus, unknown values have to be varied until the differences in output data are sufficiently small.

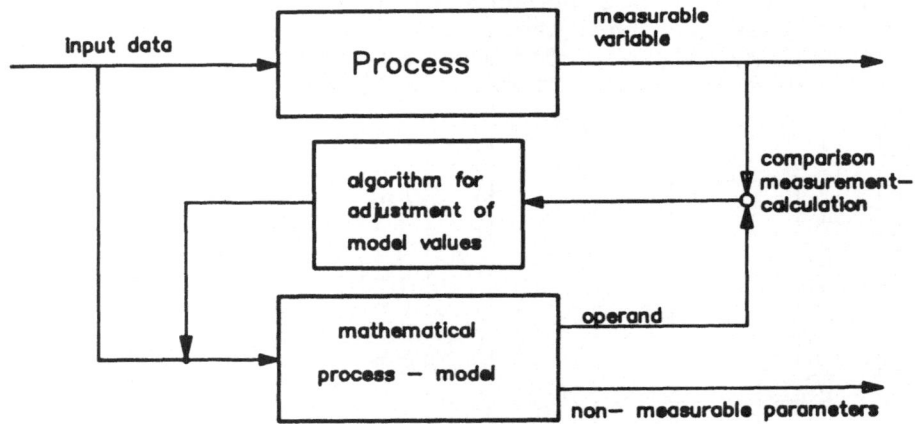

Fig.3: Numerical Identification

Although accordance between measured and calculated quantities cannot be expected, we start the identification with
values obtained in simple tests such as upsetting tests
/2,3/. During the next step, the influences of inaccurately
known material- and process parameters a_i on the final
shape of the specimens are determined. In our case of
dynamically balanced specimens we make use of the dimensionless sensitivity function $S_i(x)$:

$$S_i(x) = k \; \frac{\partial R}{\partial a_k} \; \frac{\partial a_k}{\partial a_i}$$

where   the summation convention is applied

$$k = k \; (x, a_i) = 1000 \; \frac{a_i}{R}$$

$R(x, a_i)$ - radii of the specimens.

Sensitivity functions $S_i(x)$ contain direct, $a_i = a_k$, as well as indirect, $a_i \neq a_k$, variations of the shape with parameters a_i in the surrounding of the parameter set dealt with. Characteristic curves of the different sensitivity functions S_i can be expected. They allow an allocation of differences between output data of real forging processes and numerical simulations to certain parameters. By additionally taking commonly known plausibility regions for parameters into account, the parameter corrections in the mathematical model are clearly defined.

3.2 Examples

This section deals with some examples of sensitivity functions.

Fig. 4 shows the influence of a change in the workpiece temperature prior to the forming operation T_0. The shape of the sensitivity function S_{T0} is influenced by the temperature T_0 through, for example, yield stress, elastic constants, heat capacities, conductance and transfer coefficients.

The influence of a variation of the time period in which the workpieces are lying in the bottom die is shown in Fig.5.

The high influence of friction on forming process results is well-known. This accords with the observation that the friction factor distribution displayed in Fig.6 is the only one, where plausible results and parameter values were obtained.

3.3 Creation of data bases

We cannot expect that every identified value is physically correct. The compensation of inaccuracies could lead to inaccurate values for single parameters.

242

The investigation of several forging processes will show
certain congruencies. Some parameters will be equal.
Changing the geometry but not the material, for example, is
a good possibility for the verification of material
parameters. Systematic investigations allow the creation of
reliable data bases.

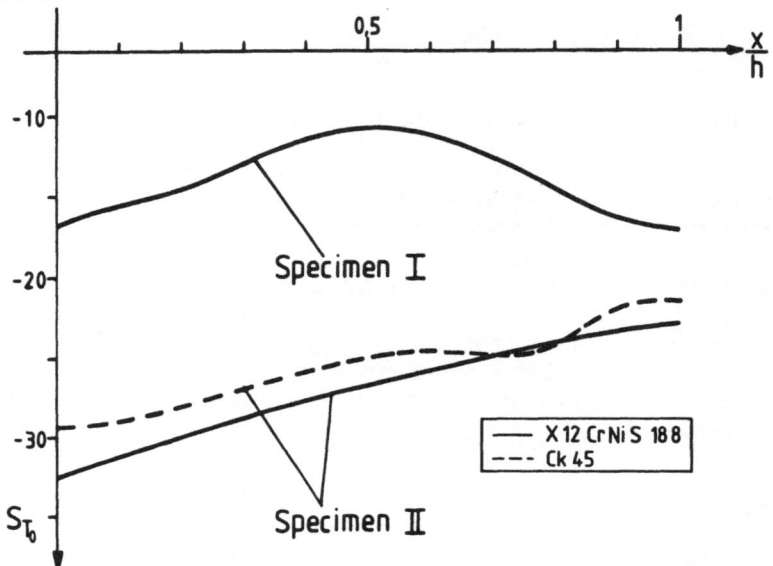

Fig.4: Influence of Initial Workpiece Temperature T_0

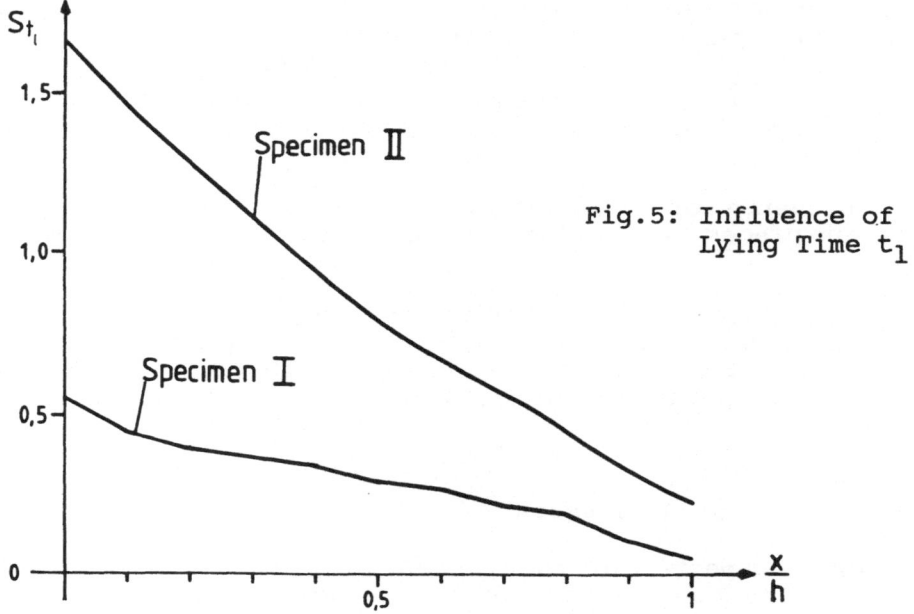

Fig.5: Influence of Lying Time t_1

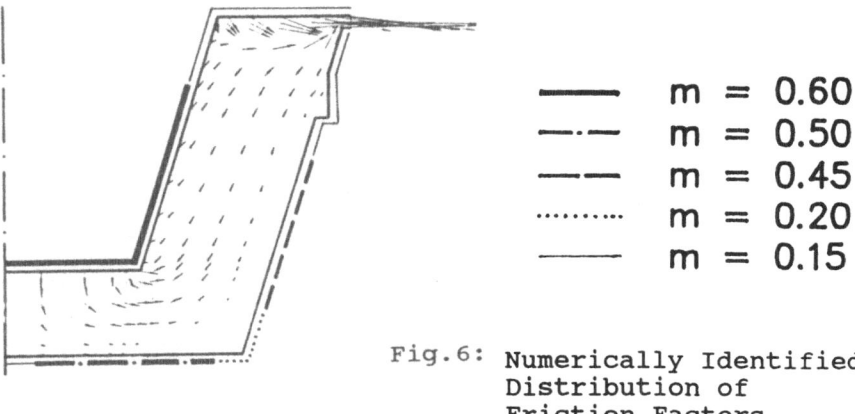

Fig.6: Numerically Identified
Distribution of
Friction Factors

4. FURTHER APPLICATIONS

Apart from sets of consistent values for non-measurable
parameters we can determine the influences of different
forging parameters. This is of great practical importance,
as the allowed deviations for each process data during a
forging process with prescribed work tolerance are thereby
calculable. Fig.7 shows the allowed percental deviations
of several process- and material parameters a_i in close
tolerance forging of specimen II.

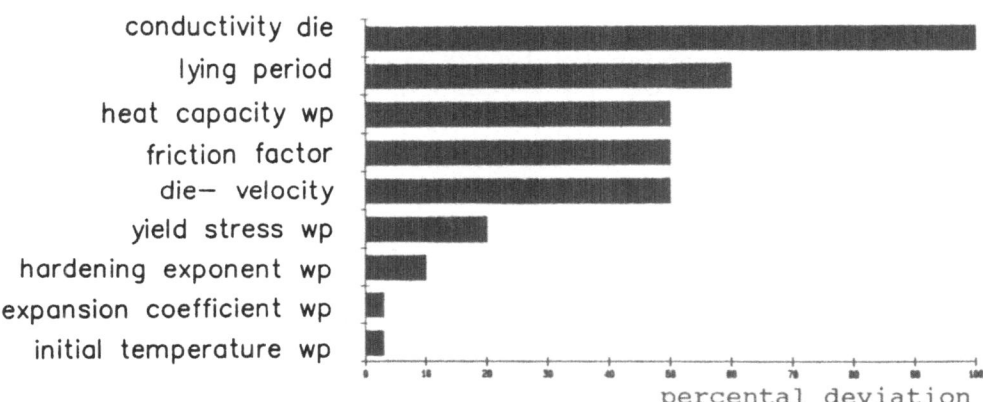

Fig.7: Allowed Deviations of Parameters a_i in Close
Tolerance Forging of Specimen II

Obviously, the identification process does not consider any
connections between micromechanics and parameter values.
But the obtained time and space-dependent distributions of
non-measurable parameters figure as an excellent foundation
for the micromechanical analysis. The distribution of

friction factors in Fig.6, for example, leads to the general conclusion that friction factors increase with relative motion and duration of contact. Reliable friction laws will have to take both influences into account.

5. CONCLUSIONS

In this paper, a procedure for the identification of inaccurately known forging parameters has been presented. Results of finite element simulations and precise measurements had to coincide. As shown in some examples, characteristic influences of different parameters on final shapes of specimens allowed the definition of corrections for parameters in the mathematical model.
The physical accuracy of identified parameter values can be checked by the determination of consistent parameter sets in similar forging processes.
Additionally obtained knowledge about permissible parameter deviations in dependence on prescribed dimensional tolerances is useful for process control.
Identified parameter values can serve as an excellent foundation for the verification of micromechanical models.

6. ACKNOWLEDGEMENT

The authors wish to express their appreciation to the 'Deutsche Forschungsgemeinschaft' (DFG) for their financial support of the project Do 190/35 and to Hibbitt, Karlsson and Sorensen, Inc. for the successful cooperation.

7. REFERENCES

/1/ Weiss,U.   Numerische Simulation von Präzi-
                               sionsschmiedeprozessen mit der
                               Finite-Elemente Methode
                               Diss. Universität Hannover (1987)

/2/ Landolt, H.   Zahlenwerte und Funktionen
    Börnstein, R.   6. Aufl., Bd.4: Technik, Teil 2:
                               Stoffwerte und Verhalten v.
                               metallischen Werkstoffen,
                               Bandteil a. Berlin, Göttingen,
                               Heidelberg (1963)

/3/ Doege, E.   Fließkurvenatlas metallischer
    Meyer-Nolkemper, H.   Werkstoffe, Hanser Verlag München
    Saeed, I.   Wien(1986)

IDENTIFICATION OF DEFECT LOCATIONS IN METAL FORMING USING A PERSONAL-COMPUTER-ORIENTED FINITE ELEMENT METHOD

N.L. DUNG
Technical University of Hamburg-Harburg,
MT II - Structural Mechanics Division,
Hamburg, FR Germany

The aim of this paper is the investigation of the ductile fracture, which can occur in the formed part during the cold forming processes, in an economical manner. A fracture criterion, based on the McClintock model of void growth and the critical value of plastic strain, is presented to identify the defect locations in the formed part. This criterion is compatible with the results of the rigid-plastic finite element method modelled for simulation of steady and unsteady forming processes on a personal computer (ATARI ST, IBM AT,...). The defect locations can be shown effectively by means of graphic display of such a low-cost computer to compare with the experimentally predicted results.

1. Introduction

In the forming processes as forging, the onset of cracking is a major limitation. Therefore, the ductile fracture of the forged metal must be avoided. The fracture initiation of metals can be investigated successfully if the deformation behaviour, temporal and local distributions of the stress and strain are known. There are many fracture criteria which deal with the defect locations in forming processes. But a fracture criterion, which takes into account the nucleation of voids as well as the void growth and coalescence is not yet well developed at present. The environmental and microstructural variables that influence fracture have not been studied fully. Most of the criteria dealed with the material behaviour beginning from the stage of deformation, at which the microscopic voids or cracks are already nucleated, except for the Gurson model /9/. They can be divided into two categories: void growth models (McClintock, Rice and Tracey, Gurson,. .../1,2,4,8,9,11/) and empirical theories (Freundenthal, Cockcroft and Latham, ... /1-4/) These criteria were used successfully for one process or class of process, but a few of them could detect both the fracture sites and strains in general cases adequately.

In this paper, a fracture criterion, based on the McClintock model of void growth and the critical value of plastic strain, is introduced to investigate the fracture initiation in cold forming processes. The deve-

245

J. L. Chenot and E. Oñate (eds.), Modelling of Metal Forming Processes, 244–252.
© 1988 by Kluwer Academic Publishers.

loped fracture criterion is implemented as a module in the rigid-plastic finite element program for simulating the forming processes on large computers (CYBER, PRIME,...) and personal computers (ATARI, IBM,...) The identification of defect locations in the axisymmetric extrusion is presented using this program FARM (Finite Element Analysis of Rigid-Plastic Metal Forming) /5-7/.

2. Fracture Criterion

From all the metallographic evidence, the sequence of events that leads to ductile fracture can be described as shown in Fig. 1 /13/.

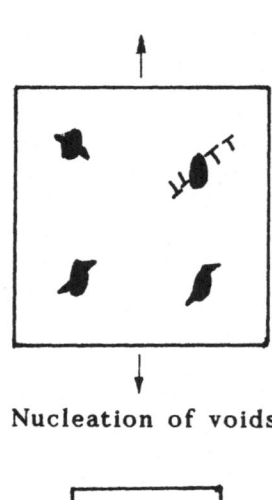

a) Nucleation of voids

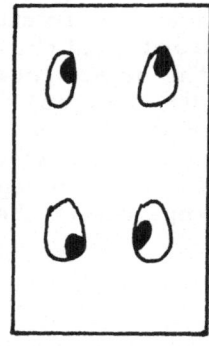

b) Growth of voids

c) Critical shape of voids and concentrated plastic defor- mation

d) Separation (Cracking)

Fig. 1: The onset of crack in the material.

Among the fracture criteria, the model due to McClintock could describe the shape change of voids in the stages of Fig. 1b and c very well. But the model of McClintock overestimates the fracture strains, because it assumes a void coalescence at fracture, i.e. the ligaments between the neighbouring voids have reduced to zero width. In spite of that, this model takes into account the effect of hydrostatic stress and is computed easily using the FE results for stress and strain /1,14/.

It is the purpose of the present work to adopt the McClintock model to establish a fracture criterion for the stages after nucleation. A closed-form expression for the damage accumulation, including the eccentric effect, under a triaxial stress system is given as:

$$d(\ln F_{ca}) = \left(\frac{\sqrt{3}}{2(1-n)} \sinh\left\{ \frac{\sqrt{3}(1-n)}{2} \frac{\sigma_a + \sigma_b}{\bar{\sigma}} \right\} + \frac{3}{4} \frac{\sigma_a - \sigma_b}{\bar{\sigma}} \right) d\bar{\varepsilon} \quad . \quad (1)$$

for a cylindrical hole with circular and elliptical cross-section in the plastic material ($\bar{\sigma} = K\bar{\varepsilon}^n$). The Eq. (1) expresses the void growth in the transverse direction a of the hole with its axial direction c. The principal maximal and minimal transverse stress components along the major and minor axes are σ_a and σ_b (Fig. 2). This interpretation is different from the original assumption of McClintock /11/. By neglecting the transient effect associated with the shape change, he used the Eq. (1) for hole growing in the b direction. But the study of the microstructure in tensile tests confirms our interpretation.

It follows that the damage accumulated during the forming process can be calculated as:

$$A_{ca} = \ln F_{ca} = \int_0^{\bar{\varepsilon}} f(\bar{\varepsilon}) \, d\bar{\varepsilon} \quad . \quad (2)$$

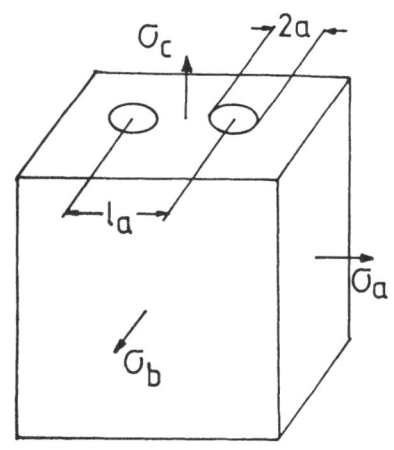

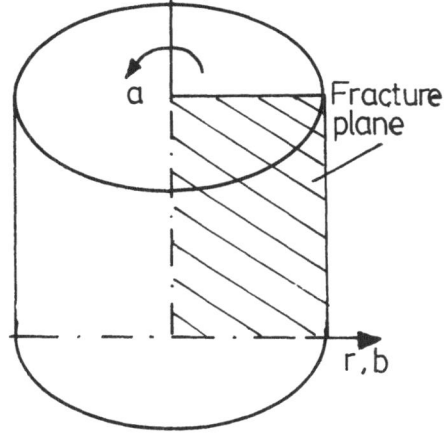

a) The cylindrical holes   b) The fracture plane

Fig. 2: The model of void growth assumed for the fracture criterion.

At fracture the critical value of the accumulated damage is reached:

$$A_{ca} = \ln F_{ca}^{f} = \int_{0}^{\bar{\varepsilon}^{f}} f(\bar{\varepsilon}) \, d\bar{\varepsilon} = A_{ca}^{*f} \qquad . \tag{3}$$

At the moment the plastic deformation is concentrated into narrow regions between neighbouring voids. Brown and Embury /1/ indicated that the coalescence of these voids occurs approximately when the size of the voids due to growth along the tensile axis equals the inter particle spacing. The plastic constraint is removed facilitating voids to link by the internal necking of the ligaments between the voids.

Freundenthal and Cockcroft & Latham /2,3/ have postulated that fracture will occur when the total plastic work:

$$\int_{0}^{\bar{\varepsilon}^{f}} \bar{\sigma} \, d\bar{\varepsilon} = C \qquad . \tag{4}$$

C is a material dependent critical value. The criterion (4) can be interpreted as a critical value of the equivalent plastic strain at fracture.

Based on the above assumptions, a criterion for ductile fracture can be defined as:

"The ductile fracture of the material will occur during the forming process first at the position in the deforming solid where the accumulated damage and also the equivalent strain reach their critical value, i.e. at the position with

$$A_{ca} \geq A_{ca}^{*f} \qquad , \tag{5}$$

and

$$\bar{\varepsilon} \geq \bar{\varepsilon}^{*f} \qquad . \tag{6}$$

Thus for the solid as a whole, the fracture plane is the cb-plane (Fig. 2)".

The critical value A_{ca}^{*f} of a material depends on the stress system applied. The critical value $\bar{\varepsilon}^{*f}$ is the fracture strain obtained from tensile test either of a strip or of a rod due to the stress system. If the stress system is triaxial, the fracture strain of tensile test of a rod is taken.

The model of void growth requires that the void should be open, i.e. the accumulated damage gets positive to indicate the void growing. To identify the defect locations, both conditions (5) and (6) must be fulfilled at fracture. However, in some simple cases, the weak form of the developed fracture criterion is suitable for predicting the fracture site and strain. Namely, if only the condition (6) is strictly considered. Crack is then expected to occur at the position in the solid where the accumulated damage is highest and the strain $\bar{\varepsilon}^{*f}$ is reached.

4. Personal-Computer-Oriented FEM

The FEM is formulated on the modified variational principle of Markov /5/ which has an additional term with Lagrangian multiplier for fulfillment of the incompressibility of the rigid-plastic material. The friction stress τ is computed using the relation $\tau = m\bar{\sigma}/\sqrt{3}$ ($\bar{\sigma}$: yield stress).

249

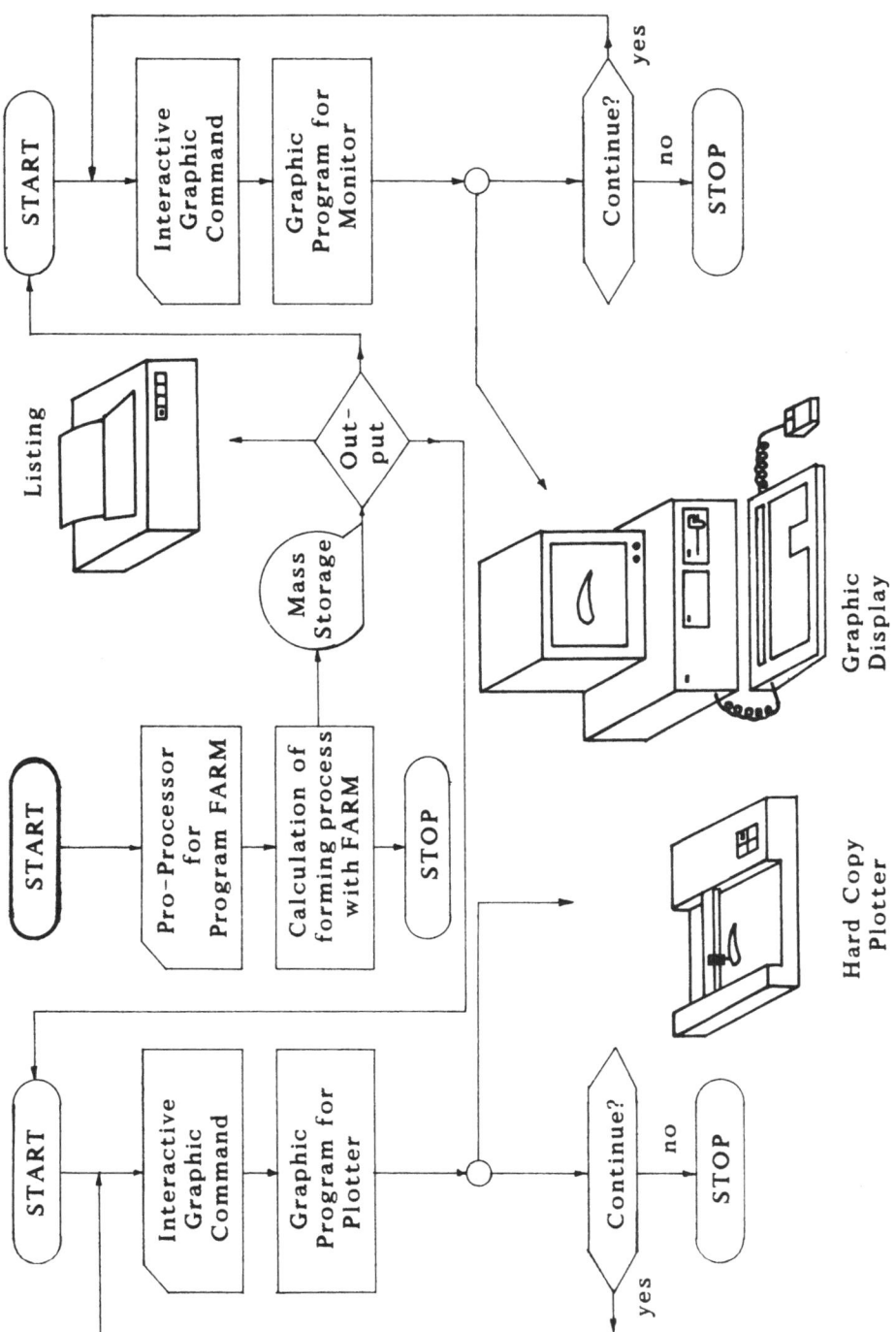

Fig. 3: FE- Simulation of metal forming processes using a personal system.

The unsteady forming process is analyzed step-by-step using many small steady deformation steps. In each step, the contact problem (normal and friction contact) is checked. Therefore, in the nonlinear calculation, there are many numerical interim solutions which repeat many times in the iteration and in the incremental procedure. In order to reduce the computing time, simplified algorithms are used in form of: linear element types, single point integration for the volume integrals, Crout method of Gauss elimination adapted with an effective skyline storage method, ...

Based on these procedures, a software package FARM has been developed. The program is written in FORTRAN 4/77 and so constructed that it can be installed easily on a personal computer with 1MBytes RAM (Random Access Memory). For our purpose, the low-cost personal computer ATARI 1040 STF is chosen. The personal system has: Computer with 1 MBytes RAM, colour monitor, floppy 3.5"/ 720 KB, hard disk 20 MB, dot printer and plotter /7/.

The graphics system GEM (Graphics Environment Manager) of this PC is applied to display the computed results (FE-mesh, distributions of stress and strain, material flow,...). The advantages of the PC are that it can be operated by the individual user at any location and without operating costs. So, the simulation of the forming process can be employed in the lab at the same time with the experiment to optimize that. The graphic display of the distributions of stress and strain in step-by-step manner gives the user very good understanding of the whole process. If the possible defect location is identified, the graphic display shows the fracture site in the formed part effectively. An efficient FE-simulation should have the flow chart as shown in Fig. 3. In the last few years, there were some works on the application of PC for simulation of metal forming processes /10,12/. The computing times, however, were found to be extremely high because of their complex FE modelling in comparison with our method.

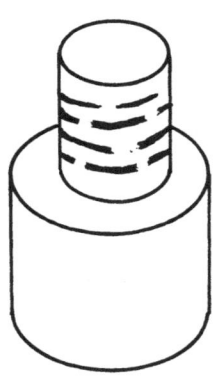

a) Christmas tree

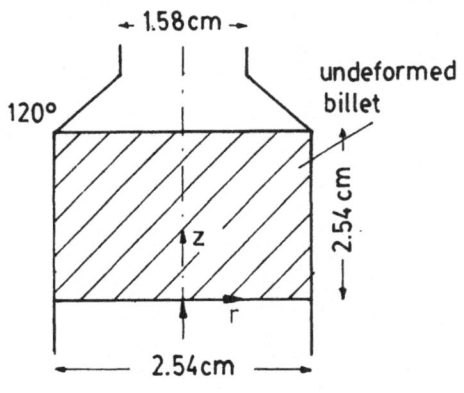

b) Extrusion test

Fig. 4: The ductile fracture in cold extrusion of circular rod of brass.

251

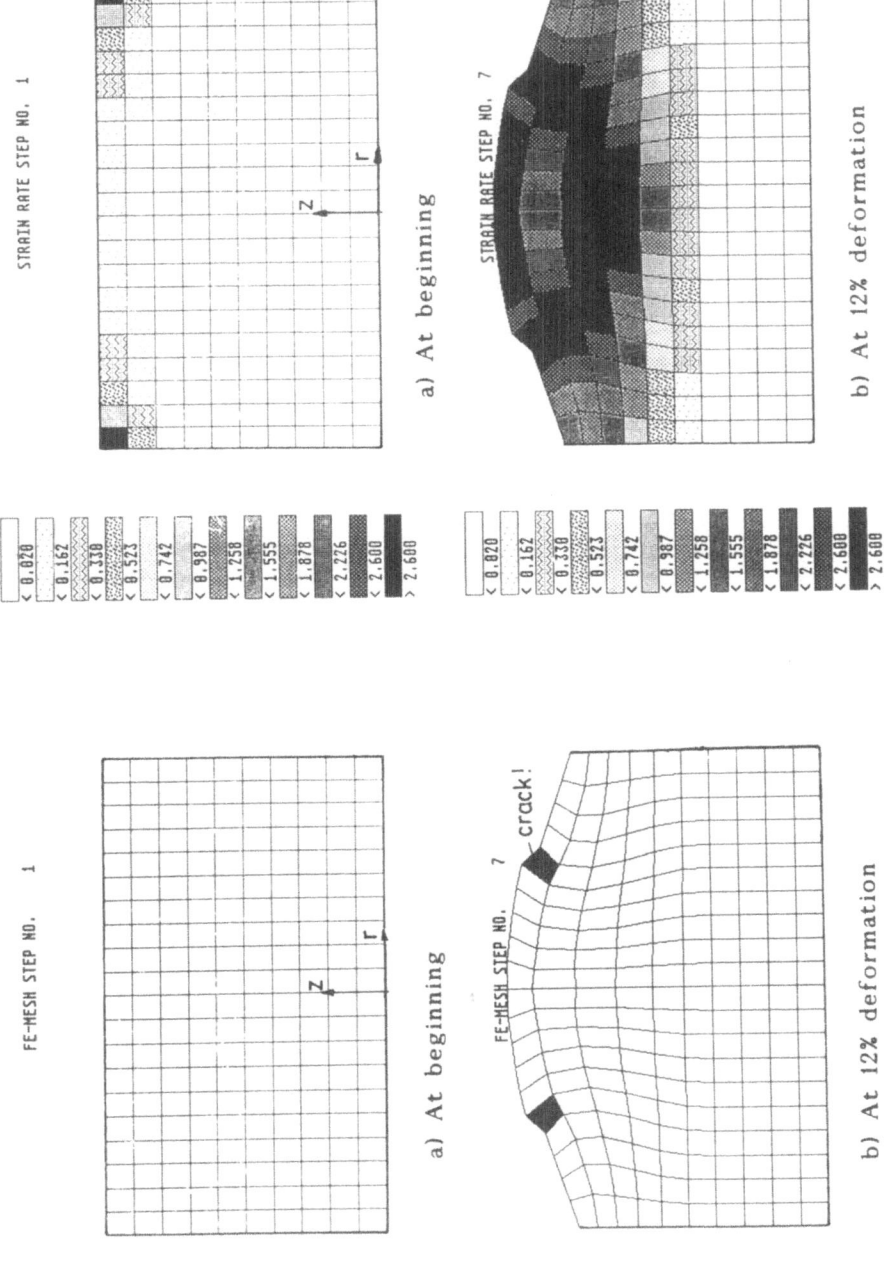

STRAIN RATE STEP NO. 1

a) At beginning

STRAIN RATE STEP NO. 7

b) At 12% deformation

Fig. 6: The distributions of the strain rate.

FE-MESH STEP NO. 1

a) At beginning

FE-MESH STEP NO. 7

crack!

b) At 12% deformation

Fig. 5: The FE-mesh in extrusion of circular rod of 60-40 brass.

4. Numerical Results

The FE-program with a module for fracture criterion is then used to investigate the onset of crack in form of the so-called Christmas tree in cold extrusion of circular rod of 60-40 brass (Fig. 4). Because of axisymmetry only a half of the rod is modelled using 100 linear quadrilateral elements. Fracture was found to occur experimentally after approximately 10% of the length of the undeformed bar had been extruded past the reduction die entrance /2/.

Fitting the critical fracture strain 0.97, obtained from the tensile test of rod, in the procedure for theoretical fracture identification, the crack is found to occur on the free surface of the rod at the position, marked with black colour, after 12% deformation (Fig. 5b). The calculation also shows that the accumulated damage is negative everywhere, except for this black region. The onset of crack at this position causes then the Christmas tree effect on the extruded bar surface. Fig. 6 illustrates the hard copies of graphic display showing the distributions of strain rate at the beginning and at 12% deformation. The white region represents the quasi-rigid material behaviour, while the dark regions mean the plastic deformed zones.

In the previous work /6/, the developed fracture criterion was also suitable for prediction of the fracture sites and strains in uniaxial compression samples. The computing time was around 4 hours for forging (0-60% reduction) and extrusion (0-16% deformation), i.e time from starting the job to obtaining the graphic display of the results.

5. References and Acknowledgements

1. Chandrasekaran, N.: Ph.D. Thesis, McMaster University, Canada (1985).
2. Clift, S.E.: Ph. D. Thesis, University of Birmingham, UK (1986).
3. Cockcroft, M.G. and Latham, D.J.: J. Inst. Metals 96 (1968), 33-39.
4. Dodd, B. and Bai, Y.L.: Ductile Fracture and Ductility. Academic Press (1987).
5. Dung, N.L.: Fortschr.-Ber. VDI-Z. Reihe 2 Nr. 6 (1981).
6. Dung, N.L.: Identification of Defect Locations in Uniaxial Compression Samples. To be published in Mech. Res. Comm. (1988).
7. Dung, N.L.: FARM - Version for Personal Computer ATARI ST - User's Manual. Technical University of Hamburg-Harburg, FR Germany (1986).
8. Gelin, J.C.: Thèse de doctorat d'Etat, Université Paris VI (1985).
9. Gurson, A.L.: Ph. D. Thesis, Brown University, USA (1975).
10. Hussin, A.A. et al.: J. Strain Analysis 21 (1986), 197-203.
11. McClintock, F.A.: J. Appl. Mech. 35 (1968), 363-371.
12. Osakada, K. and Mori, K.-I.: Annals CIRP 34 (1985), 241-244.
13. Oyane, M.: Bull. JSME 15 (1972), 1507-1513.
14. Sowerby, R. et al.: VDI-Forsch. Ing.-Wes. 51 (1985), 147-150.

This work is carried out under the grant No. I/60953 of the Volkswagenwerk Stiftung (FR Germany). Thanks are due to Professors O. Mahrenholtz (FR Germany) and R. Sowerby (Canada).

MODELLING OF METAL FORMING PROCESSES

Bramley A.N., Osman F.H, and Ghobrial M.I.

University of Bath, UK.

ABSTRACT

The design of forging processes requires considerable expertise so as to decide upon preform stages and to ensure complete cavity filling. Stepwise simulation of cavity filling in compound flow problems is demonstrated using the Upper Bound Elemental Technique(UBET). A systematic incremental procedure for the prediction of free surfaces profile using the UBET optimised velocity field is introduced. Also an attempt is made to determine the effect of workpiece diameter and flash thickness on metal flow in die corner. Theoretically predicted results are compared with those of small scale laboratory experiments.

1. Introduction

In recent years, many attempts have been made to develop computer-aided methods for forging design. They vary from empirically based packages [1], with application limited to particular shapes and materials, to sophisticated numerical techniques [2]. They require a large amount of computational time and accurate information on constitutive relations, for example, on material behaviour and the mechanism of friction. However, approximate methods, based on the principles of the theory of plasticity and characterised by flexibility in application have become attractive to the forging industry, since they are suitable for interactive usage and comparative analysis. Hence they allow for practical interaction between experience and computer based predictions. The Upper Bound Elemental Technique (UBET) is among those methods [3,4], and although it gives an approximate prediction of both loading and metal flow its application has been gaining considerable interest due to the fact that load prediction is desirable an over estimate and it has been developed for interactive use with potential application to complex shapes. However, the method has been extensively developed to simulate the forging process from start to end, and further extended, with the application of the

253

J. L. Chenot and E. Oñate (eds.), Modelling of Metal Forming Processes, 253–259.
© *1988 by Kluwer Academic Publishers.*

reverse flow algorithm, to the design of preforms [5].

Decisions on billet or preform shapes are the most difficult and critical steps in the design of a forging operation.   Apart from essential design features such as draft angles and fillets, the desgner's main concern is to ensure bulk filling of cavity and complete filling of die corners.   Bulk filling depends on the history of deformation during the process and naturally, friction and cavity profile play an important role on the instantenuous flow pattern at each stage of the deformation.   Filling of die corners on the other hand occurs at the sizing or final stages of the process and depends to a large extent on the history and variability of strain along the free boundaries of the workpiece.   corner filling however is usually achieved by the use of excessive loading and billet material.   Therefore, distribution of material in the cavity at the initial stage of the process, whether starting from a billet or a preform, is essential to the production of sound forgings with minimum   material wastage, die wear and loading requirements.

The purpose of this paper is to examine the bulk flow in double-rib forgings and the surface profile in closed die forgings.   Also, an attempt is made to compare corner filling for workpieces of different diameters and flash thickness.

2. UBET application to double-rib components

Forgings with compound flow patterns are characterised by the varying strain rate in the material undergoing deformation and the existence of neutral zones between different modes of deformation.   In order to assess the UBET metal flow algorithm in such cases the axisymmetric arrangement shown in Figure 1 is used to simulate the bulk flow in double deep cavities.   The method consists of dividing the workpiece shape into a series of standard elements connected together by surfaces of velocity discontinuity.   An optimised velocity field is then determined which minimises the rate of energy consumption and simultaneously establishes the inter-region and free boundaries velocities.   Optimised values of the boundary velocities are then used to advance through the process on an incremental basis.   The procedure is carried out automatically using an interactive computer program.

Experiments were conducted using a 750 kN mechanical press.   Billets with aspect ratio ranging from 0.7 to 2.0 were made from aluminium HE30 and preheated to 500°C.   'Metalflo 20' was used as a lubricant and the dies were coated with a thin layer of 'Powerflo 2073' in order to avoid aluminium sticking to the tooling.   The friction factor (m) was measured using the ring test and was found to be 0.1.   A final web thickness of 16 mm was obtained in each experiment and the flash aspect ratio - width/thickness - was kept constant at 2.0..

The filling characteristics for both inner and outer rib cavities are shown in Figure 2 and 3 for initial billet aspect ratio of 0.7 and 1.6 respectively. Both theoretical and experimental results show that rise in both cavites occurs simultaneously and metal flow is more predominant in the outer rib cavity for

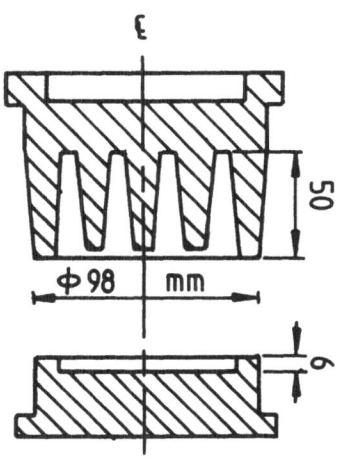

Figure 1. Double rib cavity tooling.

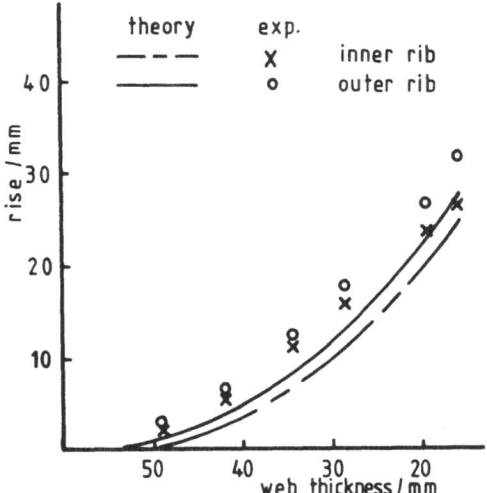

Figure 2. Cavity filling (aspect ratio = 0.7).

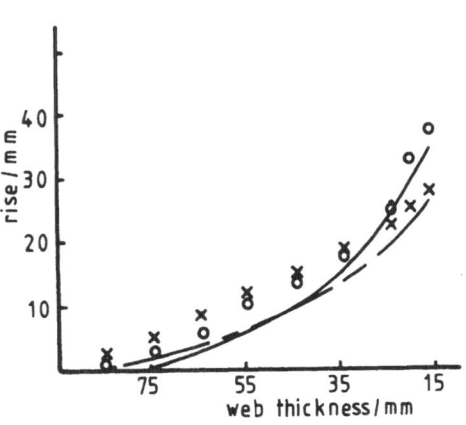

Fgiure 3. Cavity filling (aspect ratio = 1.6).

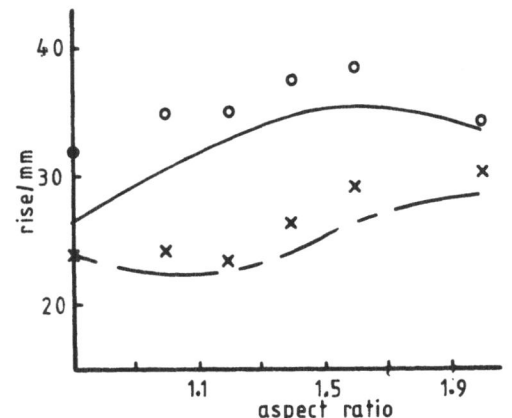

Figure 4. Effect of billet aspect ratio on filling.

initial billet aspect ratios as shown in Figure 4. It is clear from these results that the initial billet dimensions significantly influence the filling of both inner and outer cavities. Experimental results are slightly higher than those predicted by UBET, however the simulation was carried out in 10 increments and friction assumed constant along the die/ material interface.

3. Prediction of surface profile

The UBET metal flow approach for bulk deformation assumes, for simplicity, that each free boundary throughout the simulation move along a straight line with a constant average velocity according to the volume of material crossing the surface. However, in an attempt to predict the surface profile at different stages of the deformation each free boundary is divided into smaller segments each of which is advanced according to the resultant prescribed displacement. The surface profile at each increment is therefor determined by superimposition of the displacement experienced in previous increments for each segment independently. The method is applied to the configuration shown in Fgiure 5. The initial workpiece shape of this example has three free boundaries thus allowing for deformation in the axial direction into the cavity rib and boss and also in the radial direction towards the flash land.

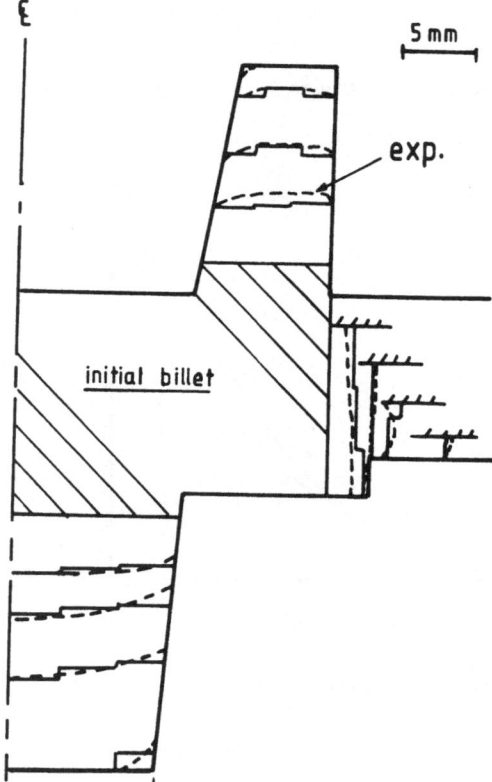

Figure 5. Surface profile in incremental deformation.

Experiment were performed on billets made of EN8 and heated to 1200°C, 'Metalflo 20' was used as a lubricant giving a friction factor of 0.3.   Six specimens were used each of which was deformed to a particular penetration. The surface profile was then traced using Taylor-Hobson Talysurf.   Figure 5 shows the theoretically predicted stepped profile and those obtained experimentally.   In order to reduce the computational time each free surface was divided into three segments and the simulation carried out using the bulk flow algorithm while the displacement of the segment was noted at the end of each increment.

4. Corner filling

In most forging operations flash is added so as to trap the metal inside the cavity and to provide sufficient pressure, at the expense of excessive loading, inside the cavity forcing the metal to fill the remaining die corners. However, at this stage the bulk of the material inside the cavity behaves as a rigid body under hydrostatic pressure and plastic deformation takes place within a diminishing volume at the vicinity of the die corners and flash land. Work carried out in that field has been directed towards the loading associated with filling of small corners in coining operations[6].

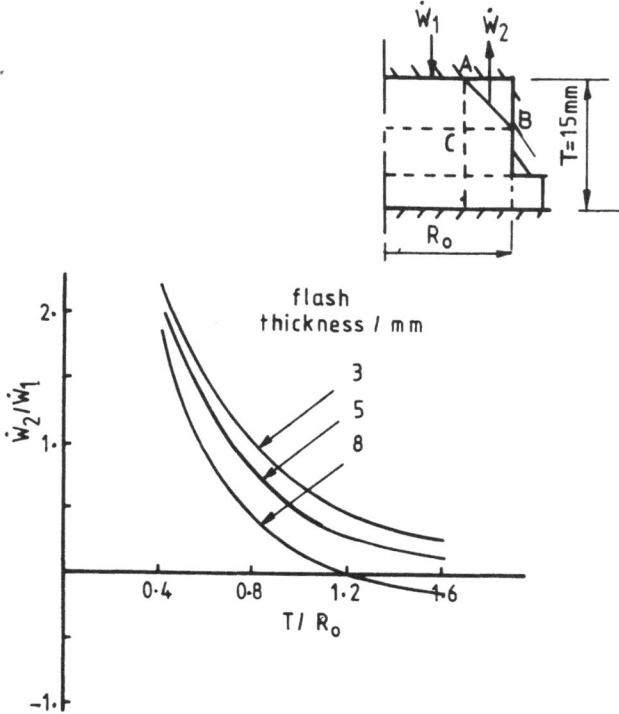

Figure 6. Die corner model

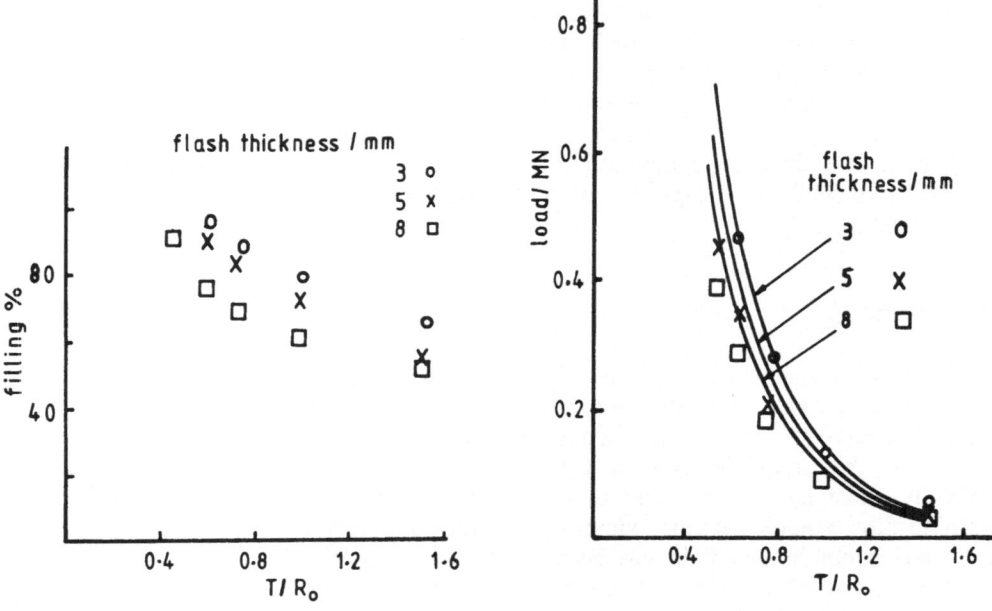

Figure 7. Effect of flash on filling

Figure 8. Effect of flash on loading.

In flashless forgings where the material is in completely closed dies knowledge of corner filling and its effect on loading is also of considerable importance in order to avoid over stressing of tooling [7].

The simple forging shown in Figure 6 was used to investigate the flow into die corners.   Cylindrical billets chamfered at 45° billets were made of aluminium HE30 and preheated to 500°C.   Experiments werd carried out on billets of different diameters.   In the analysis the workpiece was divided into 7 regions with a triangular element ABC at the die corner.   The velocity across the free surface AB was taken as a measure of filling and compared for various flash thicknesses.   The effect of flash thickness and outer forging diameter on the relative velocity (velocity across AB/die velocity) at the die corner are given in Figure 6.   Figure 7. shows the experimentally obtained filling at die corner (displaced volume/initial corner volume) while the effect on the loading requirements are given in Figure 8.   Both theory and experiment show that filling increases as the flash thickness becomes smaller and the rate of filling is greater for components with larger diameter with rapid increase of the loading requirements.

5. Conclusions

The effect of cavity profile on the degree of filling in double-rib components was examined and a systematic procedure based on the UBET method of analysis for the prediction of surface profile and corner filling was introduced. Experimental results obtained from small scale laboratory experiments gave good correlation with those predicted.

6. Acknowledgments

The authors acknowledge the provision of research facilities by the Department of Mechanical Engineering at the university of Leeds where much of the work was carried out. The science and Engineering Research Council and Doncaster Monkbridge are also thanked for the provision of a co-operative award supporting much of the work.

7. References

1. AKGERMAN N., BECKER J.R., and ALTAN T.
   Preform design in closed die forging., *Metallurgia and Metal Forming*, V 40, pp 135, 1973.
2. OH S.I., LAHOTI G.D. and ALTAN T.
   Application of rigid-plastic finite element to some metal forming operations, *J. Mech. Working Tech.*, V 6, pp 277-290, 1982.
3. 'OUDIN J. and RAVALARD Y.
   An upper bound method for computing load and flow pattern in plane strain froging processes, *Int., J., MTDR Conf.*, V21, pp 237-250, 1981.
4. OSMAN F.H.
   Computerised simulation of forging processes., Ph.D. Thesis, Leeds University, 1981.
5. OSMAN F.H., BRAMLEY A.N. and GHOBRIAL M.I.
   Forging and preform design using UBET, *1st ICTP Conf.*, pp 563-568, 1984.
6. MONAGHAN J.M. and TORRANCE A.A.
   An investigation of plan strain coining., *J. Metals Technology.*, V11, pp 20-28, 1984.
7. NEDIANI G. and DEAN T.A.
   Forging of rectangular sections in a completely closed die cavity., *Int. J. Mech. Sci.* ,V 25, P347-360, 1983.

NUMERICAL ANALYSIS OF COLD DRAWING
OF TUBES

J.M. RIGAUT[1],D. LOCHEGNIES[2],J. OUDIN[2],J.C. GELIN[3],Y. RAVALARD[2]
[1]Vallourec Industries, Direction Recherche et Développement, 51300 Vitry le François
[2]Laboratoire de Génie Mécanique, GRECO Grandes Déformations et Endommagement,
MECAMAT, Université de Valenciennes et du Hainaut Cambrésis, 59326 Valenciennes
[3]Laboratoire de Mécanique Appliquée, Université de Franche-Comté, 25030 Besançon

1.Summary

Up to now, empirical design of tubes multi-pass drawing schemes remains frequent. So, to improve the design, a new three steps strategy is developed : step one testing and grading the involved schemes set with kinematic and static element approaches ; step two testing the selected scheme with analogical experiments ; step three testing again the selected scheme by finite element approach.

In the static and kinematic element methods, plastic isotropic constitutive equations are considered for bulk properties and isotropic friction sublayer equation for interface properties. Velocities, strains and strain rates are easily calculated from kinematic element ETIRC software with regard to bulk and interface properties, tools geometry and plug position. Stresses and contact pressures are also quickly obtained from static element ETISLAB software.

For the finite element approach, viscoplastic isotropic constitutive equations and general isotropic interface equations are implemented in ASTRID software and an other prediction of strain and stress is achieved.

Examples of numerical simulation with the above methods and related softwares are given to illustrate the interest of the proposed strategy.

2.Introduction

Seamless tubes are usually obtained by hot piercing and tube rolling. Most tube blanks are cold drawn to reduce wall thickness, improve surface finish and mechanical properties. Four ways are now available for drawing : the first and simplest way is to draw through a conical die ; in the second way the inner diameter of the tube is calibrated with a plug which is held by a bar attached on the draw bench ; in a third way, a conical plug is used and the friction stress occuring keeps it in the deformation zone without bar helding ; in a fourth way, the tube is drawn on a cylindrical mandrel.

The floating plug drawing seams to be the most promising way and also the most flexible one ; however the second way remains very utilized. Most tube drawings are multi-pass ones and the definition of the sequences are commonly achieved in considering two aspects : first, the drawing stress σ_y is calculated with empirical or simplified formulae and is compared to a limiting value which depends on material ultimate elastic stress ; second, the entry-exit cross sections ratio is compared to a limiting value which is related to the mean necking stress observed in the tensile test of

261

J. L. Chenot and E. Oñate (eds.), Modelling of Metal Forming Processes, 261–268.
© *1988 by Kluwer Academic Publishers.*

a cylindrical specimen. Other informations are also taken into account such as influence of tool material, tool fitting rigidity, roughness, lubricant, drawing speed and temperature. A methodical strategy would be very useful for speedy design of the multi-pass scheme and objective data bank achievement.

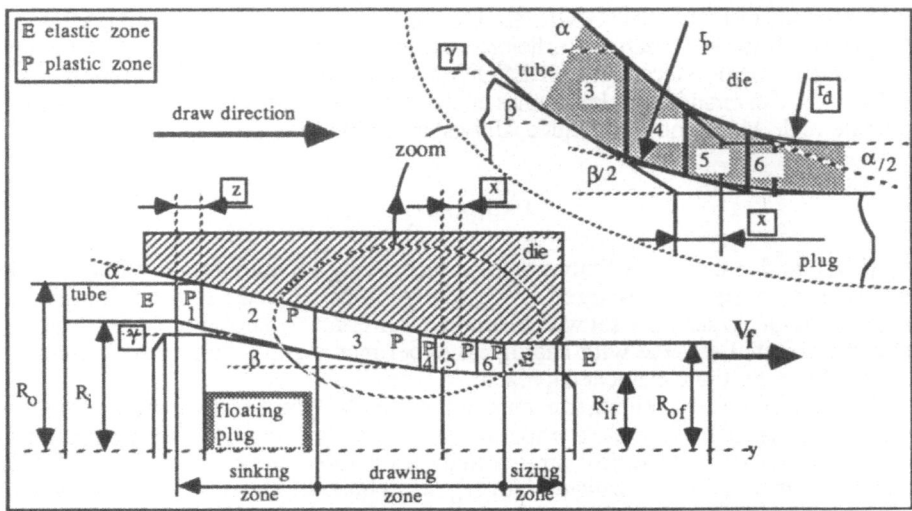

FIG. 1. Floating plug drawing : schematic representation of elements used in the kinematic element method and in the static element method.

Up to now, the main approaches are concerned with tube drawing simulation, slip line field [1] and bounding methods [2, 3]. However, these approaches are mainly available for specific drawing conditions, slip line field for quasi zero friction stress and upper bound methods for a single set of entry and exit diameters (Fig. 1), for instance :

$$\frac{R_o}{R_i} = \frac{R_{of}}{R_{if}}$$

The main results of these approaches are the semi-cone angles which correspond to the minimum of the total dissipated power functional. To facilitate multi-pass drawing design, the authors have defined a three steps procedure : first step tests the validity of a set of multi-pass schemes with kinematic and static element models and chooses the best one ; second step tests the validity of this best scheme with model material simulation ; third step tests in depth the validity with finite element models (Fig. 2).

The paper is concerned with first and third steps of the procedure : these steps involve numerical analyses using kinematic, static and finite element methods. The main principles of these methods are related in [4]. For each method, the main derivations and the results of calculations for a one pass drawing of steel tubes through conical tungsten carbide dies are given. Accurate values of strain, strain rate and stress illustrate the efficiency of the numerical simulation steps for the design procedure of the drawing pass.

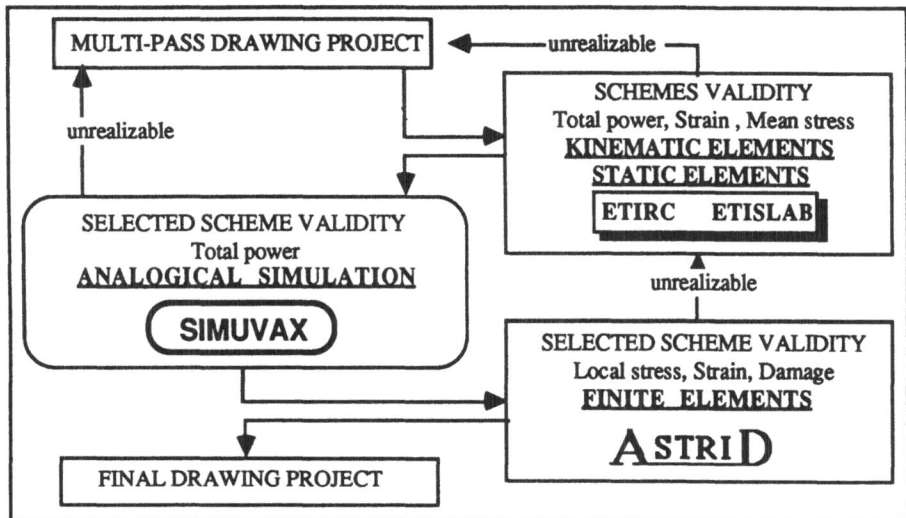

FIG. 2. Flow chart of the proposed design procedure for multi-pass drawing.

3. Mechanical Analysis

3.1. EQUILIBRIUM EQUATION

Equilibrium conditions on Cauchy stress tensor σ components, div $\sigma + f_v = 0$, are conveniently turned into the variational form

$$G^*(v,\eta) = \int_\Omega \sigma : \nabla^s \eta \, dV - \int_\Omega f_v \, \eta \, dV - \int_{\Gamma_\sigma} \bar{t} \, \eta \, dS - \int_{\Gamma_c} t^c \eta \, dS = 0,$$

where η is a kinematically admissible velocity field, Ω the deformed volume of the tube,

f_v the body force vector, $\bar{t}$ the prescribed stress vector on the surface part Γ_σ,

$t^c = \sigma \, n^c$ the stress vector imposed on tube-die and plug-die contact surface Γ_c with the condition $t^c n^c \leq 0$.

3.2. KINEMATIC ELEMENT METHOD

As the diameter-wall thickness ratio is usually in the range from five to twelve, the boundaries between elastic and plastic deformed volumes are very close to plane surfaces which are almost normal to the y axis of the tube. The limiting lines of the plastic deformed zone in a meridian r,y plane related to die-tube contact are straight lines in the sinking zone and straight lines connected by a circular curve which radius is r_d, the others related to tube-plug contact are also straight lines connected by a r_p radius curve.

These two radius are rather high say almost the same values than the tube radius, so an efficient model consists in substituting the two curves by a tangential straight line inclined of $\alpha/2$ for the tube-die contact and of $\beta/2$ for the tube-plug contact.

Therefore, an admissible velocity field for the deformed zone is obtained by using a single standard kinematic element : its volume is bounded by two plane surfaces normal to the y axis and by two conical surfaces. The solution presented here involves six elements : elements 1 and 2 for the sinking zone, elements 3, 4, 5 and 6 for the full drawing zone (Fig. 1). The r and y velocity components available for the standard element are given in [5]. As the mean equivalent strain of a drawing pass is in the range of 0.2 to 0.6, a mean isotropic yield stress σ_0 has been considered and the functional G* becomes :

$$ G^{*}(v) = \int_{\Omega} \sigma_0 \, \dot{\overline{\varepsilon}}^{\,p} \, dV + \int_{\Sigma} k \, \|\Delta v\| \, dS - \int_{\Gamma_c} \sigma_t \, v_t \, dS, $$

where $\dot{\overline{\varepsilon}}^{\,p}$ is the equivalent plastic strain rate, k the isotropic shear yield stress,

Δv the velocity discontinuity between two adjacent kinematic elements, σ_t the friction stress, v_t the sliding velocity (tube-die and tube-plug) on the contact surfaces Γ_c. G* (v) depends on four geometrical parameters, z thickness of element 1, γ angle of the inner tube surface in the sinking zone with y axis, x position of the plug, r_p curve radius of the inner surface of the tube at the exit zone, which are calculated to minimize the functional G*. A mixed lubrication regime is frequently observed in the first pass drawing of phosphated and lubricated steel tubes with tungsten carbide die and plug ; the mean value of $\|\sigma_t\|$ / k ratio is rather low from 0.04 to 0.09 and the isotropic friction constitutive equation of Tresca $\sigma_t = - f \, k \, v_t / \|v_t\|$ may be used for a speedy numerical simulation [6].

3.3. STATIC ELEMENT METHOD

The above described elements are now used to express equilibrium conditions and to obtain mean values of stress components available at coordinate y.

Two types of static elements are defined : the type 1 elements are used in the sinking zone and the type 2 in full drawing zone. For the type 1 element, mean stress $\sigma_y(y)$ and $\sigma_\theta(y)$ are involved whilst σ_r is in the form

$$ \sigma_r(r,y) = \sigma_\theta(y) \, \frac{r - r_i}{r} $$

with r_i inner radius of the element at coordinate y. For the type 2, mean stress $\sigma_r(y)$, $\sigma_y(y)$ and $\sigma_\theta(y)$ are used in the equilibrium equations. Elements shear stress are lower than friction stress and can be neglected in the equations. Stresses $\sigma_r(r,y)$, $\sigma_\theta(y)$, $\sigma_y(y)$ and $\sigma_{nd}(y)$ in the sinking zone and $\sigma_r(y)$, $\sigma_\theta(y)$, $\sigma_y(y)$, $\sigma_{nd}(y)$ and $\sigma_{np}(y)$ in full drawing zone, are obtained with element equilibrium conditions, limiting conditions on the internal and external tube surface, and Tresca yield isotropic criterion. The contact pressures are calculated from $\sigma_{nd} = - \sigma_r + \sigma_{td} \, \tan \alpha_1$ and $\sigma_{np} = - \sigma_r - \sigma_{tp} \, \tan \beta_1$ and the mean friction stresses σ_{td} and σ_{tp} are deduced when the isotropic friction factor f is known. Other constitutive equations could be also used in this approach.

The kinematic and static element models are implemented in ETIRC and ETISLAB softwares written in FORTRAN 77 in a version for Apollo stations. It takes a maximum of 20 mn CPU to have a good prediction of plastic deformed zone, velocities, strain rates, strains, bulk and interface dissipated powers, mean stresses and contact pressures. So, these approaches and softwares seam to be at the moment the most appropriate means for the first step of the proposed design procedure.

3.4. FINITE ELEMENT METHOD

In order to analyse further on high tube drawing speed, a viscoplastic constitutive equation has been considered for the finite element discretization [7].

Let $C\left(\theta, \bar{\varepsilon}^{\,p}\right)$ the billet consistency which is temperature dependent and strain sensitive, m is the isotropic strain rate coefficient of the tube material. In this case, the deviatoric Cauchy stress tensor s is expressed by the equation

$$s = \frac{2}{3}\, C\left(\theta, \bar{\varepsilon}^{\,p}\right)\, \left(\dot{\bar{\varepsilon}}^{\,p}\right)^{m-1}\, \dot{\varepsilon}\ .$$

The problem is described by the functional

$$G^{o}(v, p, \dot{e}) = \int_{\Omega} W\,(\dot{\varepsilon} + \frac{1}{3}\,\dot{e}\,I\,)\ dV + \int_{\Omega} p\,(\,\mathrm{tr}\,\nabla v - \dot{e}\,)\ dV - \int_{\Gamma_{\sigma}} \bar{t}\ v\ dS\ .$$

Here $\mathrm{tr}\dot{\varepsilon} = 0$, $\dot{e}$ is the dilatancy rate and p the hydrostatic stress.

This three field functional is used to compute velocity field, strain and stress. Its derivation with respect to v, p and e leads to finite element discretization. The mesh used here is shown in Fig. 3. The non linear system is solved by Newton-Raphson procedure and the introduction of a friction law in the functional G^{o} is related to the friction tangent operator needed for the resolution. This model is implemented in the ASTRID software available in versions VAX and APOLLO.

4. Numerical Results

Numerical simulations of 1038 steel tube drawing have been performed by using the three above described approaches implemented in ETIRC (Kinematic Element Method), ETISLAB (Static Element Method) and ASTRID (Finite Element Method). The strain stress relation was taken isotropic and weakly viscoplastic during the pass with a m coefficient of 0.01. The mean isotropic yield stress was 700 N.mm^{-2} and the mean friction factor during the pass was 0.05. The dimensions were : ϕ 27 mm x 2.37 mm for the blank tube, ϕ 21 mm x 2 mm for the drawn tube after one pass at 0.6 m.s^{-1}. The inclined part of the die is a conical surface inclined of 15° to the y axis and the radius of curvature r_d equals 30 mm.

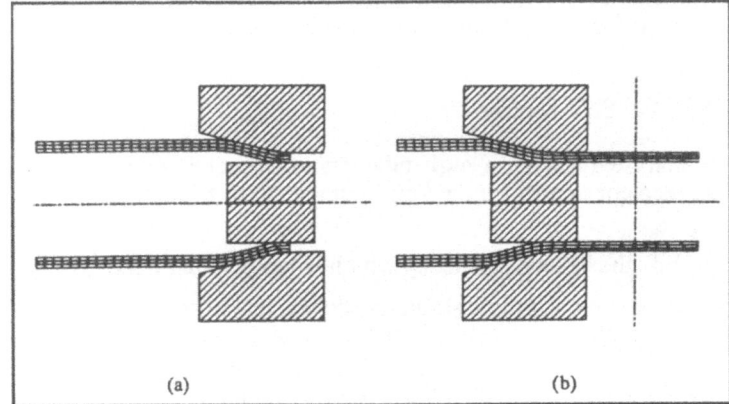

FIG. 3. Finite element model :
       (a) 174 bilinear elements initial mesh ;
       (b) deformed mesh.

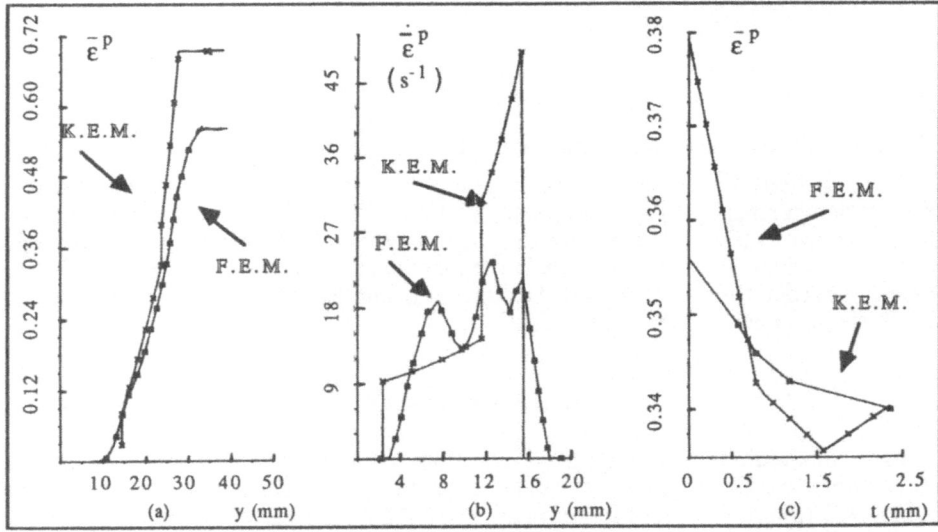

FIG. 4. Local strain and strain rates evolutions obtained by kinematic element and by finite element methods for 1038 steel phosphated tube drawn at 0.6 m.s^{-1} speed from ϕ 27 mm x 2.37 mm to ϕ 21 mm x 2 mm, with a viscoplastic coefficient m equal to 0.01 and an isotropic friction factor f equal to 0.05 :
       (a) equivalent strain along a mean flow line ;
       (b) equivalent strain rate along a mean flow line ;
       (c) equivalent strain profiles before tube plug contact zone.

Figs. 4 give the prediction of equivalent plastic strain and of equivalent plastic strain rate at the middle of the wall thickness of the tube during drawing. K.E.M. results are very similar to F.E.M. ones in the sinking zone and in most part of the drawing zone. As expected, K.E.M. overestimates a lot plastic strain rate at the end of the drawing because of the rather crude kinematic element discretization.

When considering now, as another illustration of the efficiency of the two numerical approaches, the equivalent strain prediction throughout the wall thickness at the boundary between sinking zone and full drawing zone, it appears that the strain gradient has the same predicted form with K.E.M. and F.E.M.. In this case, K.E.M. underestimates a little bit the gradient.

Meanwhile the strain and strain rate are of importance for a good drawing schemes design, it remains necessary to have speedy predictions of stress, drawing force and floating plug position. Fig. 5-a gives the evolutions of axial stress σ_y at the middle of the wall thickness both predicted from static element method and from finite element method : the evolutions are very similar and the mean final values are almost the same. The accuracy of the new static element for the sinking zone is confirmed : the radial stress varies from 10 to 90 N.mm^{-2} in the sinking zone and Fig. 5-b shows that the S.E.M. prediction is correct.

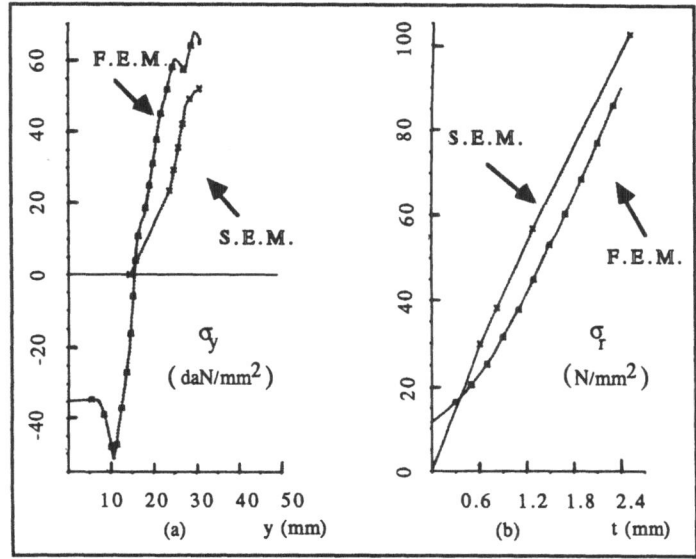

FIG. 5 - Local stress evolutions obtained by the static element and finite element methods for 1038 steel phosphated tube drawn at 0.6 m.s^{-1} speed from ϕ 27 mm x 2.37 mm to ϕ 21 mm x 2 mm, with a viscoplastic coefficient m equal to 0.01and an isotropic friction factor f equal to 0.05 :
        (a) axial stress along a mean flow line ;
        (b) radial stress profiles in a sinking zone drawing zone boundary.

5. Conclusions

Tubes specifications are more and more numerous and the design of drawing schemes must be achieved as quickly as possible. Therefore, efficient design of the schemes may be realized by a new three steps strategy :
(i) first, select a drawing scheme by velocities, strain, stress and forces analysis with ETIRC (Kinematic Element Method) and ETISLAB (Static Element Method) softwares ;
(ii) second, test the above selected scheme with model material technic ;
(iii) third, confirm the results and improve a little bit with ASTRID software (Finite Element Method).
The efficiency of the methods involved in strategy is demonstrated by the numerical simulation of a one pass drawing. Strain, strain rate, stress predicted by K.E.M. and S.E.M. are very close to those predicted by F.E.M.

Acknowledgements

The authors would like to thank Nord Pas de Calais Region, C.N.R.S., Ministry of Research and Higher Education and Vallourec Industries for supports.

References

[1] COLLINS I.F., WILLIAMS B.K., Slipline fields for axisymmetric tube drawing, Int. J. Mech. Sci., Vol. 27, N°4, 1985, pp. 225-233.

[2] BRAMLEY A.N., SMITH D.J., Tube drawing with a floating plug, Metals Technology, July 1976, pp. 322-331.

[3] AVITZUR B., Drawing of heavy wall tubing with a floating plug, J. Applied Metal Working, Vol. 2, N° 4, January 1983, pp. 259-263.

[4] GELIN J.C., OUDIN J., PICART P., RAVALARD Y., Computation of plastic flows in metal forming processes, New metallic materials and new fabrication processes, Edited by MASOUNAVE J., HAMEL F. G. and BATHIAS C., 1987, National Research Council, Canada.

[5] RIGAUT J.M., OUDIN J., RAVALARD Y., GELIN J.C., Modélisation de l'étirage de tubes par la méthode des éléments cinématiques, Congrès Français de Mécanique, Nantes, 1987.

[6] OUDIN J., RIGAUT J.M., GELIN J.C., RAVALARD Y., Approches expérimentales et numériques des conditions de contact et de frottement, Ecole d'été Matériaux Mise en Forme Pièces formées, Saint Pierre d'Oléron, 1987.

[7] LOCHEGNIES D., Résolution numérique par la méthode des éléments finis des problèmes viscoplastiques : applications à l'étirage et au forgeage, Thèse de doctorat, Université de Valenciennes et du Hainaut Cambrésis, 1988.

PART 5

HOT AND COLD ROLLING

A THREE DIMENSIONAL THERMOMECHANICAL ANALYSIS OF STEADY FLOWS IN HOT FORMING PROCESSES. APPLICATION TO HOT FLAT ROLLING AND HOT SHAPE ROLLING.

C. BERTRAND-CORSINI (*), C. DAVID (*), A. BERN (*)
P. MONTMITONNET (*), J.L. CHENOT (*), P. BUESSLER (**), F. FAU (**)

(*)   CEMEF, ENSMP, SOPHIA ANTIPOLIS, FRANCE
(**)   IRSID-CIREP, MAIZIERES-LES-METZ, FRANCE

ABSTRACT:Few studies exist which adequately treat the forming of shape rolling, primarily because the process is so difficult to model and control. So, several years ago, the CEMEF developed a computer program, based on the Norton-Hoff model of viscoplastic behavior, to simulate three dimensional flow during the hot rolling process. Since then, the finite element techniques utilized in the program have been modified to account for the geometric deformations, thermal effects, and stresses occurring within the product. In this paper, we review the aforementioned techniques and discuss their successful application to flat rolling and hot shape rolling processes.

1.Introduction

As the industry of hot rolling gets on, there appears an increasing demand for more and more precise models in view of a better understanding and control of the process. Among the many problems involved in hot rolling, two of the most important pertain to thermomechanics. First, the final shape of the rolled stock: end shapes, shape of lateral surfaces (spread) depend on the three dimensional metal flow, whereas profile is linked to the contact stresses. The second point is material damage and metallurgical evolution, related to temperature, strain and strain rate, stresses.These are the quantities to calculate. The development of the Finite Element Method, together with increasingly powerful computers, has permitted these requirements to be met /1-9/. This paper is devoted to the description of the principles and applications of a thermomechanical model of three dimensional metal flow in hot rolling of complex shapes.

2. mechanical model

As it has been described at length in other texts /10-12/, it will only be briefly summarised here. As geometry is the main interest, the flow formulation is used. The material obeys NORTON-HOFF equations:

$$s = 2K(\sqrt{3}\,\dot{\bar{\varepsilon}})^{m-1}\,\dot{\varepsilon}$$

where **s** is the deviatoric stress, $\dot{\varepsilon}$ the strain rate tensor. Consistency K and strain rate sensitivity m depend on temperature T and equivalent strain $\bar{\varepsilon}$. As for friction law, a similar expression is chosen ("NORTON friction law"):

J. L. Chenot and E. Oñate (eds.), Modelling of Metal Forming Processes, 271–279.
© *1988 by Kluwer Academic Publishers.*

$$\tau = -\alpha K \, \|\Delta V\|^{p-1} \, \Delta V$$

where τ is the friction stress, ΔV the sliding velocity; α and p are friction coefficients. Under these hypotheses, the velocity field V satisfying the equilibrium equations minimizes the functional :

$$\Phi = \int_\Omega K(\dot{\epsilon}\sqrt{3})^{m+1}/(m+1) \, d\Omega + \int_{\partial\Omega_f} \alpha K |\Delta V|^{p+1}/(p+1) \, dS - \int_{\partial\Omega_d} F^d . V \, dS$$
$$+ \rho' \int_\Omega K (\mathrm{div} V)^2 d\Omega$$

where Ω is the domain occupied by the deforming metal and $\partial\Omega$ its boundary ($\partial\Omega_f$ the contact area, $\partial\Omega_d$ where external forces F^d are prescribed); the last term ensures incompressibility. Spatial discretization makes use of 8-node hexahedral isoparametric elements. Minimization of Φ is performed by NEWTON-RAPHSON method, giving the velocity field V.

In this paper, only applications to steady state problems (neglecting end effects) are presented. Hence the problem is solved in a stationary way (one step only). However, due to spread, domain Ω is not known at first. Hence a method has been derived for the determination of the shape of the lateral free surface /12,13/. It consists in minimizing the cost functional J (flux of matter through the surface, which should be zero for a stationary free surface):

$$J = \int_{\partial\Omega} (V.n)^2 \, dS$$

n is the normal to the boundary, V minimizes Φ. J is a function of the position of surface nodal points. A NEWTON-RAPHSON method consisting in moving these points is used to minimize J iteratively, starting from an estimated initial domain.

Stress computation is performed as a post processor consisting in:

-computing a smooth nodal $\dot{\epsilon}$ /14/
-computing s from NORTON-HOFF law
-then the pressure field best balancing s is found by minimizing /11,15/:

$$f(p) = \{ \int_\Omega (s_{ij,j} - p_{,i}) N \, d\Omega \}^2$$

where N are the interpolation functions.

3. thermal-rheological coupling and strain-consistency coupling

Here again, the steady state thermal problem is addressed. Moreover, it is easily shown that conduction is negligible with regard to convection /10/. Finally, the following equation is solved:

$$\rho c V \, \mathrm{Grad} T = K(\sqrt{3} \, \dot{\epsilon})^{m+1}$$

where V and $\dot{\bar{\varepsilon}}$ derive from flow calculation. ρ is the specific gravity, c the specific heat. This equation is integrated along streamlines, as is

$$\bar{\varepsilon} = \int \dot{\bar{\varepsilon}}\, dt$$

As for boundary conditions, an initial temperature map (ahead of the entry plane) in a cross section is needed. Lateral surfaces are under zero flux conditions. A contact temperature is computed /10/, accounting both for frictional heating and conductive losses towards the cool rolls, through a thermal resistance. The results of these integrations are T and $\bar{\varepsilon}$ maps, which are used to update values of K(x,y,z). A coupled computation is thus performed.

4. Application to hot rolling

Two series of tests have been carried out at the "Laboratoire d'Elaboration et de Transformation des Matériaux" (LETRAM) of the "Centre d'Etudes Nucléaires" (CEA) at Saclay (France).The aim of these tests was to obtain a set of geometrical, mechanical and thermal data in order to prove that our thermomechanical model of rolling is able to simulate the hot rolling process.

4.1. HOT SHAPE ROLLING APPLICATION

4.1.1. Description of the experiment

The tests have been carried out on a full scale hot rolling mill. We used 254 mm diameter rolls with 24 mm diameter round grooves. Roll separation E (see figure 1) was 2 mm, 5 mm or 10 mm, giving different reductions. Moreover, two billet widths (30 mm and 40 mm) were used giving a total of six experiments. Rolling speed was 30 m/mn (3.94 rad/s). Figure 1 shows the entry table with its guiding system ensuring symmetrical rolling. The rolling mill is equipped with two force gauges on the upper roll. Roll torque and billet speed are also recorded. Furnace temperature was 1100°C. The furnace was situated near the rolling mill in order to minimize the time of transfer of the billet on the exit from furnace to the entry table (a few seconds).

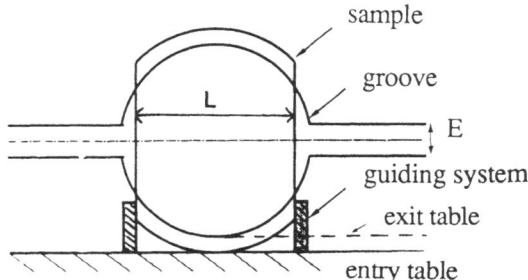

L: width of 30 mm or 40 mm
Figure 1 : Description of the groove and of the guiding system of the billet

274

The aim of these tests is to study the influence of the initial geometry and of the reduction ratio on the mechanical and geometrical parameters

4.1.2 Experimental results

The experimental forces and torques have been recorded (figure 2).Note the good reproducibility of the experimental results.

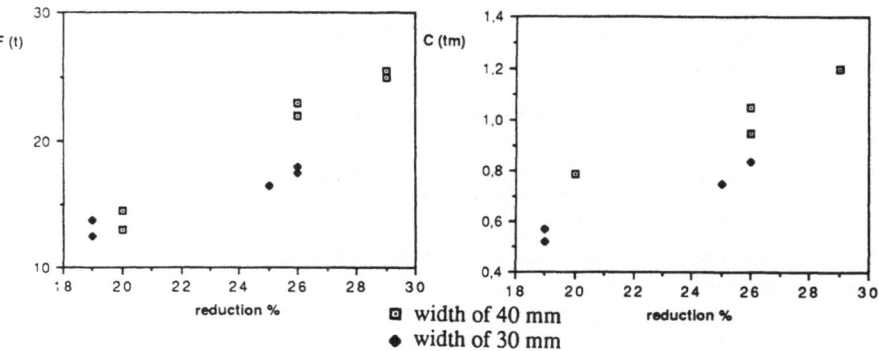

Figure 2 : Experimental rolling force and rolling torque

A section of each billet has been cut to control the filling of the groove. We present the shape of the different profile for each case on the figure 3. Spread is depending on the reduction ratio. The 'diabolo' shape is obtained every time except in the case with the initial width of 40 mm and the highest reduction ratio, where bulging is obtained.

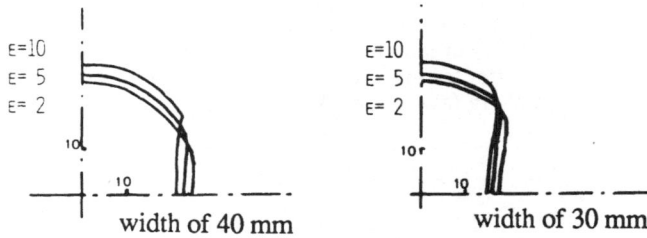

Figure 3 : The shape of the different profiles of a quarter of the billet

4.1.3 Finite element results

The aforementioned tests have been modelled with our finite element model of shape rolling. First a rheological law is determined in order to describe the behavior of the chosen steel. A finite difference thermal model gives an initial map of the temperature in the billet accounting for the time of transfer of the billet from the furnace to the roll mill.

Flow stress σ_0 is calculated as a function of temperature T, equivalent strain $\bar{\varepsilon}$ and equivalent strain rate $\dot{\bar{\varepsilon}}$ (experimental data provided by IRSID). An average strain is taken: $\bar{\varepsilon}$ = 1/2 ln S0/S1, where S0 is the section in the entry and S1 the section in the exit of the roll gap. In this paragraph temperature is supposed constant along the streamline (approximated by a line of nodes along rolling direction). Experimental rheological curves are fitted by: $\sigma 0$ (T, $\bar{\varepsilon}$, $\dot{\bar{\varepsilon}}$) = b exp(a/T) = K0($\bar{\varepsilon}$) $\dot{\bar{\varepsilon}}^{m}$ / $\sqrt{3}^{(m+1)}$ exp(a/T) .

We find m = 0,114 at T = 1100°C. The values of a and K0 at $\bar{\varepsilon}$ =1/2 ln S0/S1 are reported in table 1 for the six configurations, named after the value of E and the initial billet width L.

The parameters of the friction law are : p = m = 0.114 and α = 1.

The same two meshes have been used for all six configurations: 208 nodes (13x4x4) and 260 nodes (13x4x5). The rolling force and the roll torque are presented on figure 4 for the series with initial width 40 mm.

test	E10L30	E10L40	E05L30	E05L40	E02L30	E02L40
K0 x 10 6	7.43	7.65	7.38	7.38	7.33	7.38
a	1802.72	1751.8	1895.58	1917.18	1998.11	1968.1

Table 1 : Values of K0 and a in the expression of the consistency law

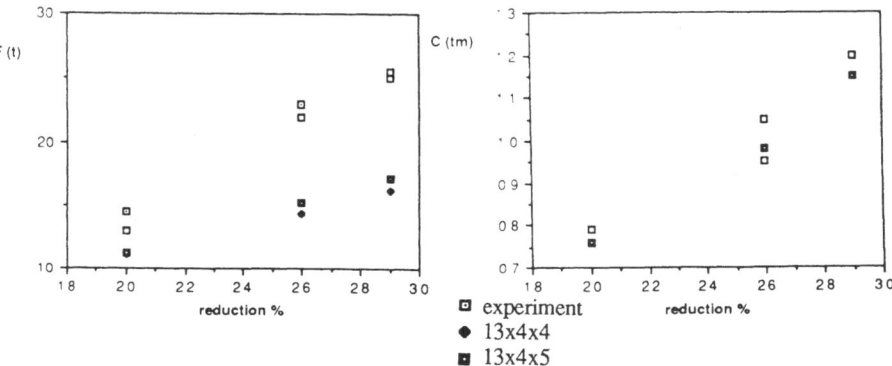

Figure 4 : The rolling force and the rolling torque for the initial width of 40 mm

The computed rolling force is systematically underestimated. A first explanation may be found in coarseness of meshing. Although the meshes are rather similar, the finer one gives a somewhat better rolling force calculation. A second phenomenon may be involved: roll contact cools and hardens the surface of the metal, thus increasing the rolling force. This thermomechanical coupling has not been accounted for in this example. Figure 4 shows that we obtain a very good approximation of the rolling torque. The maximum relative error obtained is 7%.

276

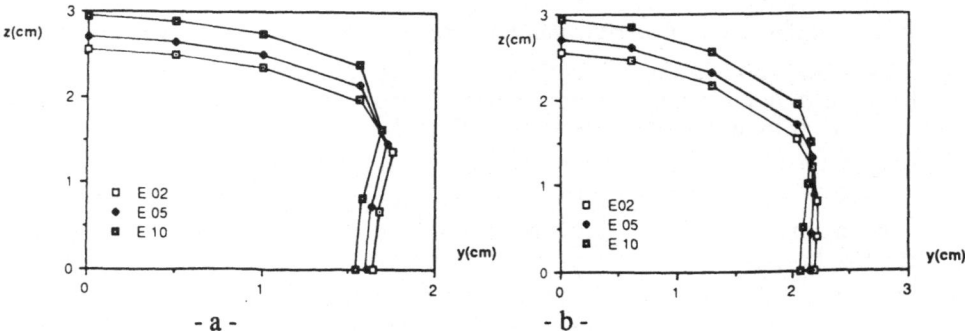

- a -   - b -

Figure 5 : Sections in the exit of the roll gap.
         a - Initial width 30 mm ;13x4x4 nodes   b - initial width 40 mm ;13x4x5 nodes

For each configuration, free surface calculations have permitted to determine the lateral side of the deformed billets. A quarter of the exit sections of the deformed meshes are presented on the figure 5.
First , we can see that qualitatively, the profiles obtained are similar in every case to the experimental profiles presented on figure 3: same 'diabolo' shape for every configuration and quite limited for the test E02L40. So a good representation of the lateral side of the rolled billet is obtained. However, the finer mesh gives us a better estimation of the free side. The spread is located near the groove.
Quantitatively, the experimental shape and the computed one have been superposed on figure 6 for the test E10L40. This figure confirms the good results obtained. The maximum spread seems to be slightly underestimated, as is the contact surface.But a good approximation of the contact surface is dependant on accounting for thermal effects. Examples of coupled thermomechanical resolutions are presented in paragraph 4.2

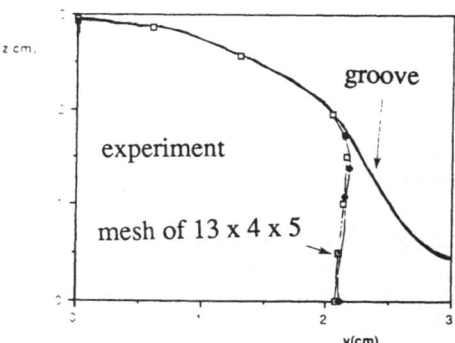

Figure 6 : Comparison between experimental and computed results. Test E10L40

A quarter of the deformed meshes for the test E02L40 and the test E02L30 are presented in figure 7.

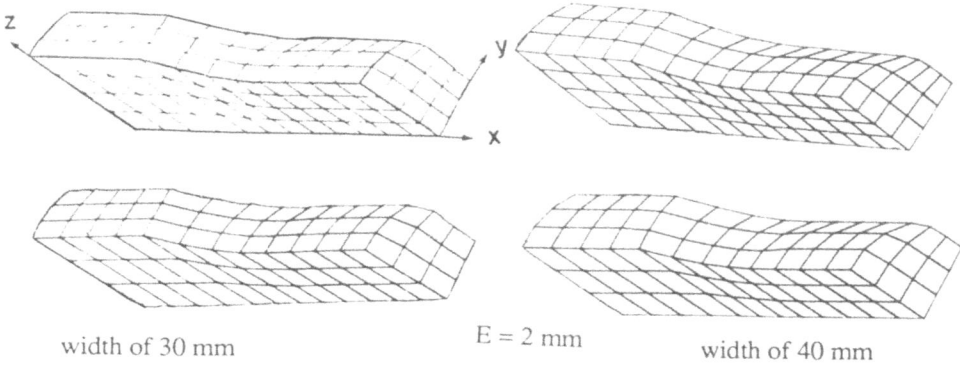

width of 30 mm   E = 2 mm   width of 40 mm

Figure 7 : Deformed meshes. Tests E02L40 and E02L30

4.2 HOT FLAT ROLLING OF SQUARE BILLETS : THERMAL EFFECTS

The experimental set up is similar to paragraph 4.1, except that non grooved rolls are used. Billets are 50x50x350mm. Furnace temperatures are 850°C, 1000°C, 1150°C. Reductions of 10%,20%,30%,40% have been chosen.Computations were carried out with the complete $K(T,\bar{\varepsilon})$ law, and thermomechanical coupling as described in 3:

$K(T,\bar{\varepsilon}) = (166830/T-96.3)\exp\{ (99/T-0.03)/(\bar{\varepsilon}+23.82/T+0.01)\}$ (MPa) (T in °C)

friction : p=m=0.114 and α=1.

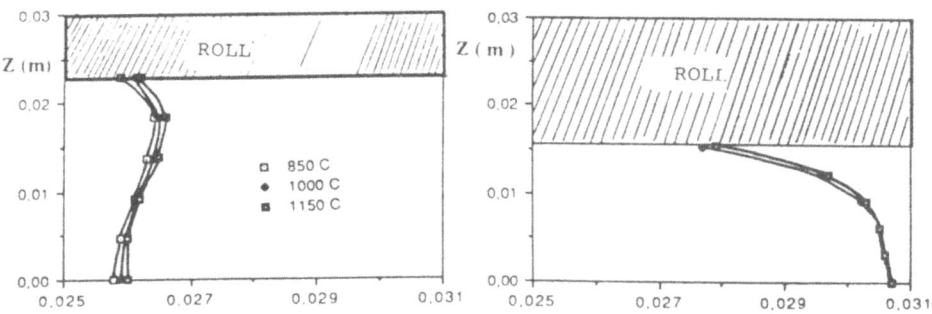

figure 8 : computed lateral shape; left: reduction 10%; right: reduction 40%
            initial half width: 0.025 m

Thermal parameters: billet effusivity b_b=10000 Jm^{-2} K^{-1} s$^{-1/2}$,roll effusivity b_r=10000
roll core temperature 50°C , contact thermal conductance 20000 I.S.
billet volumic specific heat $\rho c = 4\ 10^6$ J K^{-1} m^{-3}

Figure 8 shows the influence of temperature and reduction on computed lateral surface shape . as previously, low reduction leads to concave ("diabolo") and high reductions to convex ("bulged") shapes. Increasing T tends to slightly increase spread, but the effect is second order. A good prediction of spread has been obtained. Comparison of experimental and computed rolling force is shown in figure 9. Accounting for the coarseness of meshing (see figure 10), the prediction of rolling force (and contact stresses) is rather satisfactory.

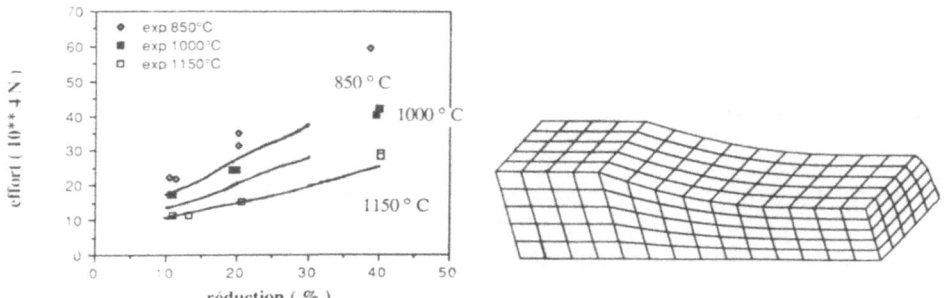

figure 9 : comparison of computed (full line)   figure 10 : example of deformed mesh
and experimental (points) forces   (40% reduction)

5. Conclusion

At the present time, modelling of most hot rolling operations is possible with inclusion of all important effects (complex material behavior, thermal flow). Apart from the cases exposed in this paper, application to more complex profiles (rails) or tubes has proved possible. End shapes can be modelled by a transient flow version of the code (explicit time integration) /15,16/ but require very fine meshes and time steps, and hence are highly CPU time consuming. A preliminary application to the prediction of material damage and metallurgical evolution has also been performed /17/.

6. Acknowledgement

Thanks are due to IRSID-CIREP for financial support and for providing the flow stress data, and to CEA-LETRAM for performing the rolling experiments described in this paper.

7. References

/1/ G.C. CORNFIELD, R.H. JOHNSON: Theoretical prediction of metal flow in hot rolling including the effect of various temperature distributions. J. Iron Steel Inst. 211 (1973) 567

279

/2/ C. LIU, P.HARTLEY, C.E.N. STURGESS, G.W. ROWE: Analysis of stress and strain distribution in slab rolling using an elastoplastic FEM. Proc. NUMIFORM'86 (Gothenburg,25-29 August 1986) K. MATTIASSON et al., eds. Publ. BALKEMA.

/3/ M. PIETRZYK: Rigid plastic finite element simulation of plane strain rolling with significantly non uniform flow of metal.Proc. NUMIFORM'86 (Gothenburg,25-29 August 1986) K. MATTIASSON et al., eds. Publ. BALKEMA.

/4/ J.H. BEYNON, P.R. BROWN, S.I. MIZBAN, A.R.S. PONTER, C.M. SELLARS: Inclusion of metallurgical developments in the modelling of industrial hot rolling of metals. Proc. NUMIFORM'86 (Gothenburg,25-29 August 1986) K. MATTIASSON et al., eds. Publ. BALKEMA.

/5/ K. MORI, K. OSAKADA: simulation of 3-D deformation in rolling by the FEM. Int. J. Mech. Sci. 26 (1984) 515-525

/6/ K. MORI, K. OSAKADA: Experimental and finite element analysis of hot rolling of slabs in width direction with flat rolls. J. Jap. Soc. Techn. Plasticity 23 (1982) 262

/7/ LI GUO JI, S. KOBAYASHI: Spread analysis in rolling by the rigid plastic FEM. Proc. NUMIFORM'82.(SWANSEA, 1982)J.F.T. PITTMAN et al. eds. Publ. PINERIDGE PRESS.

/8/ H.G. HUISMAN, J. HUETINK; Combined eulerian-lagrangian 3D FEM analysis of edge rolling. J. Mech. Working Tech. 11 (1985)

/9/ C. DAVID, C. BERTRAND, P. MONTMITONNET, J.L. CHENOT, P. BUESSLER: A 3D thermomechanical analysis of rolling by FEM. Proc. 4th Int. Steel Rolling Conf. (DEAUVILLE 1987) Publ. IRSID-ATS

/10/ C. DAVID: Modélisation tridimensionnelle par éléments finis de l'écoulement d'un matériau viscoplastique pour l'étude du laminage à chaud. Thesis, CEMEF-ENSMP(1987)

/11/ C. BERTRAND: Modélisation numérique par éléments finis du calibrage à chaud des métaux . Thesis, CEMEF-ENSMP (1987)

/12/ A. BERN : Contribution à la modélisation par éléments finis des surfaces libres en régime permanent. Thesis, CEMEF-ENSMP (1987)

/13/ A. BERN, C. DAVID, J.L. CHENOT, P. BUESSLER: FE simulation of the spread in hot rolling of thick slabs. Proc. 4th Int. Steel Rolling Conf. (DEAUVILLE 1987) Publ. IRSID-ATS

/14/ T. LISZKA, T. ORKISZ: The finite difference method at arbitrary grids and its application in applied mechanics. Int. J. Computers and Structures 11 (1980) 83

/15/ C. BERTRAND, C. DAVID, J.L. CHENOT, P. BUESSLER: Stresses calculation in FEM analysis of 3D shape rolling.Proc. NUMIFORM'86 (Gothenburg,25-29 August 1986) K. MATTIASSON et al., eds. Publ. BALKEMA

/16/ C. DAVID, C. BERTRAND, J.L. CHENOT, P. BUESSLER: A transient 3D FEM analysis of hot rolling of thick slabs.Proc. NUMIFORM'86 (Gothenburg,25-29 August 1986) K. MATTIASSON et al., eds. Publ. BALKEMA

/17/ O. BRANSWICK, C. DAVID, C. LEVAILLANT, J.L. CHENOT, J.P. BILLARD, D. WEBER, J.P. GUERLET: Surface defects in hot rolling of copper based alloys heavy ingots: 3D simulation including fracture criterion. Proc. Comp. Methods Pred. Mat. Proc. Defects (CACHAN,FRANCE, 1987) Publ. ELSEVIER

THERMAL-MECHANICAL MODELLING FOR HOT ROLLING:
EXPERIMENTAL SUBSTANTIATION

M.Pietrzyk*, J.G.Lenard**

* Mining and Metallurgy University, Krakow, Poland
(currently Post Doctoral Fellow at UNB)

** Department of Mechanical Engineering, UNB,
Fredericton, Canada

1. Introduction

The accuracy of predictions of mathematical models of hot rolling depends on the rigour and quality of the description of the boundary conditions. For the mechanical component these involve the distribution of surface tractions on the roll/strip interface and the shape of that interface. In the thermal portion the heat transfer coefficient in the roll gap needs to be specified. .

Experimentally determined values of that coefficient have been reported in the literature - see, for example [1,2,3,4,5]. Since the number of parameters that influence the magnitude of the heat flux at the contact surface is large, choice of the heat transfer coefficient for use in modelling is still problematic, however.

Computer simulation of transient viscoplastic metal deformation problems coupled with heat transfer analysis has been presented by several authors. Only a few publications deal with the rolling process however, which presents difficulties connected with the existence of the neutral point and with the cyclic thermal state of the work rolls. A review of some of the relevant publications has been presented in [6]. The scope of the work ranges from a simplified approach yielding a closed form solution [7], through more complicated models [8,9,10 12] and [13] to finite element solutions [14,15] and [16].

The objectives of the present project are threefold. First, a thermal-mechanical model of hot rolling has been developed. Following that, the dependence of the heat transfer coefficient on the process parameters was established experimentally. Finally, the predictive capability of the model was tested by inserting in it appropriate values of the heat transfer coefficient and comparing its predictions of temperature distribution to those measured while hot rolling slabs.

281

J. L. Chenot and E. Oñate (eds.), Modelling of Metal Forming Processes, 281–288.
© 1988 by Kluwer Academic Publishers.

282

2. Mathematical Model

The method of computation, presented in [6] is used in the present work. The model is based on the finite element method [17] and it contains two parts - the thermal and the mechanical components. The first part is connected with heat transfer and heat generation within the strip. The general quasi-harmonic equation for a two dimensional, time dependent problem is:

$$\frac{\partial}{\partial x}(k_x\frac{\partial T}{\partial x}) + \frac{\partial}{\partial y}(k_y\frac{\partial T}{\partial y}) + (Q - \rho c_p\frac{\partial T}{\partial t}) = 0 \qquad (1)$$

where T is the temperature, Q is the heat generation rate due to plastic deformation, k_x and k_y are the heat conduction coefficients, ρ is the density and c_p stands for the specific heat.

The boundary equation is specified through an energy balance and is given by:

$$k_x\frac{\partial T}{\partial x}l_x + k_y\frac{\partial T}{\partial y}l_y - \alpha(T_0 - T) - q = 0 \qquad (2)$$

where α is the interface heat transfer coefficient, l_x and l_y are the direction cosines and T_0 is the temperature of the roll.

Equation (1) is discretized and minimization with respect to nodal temperatures is performed. Assuming that the nodal temperatures are linear with respect to time the solution is obtained as [18]:

$$(2[H] + \frac{3}{\Delta t}[C]){\{T_{i+1}\}} = (-[H] + \frac{3}{\Delta t}[C]){\{T_i\}} - 3\{P\} \qquad (3)$$

where:

$$[H] = \int_v[k_x\{\frac{\partial N}{\partial x}\}\{\frac{\partial N}{\partial x}\}^T + k_y\{\frac{\partial N}{\partial y}\}\{\frac{\partial N}{\partial y}\}^T]dV - \int_s\alpha\{\overset{*}{N}\}\{\overset{*}{N}\}^TdS$$

$$\{P\} = \int_s(\alpha T_0 + q)\{\overset{*}{N}\}dS - \int_v\{N\}\{N\}^T\{Q\}dV; \qquad [C] = \int_v\{N\}c_p\rho\{N\}^TdV$$

$\{N\}$ and $\{\overset{*}{N}\}$ are the vectors of shape and surface shape functions, respectively, t is a time interval, V is the volume and S designates the boundary surface area. Further, T_i is the initial temperature field for the current iteration and T_{i+1} is the temperature field after time Δt.

The solution requires additional data which are supplied by the second component of the model, concerned with the determination of the velocity, strain rate and strain distributions. The method suggested in [19], based on an extremum principle stating that the actual solution minimizes the functional:

$$J = \int_v\sigma_i\dot{\varepsilon}_idV + \int_v\lambda\dot{\varepsilon}_idV - \int_s\{T\}\{v\}dV \qquad (4)$$

where $\dot{\varepsilon}_v$ is the volumetric strain rate, σ_i is the effective stress, $\dot{\varepsilon}_i$ is the effective strain rate and $\{v\}$ is the vector of velocities on the surface S, is employed. $\{\underline{T}\}$ stands for a traction, prescribed on part of the surface.

The material is assumed to be rigid-plastic and it obeys the von Mises yield criterion. The discretization of the functional (4) leads to the non-linear stiffness equations which are solved by a Newton-Raphson technique. The frictional boundary conditions are introduced as functions of the slip velocity [20].

Two sets of equations are formulated in the present paper, solved simultaneously and the results are iterated between the two solutions. The flow strength is given in [21] in terms of the temperature, carbon content, strain and strain rate. The length of the contact is calculated from the modified Hitchcock formula.

3. Experiments

The details of the experimental program have been given in [22]. Briefly, low carbon steel slabs were rolled at various temperatures and rolling speeds to various reductions. The slabs had several thermocouples embedded in them and by monitoring their output during rolling the time-temperature profile for each was determined.

4. Results and Discussion

The results of predictive calculations are compared with experimental data in this section. Finite element meshes and boundary conditions used in the calculations are given in [6,23]. All physical properties are introduced as functions of the temperature [24,25]. Empirical relations for the heat conduction coefficient, specific heat and the density of the steel are given in [6].

One of the most important factors contributing to the accuracy of the solution is the heat transfer coefficient between the strip and the roll. In the present work the value of 4800 W/m²K [22] is used. In order to assess its effect on the predictions, calculations for three various coefficients were carried out - see Figure 1. It is noted that increasing to 7000 W/m²K overestimates the cooling rate of the strip while a = 3000 W/m² K results in slower cooling. After exit the temperature is distributed more uniformly and predictions for various heat transfer coefficients are close to each other. This situation is due to the short time of contact between the roll and the strip, which was 0.6 sec. in the case considered. In hot strip mills the effect of the heat transfer coefficient may be cumulative and significant differences may be obtained after the last stand.

In Fig.2 the results of calculations for a = 4800 W/m²K and a = 70000 W/m²K are compared with measurements.

Temperatures at the strip surface (dotted line) and at a point 2 mm below are considered (solid line). It is noted that the cooling rate for a = 70000 W/m²K exceeds several times the experimental

284

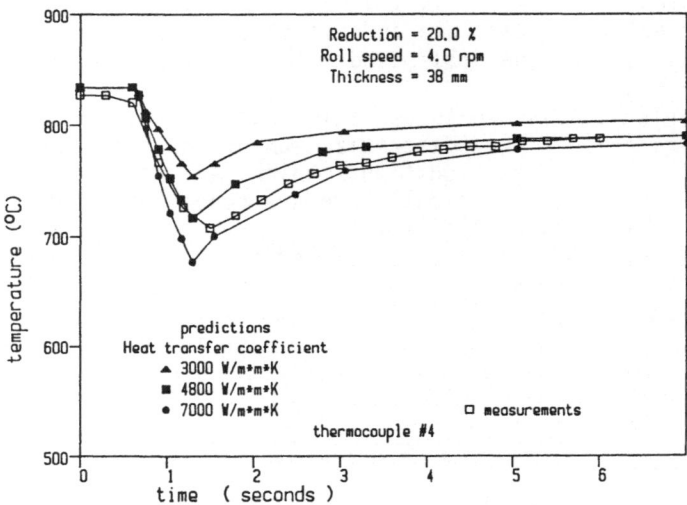

Fig.1. The effect of the heat transfer coefficient on the temperature variation during rolling.

rate. In the present model the roll temperature was assumed to remain constant. Since that can increase in the pass by several hundred degrees, the calculations for a = 70000 W/m²K were repeated with

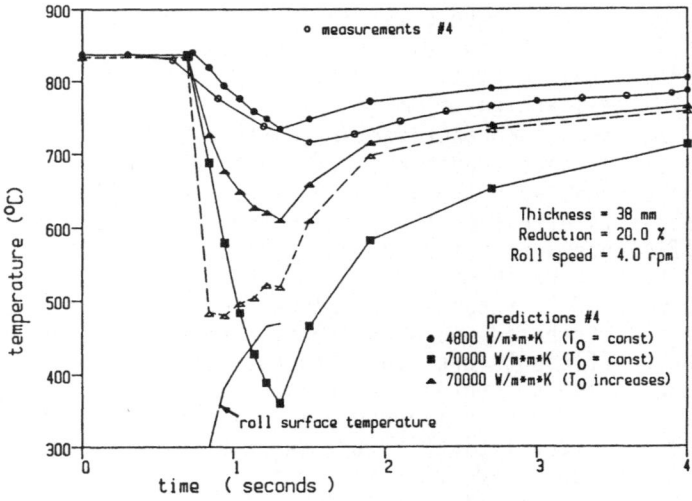

Fig.2. The effect of the heat transfer coefficient and roll surface temperature on the predictions of the model.

varying roll temperature. Results were much closer to the experimental data this time but the cooling rate of the strip was still overestimated. The present tests were carried out on a laboratory mill without lubrication. As given in [1] introduction of lubricants may increase the heat transfer coefficient several times. The presence of scale on the surface is also of importance.

In all further calculations a = 4800 W/m²K is used. Results are presented in Figs.3,4,5 and 6.

The time-temperature profiles during rolling at the centre and in one corner of a 19 mm thick slab are given in Fig.3, assuming a uniform initial distribution of 870 °C. It is observed that the differences between predictions and measurements nowhere exceed 3.5% of the measured values and are mostly much less than that, indicating the exceptional predictive capability of the model.

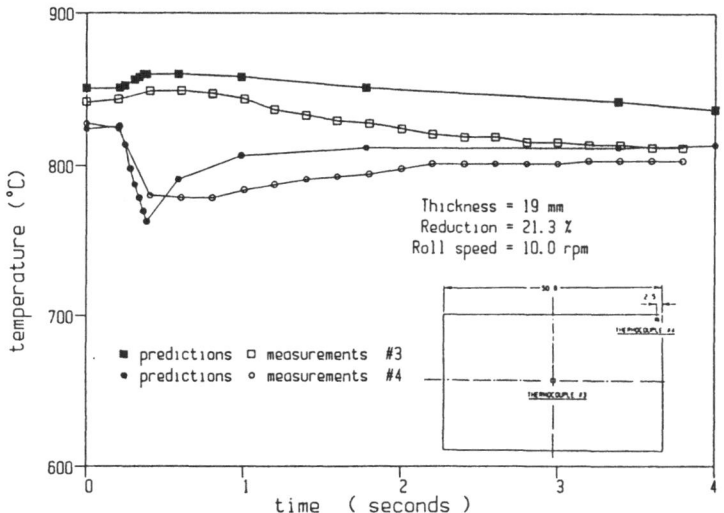

Fig.3. Measured and calculated temperatures during rolling of a 19 mm thick slab.

Fig.4 shows the lines of constant temperature on a plane parallel to the direction of rolling. As expected, the coolest part of the strip is the one near the contact zone and sharp temperature gradients toward the strip centre can be observed. Both analysis and experiment indicate that heating as a result of plastic work done, if not excessive in the laboratory situation, may be very significant.

The temperature distribution in a plane perpendicular to the direction of rolling and located at exit is shown in Fig.5. The hot centre and the cool surface are again observed. As well, the different rates of cooling across and through the strip are also indicated.

The variation of the effective strain rate within the rolled

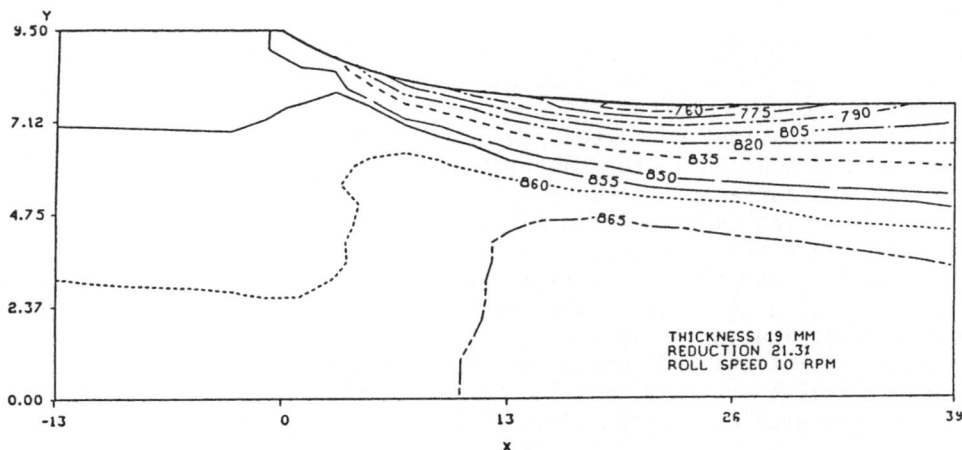

Fig.4. Temperature distribution in the plane perpendicular to the
      roll axis.

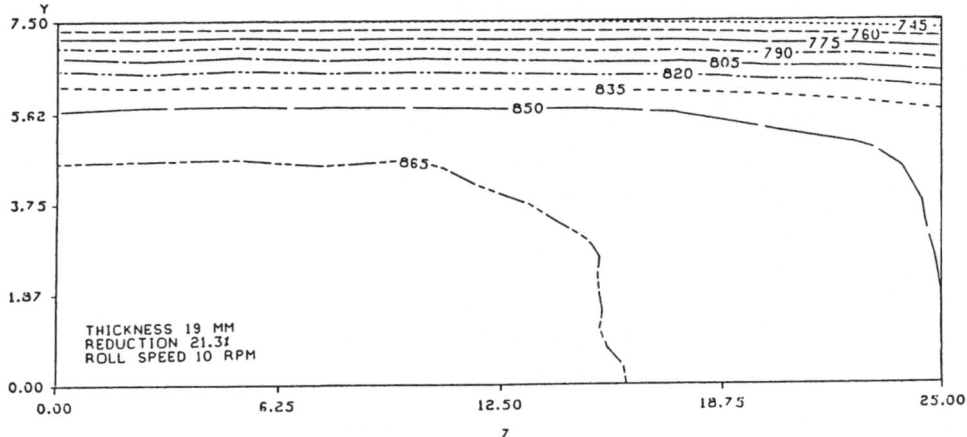

Fig.5. Temperature distribution in the exit plane.

strip is given in Fig.6.   Concentration of the strain rate close to
entry and in the centre of the deformation zone can also be observed.

5. Conclusions

The predictions of a thermal-mechanical model of the hot flat rolling
process were compared with temperatures measured during hot rolling of
slabs. The effect of the assumed heat transfer coefficient on the
predictions was also investigated. It was shown that numbers published
in the literature vary significantly and they cannot be employed
directly in the mathematical models.   The method followed in the
determination of the heat transfer coefficient used in the calculations
should correspond closely to the assumptions which are introduced in

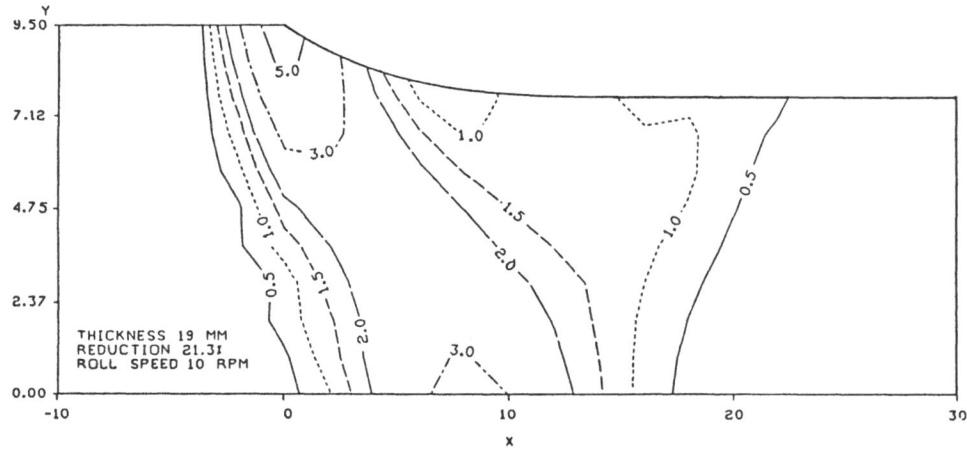

Fig.6. Lines of constant effective strain rate during rolling of a 19 mm thick slab.

the model. In particular, the problem of heating of the roll has to be taken into account. Indeed, due to a very short time of contact the effect of the heat transfer coefficient on the average temperature after one pass was small, but a very strong influence of this coefficient on the temperature variations in the contact zone was observed. Moreover, in continuous rolling the effect of the heat transfer coefficient may be cumulative and significant differences can appear after the last stand.

The differences between the analytical and experimental results were found to be consistently less than 5%, showing the capabilities of the model, especially when heat transfer coefficients, developed experimentally, were introduced.

References

1. Devadas C.,Samarasekera I.V., Ironmaking and Steelmaking,13,1986, 311-321.
2. Murata K. et al., Trans.ISIJ,24,1984,B309.
3. Stevens P.G.,Ivens K.P.,Harper P,J., Iron Steel Inst.,209,1971, 1-11.
4. Harding R.A., PhD dissertation, Univ.of Sheffield,1970.
5. Preisendanz H.,Schuler P.,Koschnitzke K.,Arch.Eisenhuttenwes.,38, 1967,205-213.

288

6. Pietrzyk M.,Lenard J.G., Proc.25th Nat.Heat Transfer Conf.,Houston, 1988.
7. Lee P.W.,Sims R.B.,Wright H., J.Iron Steel Inst.,201,1963,270-279.
8. Wilmotte S. et al., CRM,1973,no.36,35-44.
9. Bryant G.F.,Heselton M.O., Metals Technol.,9,1982,469-477.
10. Bryant G.F.,Chiu T.S.L., Metals Technol.,9,1982,478-484.
11. Lahoti G.D.,Shah S.N.,Altan T., Trans.ASME,J.Eng.Ind.,100,1978, 159-166.
12. Sheppard T.,Wright D.S., Metals Technol.,7,1980,274-281
13. Tseng A.A., Trans.ASME,J.Heat Transf.,106,1984,512-517.
14. Silvonen A.,Malinen A.,Korhonen A.S.,Scand.J.Metal.,16,1987,103-108.
15. Thompson E.G.,Berman H.M., Numerical Analysis of Forming Processes, Edited by J.F.T.Pittman,O.C.Zienkiewicz,R.D.Wood,J.M.Alexander, John Wiley & Sons, 1984.
16. Zienkiewicz O.C.,Onate E.,Heinrich J.C.,Int.J.Num.Meth.Eng.,17, 1981,1497-1514.
17. Zienkiewicz O.C., The Finite Element Method in Engineering Science, McGraw Hill, London 1971.
18. Glowacki M. et al., Hutnik,55,1988,(in press).
19. Lee H.C.,Kobayashi S., Trans ASME,J.Eng.Ind.,95,1973,865-873.
20. Chen C.C.,Kobayashi S., Appl.Num.Meth.Form.Proc.,ASME,ADM,28,1978, 163.
21. Shida S., Hitachi Res.Lab.Report,1974,1-9.
22. Karagiozis A.N.,Lenard J.G.,Trans.ASME,J.Eng.Mat.Techn.,110,1988, 17-21.
23. Pietrzyk M., Hutnik,49,1982,213-216.
24. Touloukian Y.S. et al., Thermophysical Properties of Matter,Vol 1, IFI Plenum, Washington-New York 1970.
25. Touloukian Y.S.,Buyko E.H., Thermophysical Properties of Matter, Vol 4,IFI Plenum, Washington-New York 1970.

Acknowledgements

The financial assistance of the Natural Sciences and Engineering Research Council of Canada and NATO is gratefully acknowledged. Figures 1,3,4,5 and 6 have been reproduced from "Experimental Substantiation of Modelling Heat Transfer in Hot Flat Rolling" by Pietrzyk and Lenard, published in a special Symposium Volume of ASME (in press).

SIMULATION OF ROLLING BY PLASTICINE

COMPARISON BETWEEN THE FLOW OF STEEL AND THE FLOW OF PLASTICINE
OR WAXES BY ACTUAL ROLLING AND FINITE ELEMENT MODELS

AUTHORS: F. FAU*, P. BUESSLER*, C.H. QUAN*, C. BERTRAND**
          * IRSID-CIREP, MAIZIERES-LES-METZ, FRANCE
          ** CEMEF, SOPHIA ANTIPOLIS, FRANCE

Abstract

The IRSID-CIREP research center utilizes plasticine and waxes to simulate the hot rolling of steel. In the case of width reduction of slabs in grooved rolls, the flows of steel and plasticine are much different. By rolling trials on steel, plasticine and waxes and by numerical simulations with a finite element model developped by the CEMEF research center, we studied the reasons of this bad simulation : the different flow stress coefficients n and m of steel, waxes and plasticine, the different friction coefficients, the bad simulation of temperature gradients. The relative influence of these parameters has been clarified.

1. Introduction

The IRSID-CIREP is the process research centre of the USINOR-SACILOR french steel group concerning the steel making processes and the rolling of steel. Its rolling and forming department utilizes since 1971 the simulation to study the hot rolling of steel. It is equipped with a plasticine simulation laboratory and a hot steel pilot mill (AXEL). After 15 years of simulation we have simulated the rolling of most of the products rolled by SACILOR. In order to improve the accuracy of simulation we have been obliged to make comprehensive studies concerning the theory and the practice of simulation.

2. Plasticine simulation laboratory

This laboratory was created in 1970. Presently, the following equipment exists for the preparation of plasticine : 2 rolling mills for the preparation of slabs, a compacting press, an extrusion press and a wire slicing table. Plasticine blocks are prepared in the following way : after rolling, homogenous slabs of 5 mm thickness are obtained. Afterwards, these slabs are piled and pressed during several hours. The blocks thus obtained may

289

J. L. Chenot and E. Oñate (eds.), Modelling of Metal Forming Processes, 289–296.
© *1988 by Kluwer Academic Publishers.*

be directly cut to dimension by wire slicing, or may be extruded through a die to obtain bars of various cross sections (billets, slabs, intermediate blanks).

The rolling tests may be achieved on 4 simulation mills : 2 universal stands, a double duo mill with 2 parallel stands and a two stands tandem duo mill for the study of interstand tension (fig. 1).

Simulation is used in four major fields :
- Flow studies and roll pass design. We have, in this way, prepared roll pass designs for H beams, Z and Larssen sheet piles, rails, angles, U shape sections and round sections
- Optimization studies of rolling schedules for a better use of the existing facilities and for the reduction of end crops, especially in the case of blooming mills, slabbing mills or hot strip mill roughing stands
- Studies of surface defects (opening or closing of cracks), by following the evolution of surface inprints, and by measuring normal stresses in the roll gap. For some studies we track the evolution of artificial defects. This has been achieved for rails, slabs, rounds and billets.
- Studies of internal flows, of the evolution of segregations which can be traced with plasticine wires or by using blocks with layers of different colours.

3. AXEL Steel Pilot Mill

The AXEL steel pilot mill was built in 1984. It consists of :
- a gas reheating furnace, for heating products up to 2 meters long at 1 250° C
- a duo stand
- an universal stand which can also be used as a duo stand (universal-duo stand)
- a cold saw.

This mill is used for the development of new processes, for the establishment of roll pass designs and for the development of rolling models, for example : thermomecanical models for the rolling of H beams in a universal stand, spreading models for rounds. We have also studied on this mill the feasability of rolling composite materials..

4. Similitude Laws

In order to simulate in a proper way the rolling of hot steel with plasticine, similitude laws are to be observed. The theorical basis of simulation, as well as their application to the forming of metals have already been studied in a comprehensive way (1, 2).

The three fundamental laws of similitude, in the case of plasticine used to simulate the hot rolling of steel are :
- the dimensions of the products and of the rolls must be similar

- the friction coefficients between the rolls and the product must be equal
- the flow stresses at any point of the steel or plasticine product must be similar.

The first condition requires that the simulation is made at a given scale.

The second condition can be achieved by the selection of an appropriate lubricant or roll roughness.

The last condition, relating to the flow stresses is the one which is the most difficult to be met.

The flow stress of hot steel can be described by the following law :

$$\sigma_0 = A \; \overset{-n}{\varepsilon} \; \overset{\pm m}{\dot\varepsilon}$$

The A, n and m coefficients depend on the temperature. The flow stress varies therefore with the temperature gradient. It is often difficult, and even impossible to simulate this flow stress gradient with plasticine.

If the temperature gradients are sufficiently low to be neglected, the simulation of the flow stress is equivalent to the equality of the n and m coefficients between steel and simulation material.

In practice, the three similitude laws are very constraining and they are observed only approximately. Nevertheless, in the majority of situations, simulation with plasticine is useful. There are however examples of industrial rolling where the flow of plasticine is far different from that of hot steel. We shall now study on two examples, the width reduction of slabs and the slab edging, the reasons of these poor simulations, and the improvements that can be achieved.

5. Width Reduction of Slabs and Slab Edging

We consider two types of rolling : the width reduction of slabs and slab edging, which both induce flows of the same type.

In order to obtain a blank for wide flange beams from a slab, we can apply the slab edging method : the small sides of the slab undergo a knife pass in the first groove, and afterwards they are rolled in the two following grooves in such a way as to obtain important bulges (fig. 2). We discovered that the bulges are much more pronounced with plasticine than with steel (fig. 3).

Width reduction of slabs in a hot strip mill is obtained in a vertical stand (edger) and in a horizontal stand (fig. 4). Width reduction in the vertical stand results in the formation of bulges near the edgers. The next pass in the horizontal stand has the purpose to eliminate these bulges. The flattening of these bulges results in an increased widening of the slab, which increases with the size of the bulges. The efficiency of the width reduction process is thus defined as the ratio between the net reduction after flattening and the width reduction in the vertical stand :

$$\eta = \frac{W_0 - W_2}{W_0 - W_1}$$

This rate increases when the volume of bulges decreases.

When rolling steel, this ratio is near to 0.70. The same test with plasticine results in a ratio equal to only 0.50. The difference is due to the fact that the plasticine bulges are larger than the steel ones and therefore lead to an important re-widening.

Let us now consider the reasons for such a poor simulation. There can be three causes :
-   a poor simulation of friction
-   a lack of simulation of temperature gradients
-   a poor simulation of rheology.

In order to quantify the influence of friction, we carried out plasticine slab edging tests with three lubricants :
-   an "aquasonic" gel, with a Tresca friction coefficient $\bar{m} = 0.05$
-   talcum powder, with $\bar{m} = 0.7$, which is near to the friction of hot rolled steel
-   calcium carbonate powder, with $\bar{m} = 0.9$.

The results show a slight influence of friction, which cannot explain the differences observed between steel and plasticine (fig. 5).

The influence of temperature gradients was examined in carrying out slab width reductions on the steel pilot mill. The slab is allowed to cool during variable times and from different furnace exit temperatures prior to its rolling, in order to create different temperature gradients. The efficiency ratio varies only slightly with the temperature gradients and these variations seem too low to explain the differences between steel and plasticine (fig. 6).

The last possible explanation is the difference between the n and m coefficients of plasticine and steel.

The flow stresses of hot steels were measured by means of torsion tests, and the flow stresses of plasticine by means of cylinder compression tests.

The n and m coefficients of plasticine are equal to 0.01 and 0.05 and are lower than those of hot steel which on the average are equal to 0.25 and 0.17. So, when it is deformed, steel hardens much more than plasticine.

In the two cases of width reduction of slabs and slab edging the deformation remains localized at the contact of the rolls, where it first started. Plasticine, a material which is subject to only a slight hardening, is able to distribute this deformation to the center of the slab to only a limited degree. The deformation remains localized at the level of the rolls and leads to the formation of pronounced bulges. By contrast, steel hardens under deformation at the contact with the rolls, which prevents the formation of important bulges.

This explanation is confirmed by the finite element modelling of width reduction of slabs carried out by KOKADO, HATTA and TAKUDA which shows that the bulges do indeed diminish when n or m increases (3). We drew the same conclusions by utilizing the finite element model developped by the CEMEF research center (4).

We have also confirmed this explanation by another test : we have compared the flows in slab edging of two simulation materials : plasticine and a simulation wax manufactured by the VALENCIENNES University (5, 6). This wax has a rheology which is much different from that of plasticine. It presents an n coefficient which is positive at the beginning of the deformation, and negative afterwards, and a substantially higher m coefficient i.e., 0.38. The coefficients of friction for this wax and for plasticine are equal. We found that the bulges are far less pronounced for this wax than for plasticine (fig. 7).

We have proved that in these two cases of rolling the poor simulation of plasticine was due to its low n and m coefficients. Plasticine finds here its limits. In order to improve simulation, another model material must therefore be found, presenting n and m coefficients closer to those of steel.

6. Conclusion

By trials with plasticine, wax and hot steel simulation, by comparison with industrial rollings and by finite element simulations we succeeded in finding the reasons of the bad simulation of width reduction of slabs and slab edging by plasticine.

Another result of this research is the development of a better simulation material : a simulation wax.

But, despite its imperfections, plasticine is still very useful in the development of roll pass design and in the determination of flow laws during rolling. If a greater accuracy is required, then the study goes on the steel pilot mill, prior to the switching to the industrial process itself. In any case, the use of plasticine for simulation leads to large savings in terms of tests in an industrial mill and reduces the time required for overall development of new processes.

References

1. A.P. GREEN : "The use of plasticine models to simulate the plastic flow of metals". Phil. Mag.42 (1951).

2. T. WANHEIM : "The physical modelling of plastic working processes". Proc. of lst int. conf. on technology of plasticity. Tokyo 1984.

3. KOKADO, HATTA, TAKUDA : "Numerical analysis of the slab deformation in vertical rolling". Arch. Eisenhüttenwesen. 54 (1983). Nr 12 december.

4. C. BERTRAND, C. DAVID, A. BERN, P. MONTMITONNET, J.L. CHENOT, P. BUESSLER : "A three dimensional analysis of steady flow in hot forming processes. Application to hot flat rolling and hot shape rolling". Euromech Colloquium. August 29-31 1988. Sophia Antipolis, France

5. OUDIN, RAVALARD, ROMMENS : "On the contribution of waxes to the simulation of metal forming processes". 8th NAMRC. University of Missouri Rolla USA. 18-21 May 1980.

6. RAVALARD, OUDIN, RAVASSARD, ROMMENS : "Utilisation de cires équivalentes pour la simulation du forgeage à mi-chaud et chaud". 7ème Congrès F. de Mec. Sept 85. BORDEAUX (France) pp 1-6.

Figures are reprinted with permission from Physical Modeling of Metalworking Processes, edited by Emin Erman and S. L. Semiatin, 1988, The Metallurgical Society, 420 Commonwealth Drive, Warrendale, PA 15086.

Figure 1 - Plasticine universal mill

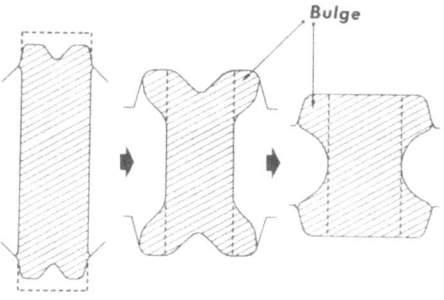

Figure 2 - Principle of slab edging

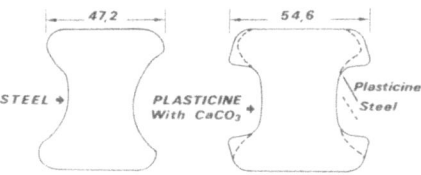

Figure 3 - Simulation of slab edging with plasticine

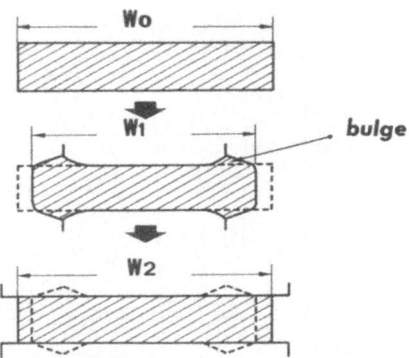

Figure 4 - Principle of the width reduction of slabs

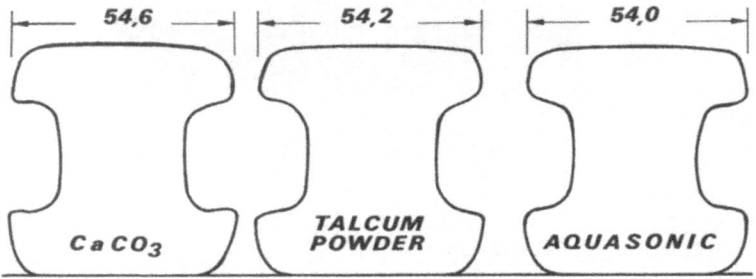

Figure 5 - Influence of friction in slab edging with plasticine

Figure 6 - Influence of temperature
gradients upon the edger efficiency

Figure 7 - Influence of the flow stress
upon the flow in slab edging

SIMULATION OF PROFILE AND FLATNESS IN THE COLD ROLLING OF STEEL STRIP

M. Malinen, A. Lankila* and A.S.Korhonen

Helsinki University of Technology, Dept of Materials Science and Engineering, Laboratory of Metal Working and Heat Treatment, Vuorimiehentie 2A, 02150 Espoo, Finland

* Rautaruukki Oy Hämeenlinna Works, 13300 Hämeenlinna, Finland

Abstract

A mathematical model for the simulation of profile and flatness in four-high rolling of steel was used to simulate the cold-rolling process in a 4-stand tandem mill. The calculated strip profiles were found to be within a few micrometres of the measured ones, and the calculated transfer stress distributions were qualitatively correct.

1. Introduction

The increasing competition in the steel industry demands better quality of the end product. In addition to accurate dimensions and mechanical properties, the customer also requires flat strip of uniform thickness. The control of flatness and transfer profile of the strip is therefore very important. The flatness defect and the non-uniformity of thickness across the strip width are connected to each other and both stem from the same two reasons. The first is that the deformation of the strip is not uniform across the width, and the other is that the roll gap during rolling is not uniform along the roll barrel.

A number of mathematical models have been developed for flatness control during the past fifteen years. Some of these models are very large and are able to describe the rolling process very accurately, though they usually demand long computing time on a large computer. Therefore they are not useful for industrial purposes, especially for cold rolling, where rolling speeds are high.

Mathematical models of profile and flatness of cold-rolled strip are usually based on modelling of the roll gap during rolling. Several parameters affect the transverse profile of the roll gap, some parameters varying slowly, others faster. These factors are presented in simplified form in fig. 1.

297

© 1988 by Kluwer Academic Publishers.

298

One of the programs which has been used in industry is the computer program **CROWN**, which has been developed in the MEFOS Metal Working Research Plant in Luleå, Sweden. The program was developed in co-operation between several Finnish and Swedish rolling mills. It is a program-package, which includes several numerical models. These include a rolling gap "friction-hill" model including heat generation, a strip temperature model, a model for work-roll temperatures and thermal expansion and a model for the elastic deformation of the rolls.

This program was used at the Rautaruukki Oy Hämeenlinna cold rolling mill in order to understand the mechanisms that change the thickness profile and flatness in strips during rolling in four-high mills.

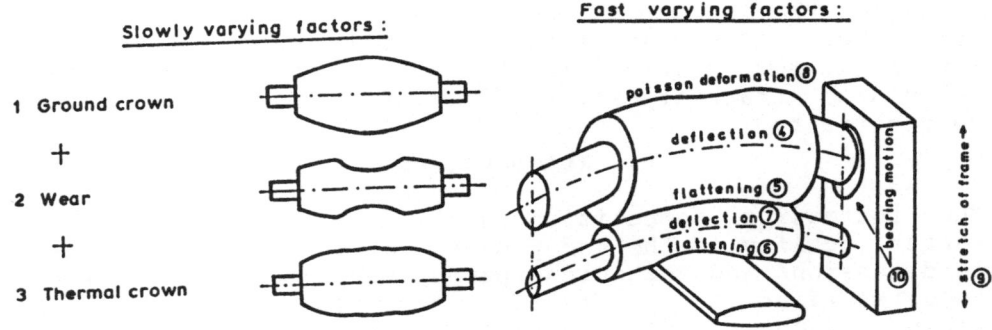

Fig. 1. Flatness model factors /1/

The purpose of this research was to compute the final profile and flatness when using a certain rolling schedule and to evaluate the influence of different parameters, control methods and equipment on profile and flatness. Finally, an aim was to define optimal values for the crown of the hot rolled strip, tandem mill settings and the shape of the work rolls.

For the verification of the simulated results, a series of industrial rolling experiments was carried out.

In the present paper the off-line version of **Crown**, **Crownoff** was used. There is also an on-line version, which is a simplified model for real-time process control.

In what follows the description of the theoretical
background of the model has been omitted, because it has
already been thoroughly described in references /2-8/

2. Crownoff modules

The program **CROWN** consists of five major program modules,
which are capable of working either alone or together.
Included are models for:

- Rolling mill elastic deformation
- Work piece temperature
- Friction energy and rolling force in cold rolling
- Work roll thermal expansion
- Work roll wear

Crown is used for the simulation of plate and strip rolling
in four-high mills. Because of its vast application area,
the program itself does not contain any mill-dependent empirical models. Almost all input variables are real physical variables or constants, that are easy to measure or
estimate. Some variables, however has to be verified by
calculations against measured data. In its present form
Crown can be used in any four-high rolling mill after some
experiments.
    In the elastic deformation module /9-11/, the rolls
are divided into segments. The force acting on each segment
causes flattening and bending of the rolls. The calculation
method is applicable to all four-high mills and can handle
rolling conditions where the rolling load is not uniformly
distributed in the roll gap, or the roll profile is affected either by roll-bending methods or by arbitrary crowning
of the rolls.
    In the thermal expansion calculation a two
dimensional temperature distribution is calculated by the
finite-difference method. Thermal coefficients necessary
for the calculation must be derived from actual measurements in the rolling mill. The actual thermal expansion is
calculated using two dimensional elasticity theory.
    During rolling, the temperature of the strip chan-
ges. This temperature change is calculated by using the
one-dimensional finite element method, which considers the
heat flow only in the thickness direction.

Rolling force and friction energy calculations are based on a model that uses a numerical slab method to calculate von Karmans equation. The "friction-hill" stress distribution is used to calculate the roll flattening. The rolling force and friction energy are calculated by summing up the result for each slab. This model is the only one which is used solely for cold rolling.

Work roll wear is calculated by an equation that integrates rolling force distribution over the length and width of the workpiece. The wear coefficient must be determined empirically in the rolling-mill.

The roll gap defines the thickness profile and the change of the relative thickness profile is used to calculate the transverse distribution of residual stresses. The lateral material flow is taken into consideration by a correction factor ξ. The stress distribution can be fed back to influence the thickness profile.

A block diagram of **Crownoff** modules is presented in figure 2.

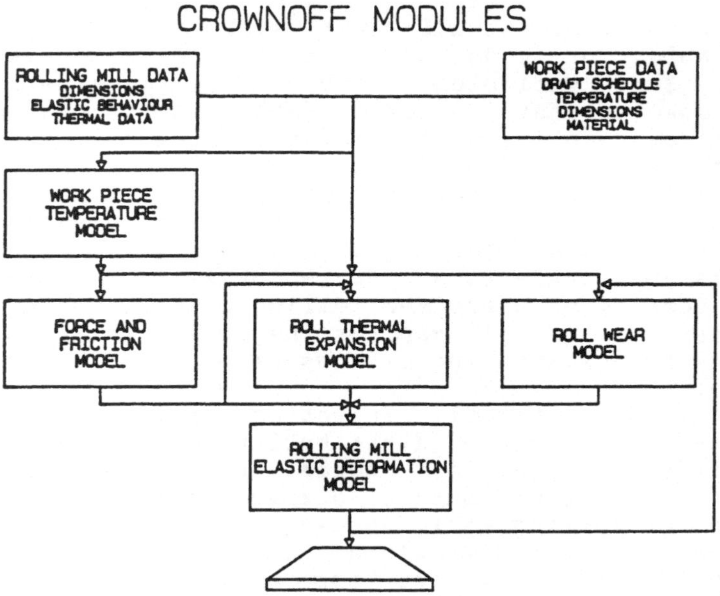

CROWNOFF MODULES

Figure 2 Block diagram: Crownoff modules /4/

3. Experimental procedure

The experiments were divided into two phases. In the first phase correct input parameters for **Crownoff** were derived from earlier rolling experiments. The input parameters included rolling mill elastic constants(E and ν for backup and work rolls etc.), thermal coefficients for water cooling and heat transfer from strip to rolls, and material parameters like shape-disturbing coefficient ξ. The work was carried out by measuring hot-band profiles as an input for the program in addition to tandem-mill parameters from the mill on-line technical reports. The calculated results were then compared mainly with measured strip output profiles. At this stage of the work, no data on strip transfer stress distribution was available. The simulations were run on two computers, on a Vax8650 and on a μVax2000. The program has been implemented on several different computers besides those of the Vax-family. The average CPU time used for a series of four coils rolled was about 40 minutes on a μVax2000.

In the second phase a major industrial rolling experiment with 29 coils was carried out. The results from **Crownoff** were used to evaluate tandem mill settings and flatness corrections. The measured hot-band profiles were fed to **Crownoff** with approximate rolling force and other tandem parameters. The calculated profiles were then compared with measured cold-band profiles. Transverse stress distributions were compared both with measurements using the on-line Stressometer and with visual observations on a hot-dip galvanizing line.

4. Results and discussion

Results from the simulations and rolling experiments are presented in figures 3 and 4. The calculated results show reasonable agreement with the actual measured data. The calculated strip crown was within a few micrometres of the actual crown. The form of the profile at the strip edge is not calculated as precisely as in the centre region. Also the numerical values of the transverse residual stresses are not correct. However, the stress pattern is correct and the visual analysis of the strip reveals flatness defects according to the calculated strip tensions. The preciseness of the calculations could be improved by adjusting the shape change coefficient ξ in the transverse direction. In practice this means that the transverse material flow is possible near the edge of the strip.

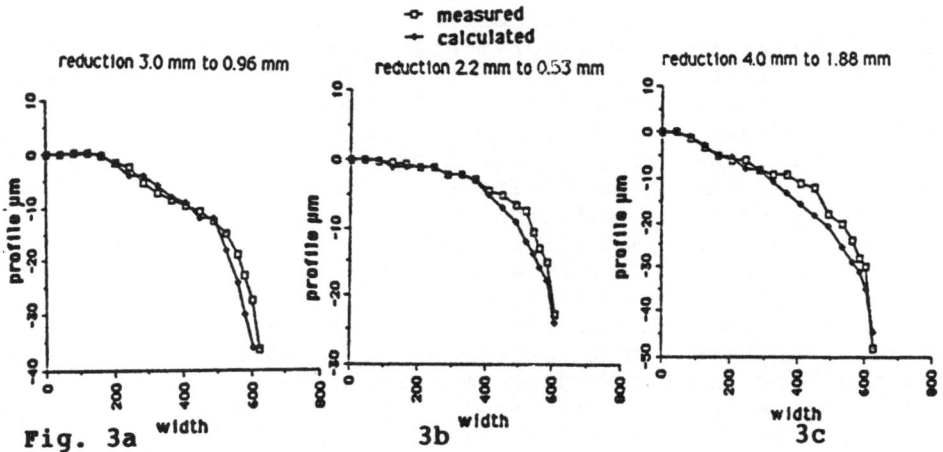

Fig.3.a,b,c Calculated and measured roll exit thickness
profiles of strips with different reductions.

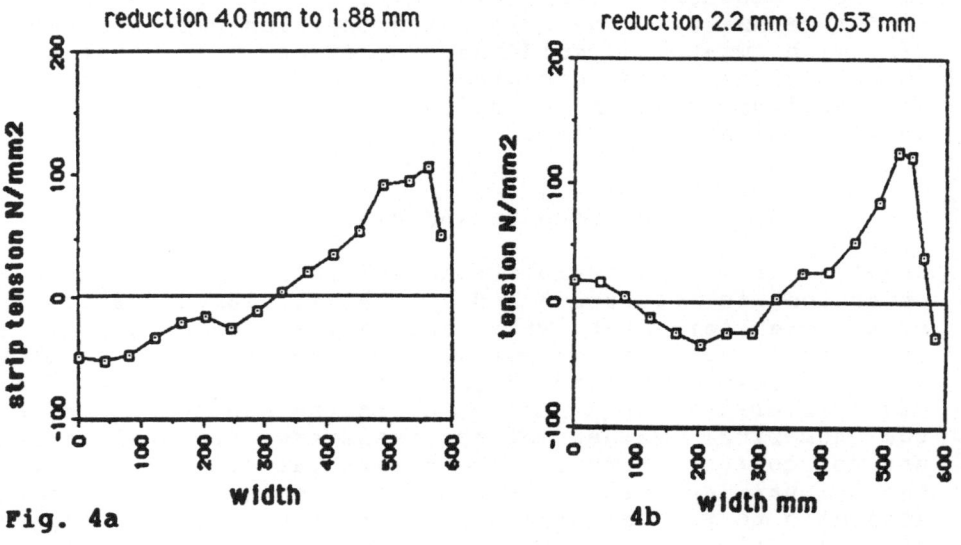

Fig. 4a,b Calculated strip transverse residual stresses.

The visually observed strips shapes of the strips in Fig.
4 were long middle for fig. 4a and long middle and long
edge for fig 4b.

References

/1/   Hollander, F. Process technology-The start of integrated process control in rolling mills, Process control in the steel industry, Ed. G.Carlsson,Mefos,Luleå 1986

/2/   Jonsson N-G and Mäntylä P., On-line control system for profile, shape and temperature in 4-high mills, Proceedings of the 27th Mechanical Working and Steel Processing Conference-Vol XXIII, Oct 27-29,1985

/3/   Wiklund, O., et al., Simulation of crown, profile and flatness of cold rolled strip by merging several physically based computer models, Proceedings of the 4th international Steel rolling Conference, Deauville, France 1.-3.6.1987,vol 2 E9

/4/   Levén, J., et al., Thermal expansion of rolls, Proceedings of the conference Mathematical models for metals and materials applications, Sutton Coldfield 12.-14.10.1987, The Institute of Metals, no.36

/5/   Johansson K. and Jonsson N-G, Handbok för användandet av programmet Crownoff, Mefos/BTF, Luleå 1984, BTF 56,(in Swedish)

/6/   Levén, J., Profil och planhet, Mefos/BTF Internal report BTF 85009, Luleå 1985 (In Swedish)

/7/   Wiklund, O., Crownoff för kallvalsning-komplettering och modifiering av dataprogrammet, Mefos/BTF Internal report BTF 84024, Luleå 1986

/8/   Wiklund, O., Crownoff för kallvalsning-slutrapport Mefos/BTF Internal report BTF 86026, Luleå 1986 (in Swedish)

/9/   Berger,B. et al, Die elastische Verformung der Walzen von Vierwalzengerüsten, Arch.Eisenhüttenwes. 47 (1976) nr. 6 s.351-356 (in German)

304

/10/   Pawelski,O et al,. A mathematical model with a
        combined analytical and numerical approach for
        predicting the thickness profile of a cold-rolled
        strip, Steel Research 56 (1985) No.6 s.327-331

/11/   Berger, B. et al, Die Beeinflussung des
        Dickenprofils von Bändern und Blechen durch
        Walzenbiegeeinrichtungen, Stahl u. Eisen 96
        (1976) Nr.8 s. 377-381 (in German)

APPLICATION OF A MATHEMATICAL MODEL
FOR THE COLD ROLLING PROCESS ON SIX-HIGH MILLS

O. PAWELSKI, W. RASP, J. RIECKMAN

MAX-PLANCK-INSTITUT.

MAX-PLANCK STRASSE 1.D 4000 DÜSSELDORF-GFR

Since the seventies, six-high mill technology has been
introduced in several hot and cold rolling plants. Its main
advantage in comparison to the classical four-high mill is
its large range of mechanical shape control. Thus, the effective roll gap is significantly influenced by work and intermediate roll bending and intermediate roll shifting. In
this way a wide range of strip widths and materials can be
rolled without fitting the crowns by roll changing. Another
advantage is the fairly small work roll diameter that reduces roll force and torque and considerably decreases the
edge drop effect. It allows the rolling of harder material
and the achievement of very thin final strip thickness.
Several papers published by manufacturers and suppliers deal
with the six-high mill technology /1,2,3/. These results are
mostly based on practical experience.

In order to gain a basic understanding of the mechanical behaviour, a mathematical model has been developed at the
Max-Planck-Institut für Eisenforschung in Düsseldorf. This
model predicts the influence of elastic and plastic deformations on strip profile in six-high, as well as in two- and
four-high cold rolling mills. Even a CVC-type mill and asymmetrical roll conditions can be considered. Calculations can
be performed for numerous passes. The outline of the model
has been previously published /4,5,6/.

1. Mathematical Model

The influence of intermediate roll shifting on the mill and
strip deformations is shown in /6/. The most recent
simulations predict the flatness of the out-going strip
according to the chosen mechanical mill adjustments. Due to
the complex elastic behaviour of the six-high mill, it is
difficult and of course not sufficient by experience, to
satisfy the increasing requirements of strip quality, i.e.
strip flatness and profile. Alone, an accurate understanding
of the mechanical mechanisms is a necessary prerequisite for
an optimization of the rolling process. The determination of
the general dependence between the mill adjustments and the
strip profile and flatness requires complex calculations.
Short computing time qualifies the model for this purpose.

J. L. Chenot and E. Oñate (eds.), Modelling of Metal Forming Processes, 305–312.
© 1988 by Kluwer Academic Publishers.

Calculations show the correlations between the setting of
the regulating units and strip flatness.

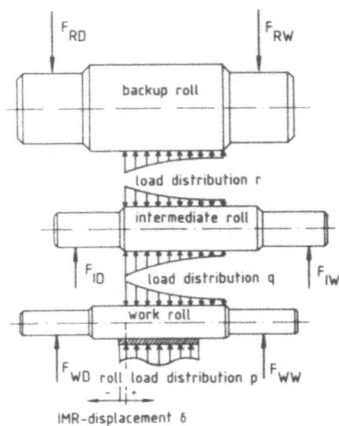

Fig. 1
Decomposition of the
loadings in the upper
half of a six-high mill

Fig. 1 illustrates the forces and load distributions taken
into account by the model. The size of the roll load distribution p depends on the interaction between the forces due
to plastic deformation of the strip and the elastic deformation of the work roll in the contact area. The roll force is
calculated by either the v. Kármán differential equation or
the theory of Bland, Ford, and Ellis. The load distributions
q and r are determined by the elastic deformations in the
contact area between two neighbouring rolls. The roll force
F_B acting on the backup roll necks is subdivided into driven
(F_{RD}) and work side (F_{RW}) in the case of asymmetrical rolling. Possible re-bending forces at intermediate or work roll
are described by the forces F_I and F_W, respectively.

The characterization of the intermediate roll position in
this mathematical model is demonstrated in Fig. 1. The left
edge of the strip is the origin of displacement δ. This displacement of the intermediate roll (IMR) in the upper half
of the mill is defined as positive if the left edge of the
roll barrel is inside the strip width, negative if it is
outside. If the displacement is equal to zero, the left roll
barrel edge is exactly above the left strip edge. Normally
the intermediate rolls are set in such a way that they are
symmetric to the longitudinal axis of the strip. This means
that the intermediate rolls are shifted in opposite axial
directions and both displacements are of the same magnitude.

2. Prediction of Strip Flatness

Strip flatness problems may occur if the rolling process
induces residual stresses in the out-going strip caused by
different elongation of the strip fibres. If these stresses
exceed the buckling stress of the strip, shape defects in
the form of waves will be seen. Due to the large ratio of
strip width to thickness, material flow in the width direction can be neglected across large parts of the strip width.
Plane strain can be assumed. The strains in the length direction are of the same magnitude as those in the thickness
direction. Uniformity in length of all in-going fibres will
be achieved for the out-going fibres if the thickness reduction is uniform across the width. This is the restriction
for positioning the mechanical regulating units when flat
strip is required.

As mentioned above, the out-going strip profile is calculated in the mathematical model by considering the elastic behaviour of the mill and the characteristics of the in-going
strip. According to the incompressibility condition, the
fibre lengths are determined from the in- and out-going
thickness distribution. For describing strip flatness, not
the absolute lengths but rather the length deviations ΔL
between the fibres are determinative. Divided by the length
of the fibre at strip centre L, the relative length
deviations $\Delta L/L$ are the criterion for strip flatness.
Simulations show the influence of different regulating units
on the $\Delta L/L$ distribution across the strip width. For easier
handling, it is advantageous to describe the distributions
by characteristic values.

$$\Delta L/L = a_2 \, \xi^2 + a_4 \, \xi^4 \qquad\qquad -1 \leqq \xi \leqq 1$$

Two coefficients a_2 and a_4 are obtained by a regression
analysis done for a fourth order parabola by using the least
squares method. $\xi = -1$ represents the left, $\xi = 1$ is the right
strip edge. Both coefficients can be displayed in a flatness
chart, Fig. 2. These points joined by characteristic lines,
allow the consequences of mill adjustment settings on strip
flatness to be estimated. For some ratios of the
characteristic values, the qualitative graphs of the $\Delta L/L$
distributions are displayed around the chart. To get
quantitative information, the chart is completed by boundary
lines. Inside these tolerance zones the difference between
the maximum and minimum value of the $\Delta L/L$ distribution
across the strip width is less than the tolerance value
according to the zone. As usual the length deviations are
specified in I-units (1 I-unit = 10 μm/m = 0.001%). It is
possible to define a critical tolerance that corresponds to
the buckling stress.

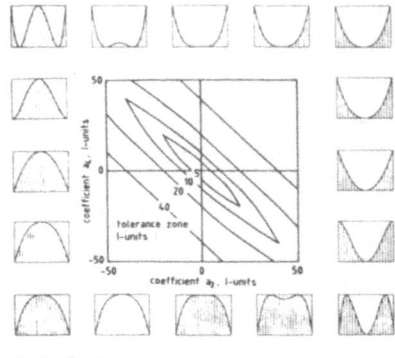

Fig. 2
Flatness chart

3. Flatness Control by Different Mill Adjustments

The simulations done for one pass on a six-high mill
(1500 mm x 300/500/1300 mm) result in the following data.
The in-going 1200 mm wide and 2.0 mm thick strip of mild
steel is reduced to 1.5 mm. The thickness distribution is
described by a second order parabola and an exponential
equation. The back tension stress is 50 N/mm^2, the forward
tension stress is 100 N/mm^2. A coefficient of friction of
0.05 is assumed. This corresponds with the calculations done
in /6/.

The Figs. 3, 4, and 5 show how the out-going strip flatness
depends on each mill adjustment separately. The work roll
bending system (range of bending force -100 kN to 300 kN)
mainly influences the coefficient a_4, whereas the intermediate roll bending system (range of bending force -200 kN to
500 kN) changes the coefficient a_2 of the $\Delta L/L$ distribution.
Due to the small work roll diameter, the deflection of the
roll caused by the work roll bending force is not trans-
mitted towards the strip centre. The $\Delta L/L$ changes are concentrated near the edges of the strip. The bending of the
intermediate roll influences the $\Delta L/L$ distribution across
the whole strip because of the larger diameter and distance
to the roll gap. According to Figs. 3 and 4 the change in
I-units of the $\Delta L/L$-deviations due to the change of bending
force, i.e. the efficiency of the mechanism, becomes
slightly smaller for both cases, when the absolute bending
force increases. The smaller diameter and the nearness to
the roll gap are the reasons for the more effective change
of the I-units, due to work roll bending rather than due to
intermediate roll bending (average value for work roll bending 0.35 I-units/kN, intermediate roll bending 0.08
I-units/kN).

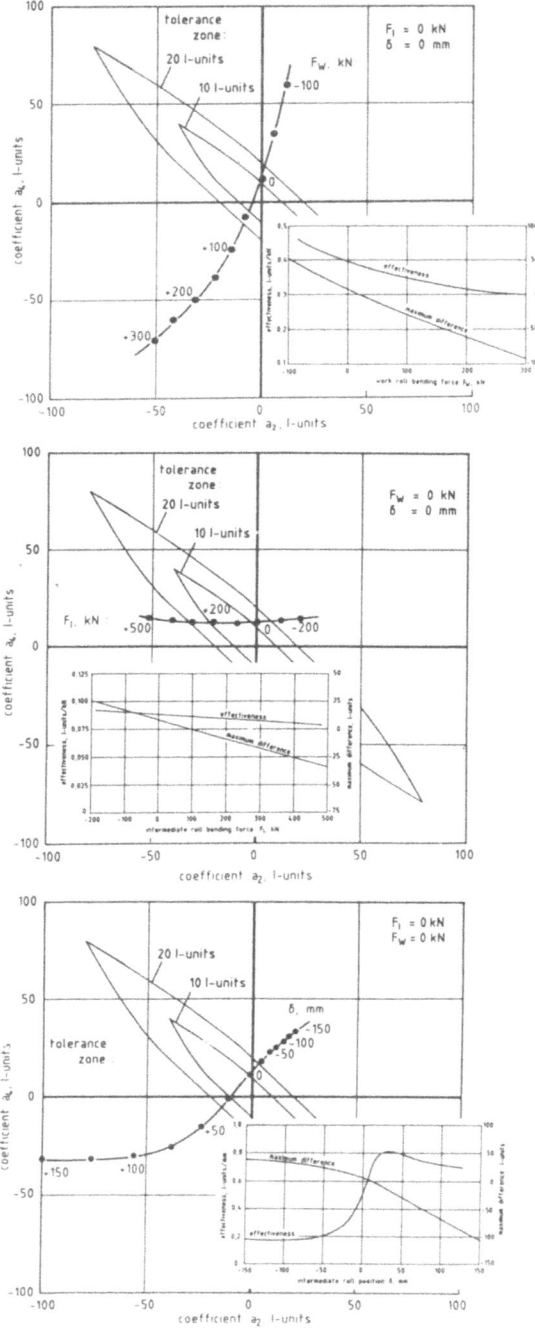

Fig. 3
Influence of work
roll bending force
F_W on strip flat-
ness

Fig. 4
Influence of inter-
mediate roll bend-
ing force F_I on
strip flatness

Fig. 5
Influence of inter-
mediate roll shift-
ing δ on strip flatness

310

The flatness chart in Fig. 5 illustrates the characteristic
line for intermediate roll shifting. The intermediate roll
is shifted from the position described by δ = -150 mm (un-
shifted roll) to δ = 150 mm. In the range from δ = -150 mm
to δ = 0 mm both coefficients change in the same way.
Further shifting beyond the δ = 0 mm position, where the intermediate roll barrel edge is exactly above the strip edge,
makes the coefficient a_2 more and more dominant. Beside the
δ = 0 mm position the characteristics of the intermediate
roll shifting are different. With non-negative δ values, the
change in I-units due to intermediate roll shifting rises
dramatically. The efficiency of shifting increases by a
factor of four. Because of this non-uniform characteristic
it seems to be disadvantageous to use the intermediate roll
shifting system for on-line flatness control, especially
when the intermediate roll position is nearby δ = 0 mm.
Additionally, the hydraulic bending systems for on-line control should be preferred because of the low speed of intermediate roll shifting (4 mm/s /3/). According to Figs. 3 and
4 the characteristic lines of the bending systems are nearly
perpendicular to each other. By using both systems, the
lines subtend a rectangular-like panel which can be
positioned in the chart by presetting the intermediate roll,
Fig. 6.

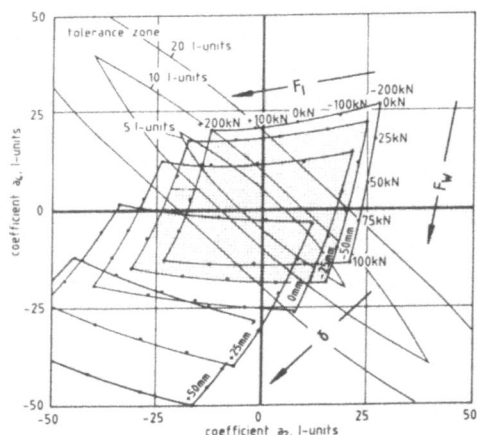

Fig. 6
Flatness control by
three regulating units:
WR bending force F_W
IMR bending force F_I
IMR roll shifting δ

The aim is to gain a maximum sectional area with the desired
tolerance zone by choosing the optimal intermediate roll
position. The edge length of the panel depends on the
available range of the bending forces.

4. Edge Drop Minimization by Optimal Mill Setting

According to Fig. 6, there are several combinations which
lead to flat strip for the given roll conditions. It is investigated how these combinations influence the edge drop of
the out-going strip. With the assumption that the edge is
mainly influenced by work roll bending and intermediate roll
shifting the characteristic lines for these mill adjustments
are given in Fig. 7. The marked combinations satisfy the requirement for flat strip. The deviations of the $\Delta L/L$ distributions are less than five I-units. The edge drop of the
strip is characterized by an edge drop value defined as the
product of the width and height of the thickness
disturbance. According to Fig. 7 the marked combinations in
Fig. 8 result in significantly different edge drop values.
They can be made smaller by using large work roll re-bending
forces in combination with less shifted intermediate rolls.

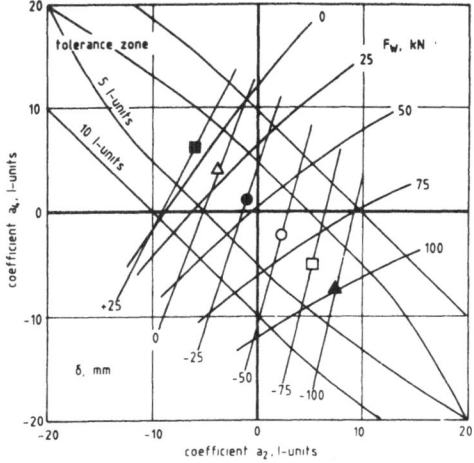

Fig. 7
Flatness control by
two regulating units:
WR bending force F_W
IMR roll shifting δ^W

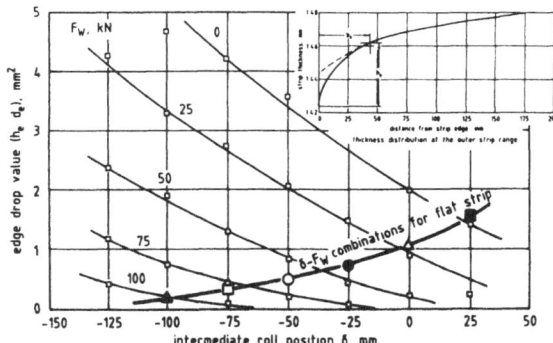

Fig. 8
Edge drop accord-
ing to mill sett-
ing

5. Summary

Calculations show that in a six-high mill the intermediate roll bending force mainly influences the coefficient a_2, the work roll bending force the coefficient a_4 of the $\Delta L/L^2$ distribution. The characteristic of the intermediate roll shifting is not uniform across the shifting range. The different effects of the bending systems and the additional intermediate roll shifting explain the enlarged range of flatness control in six-high mills compared to conventional four-high mills.

A flatness chart indicates those combinations that lead to strip flatness within a given range of tolerance. Further investigations show that these combinations can be optimized with respect to the edge drop. Thus, an optimal setting of the regulating units can reduce the thickness deviation across the strip width without disturbing the strip flatness.

6. References

/1/ Kajiwara, T. et al.: Recent Trends and New Applications of HC-Mill in Cold Rolling. Hitachi Review 28(1979)5, p. 233-238.

/2/ Kajiwara, T., N. Fujino, H. Nishi, and S. Shida: The Hitachi HC-mill - a breakthrough in strip rolling. Iron and Steel International 49(1976)4, p. 247-255.

/3/ Imai, I., S. Hirai, T. Furuya, M. Morimoto, and K. Nakajima: New 6-high mill (NHM) and automatic shape control system in cold rolling. Proc. Intern. Conf. on Steel Rolling, Tokyo 1980, p. 807-818.

/4/ Pawelski, O., and H. Teutsch: A Mathematical Model for Computing the Distribution of Loads and Thickness in the Width Direction of a Strip Rolled in Four-high Cold-rolling Mills. Engineering Fracture Mechanics 21(1985)4, p. 853-859.

/5/ Pawelski, O., W. Rasp, and H. Teutsch: A Mathematical Model with a Combined Analytical and Numerical Approach for Predicting the Thickness Profile of Cold-rolled Strip. Steel Res. 56(1985)6, p. 327-331.

/6/ Pawelski, O., W. Rasp, and J. Rieckmann: A Mathematical Model for Predicting the Influence of Elastic and Plastic Deformations on Strip Profile in Six-high Cold Rolling. Proc. Intern. Steel Rolling Conf., Deauville 1987, p. E.3.1-6.

The authors wish to thank the Deutsche Forschungsgemeinschaft (DFG) for providing the financial help to carry out this investigation.

ANALYSIS OF PLANE STRAIN COLD ROLLING USING A FLOW FUNCTION AND THE WEIGHTED RESIDUALS METHOD

M.J.M. Barata Marques and Paulo A.F. Martins
Instituto Superior Tecnico, Lisboa, Portugal

Summary

The upper bound approach used in cold rolling is based on a flow function built from the prescribed streamlines in the deforming zone. Differentiating the flow function, the velocity and strain rate fields are obtained. The roll torque is calculated from the upper bound theorem. The stress field in the deformation region is evaluated by combining the results derived from the stream function with the plasticity equations and the boundary conditions by the weighted residuals method. The roll separating load and the roll pressure distribution are then obtained. The results are presented for aluminium and steel for several geometrical and friction conditions. The method has proved to be capable of describing plane strain rolling. The results agree with experimental data and indicate a large advantage over the finite element method in computing time and computer resources.

1. Introduction

The development of theoretical methods to be applied to metalforming has provided useful information about the mechanics of the processes. In the last few years rolling has been extensively analysed mainly through the upper bound technique and the finite element method [1,2]. With the FEM, the velocity, strain rate, strain and stress fields can be obtained in the plastically deformed material. Although this method is very powerful it requires large amount of computer capabilities. The upper bound technique is based in a kinematically admissible velocity field which has to be assumed. The strain rate and strain fields can be derived and the upper bound load is calculated. The stresses can only be predicted if a numerical method is added to solve the Prandtl Reuss equations, taking into account the boundary conditions.

In this work an upper bound solution of plane strain rolling based in a flow function is proposed. The model used in the approach considers a rigid plastic material and assumes the shape of the streamlines. Coupling the upper bound technique with the weighted residuals method, the stress field in the deforming region are calcula-

313

314

ted. This approach can provide information about the mechanics of plane strain cold rolling and evaluate the strain rate, strain and stress fields and consequently the roll separation force, the roll torque and the roll pressure distribution. To check the validity of the assumptions made and to assess the advantages of the method the results obtained are compared with those determined experimentaly [3] and by the finite element method found in the literature [1].

2. Flow Function

A method of analysis based in the flow function has been developed by Marques [4] and has been applied to tube extrusion. In this model a flow function satisfying the condition of volume constancy is built from the co-ordinates of the points of the streamlines.

In a reference system of rectangular co-ordinates (r,z) a flow function is selected in the form:

$$\psi = \sum_m \sum_n A_{mn} \, r^{2m+2} \, z^{n+2} \tag{1}$$

where A_{mn} are unknown parameters. This flow function satisfies the following conditions:
- The velocity component v_r is symmetrical across the z-axis.
- The deformation boundaries along the rolls and the z-axis are streamlines.

The value of the flow function ψ along a streamline is calculated from the initial conditions in the undeformed region:

$$\psi_i = r_i b v_o \tag{2}$$

where v_o is the entrance velocity, r_i is the r-co-ordinate of the streamline and b is the stock width. Fig. 1 shows schematically the rolling process and a streamline used to evaluate the flow function.

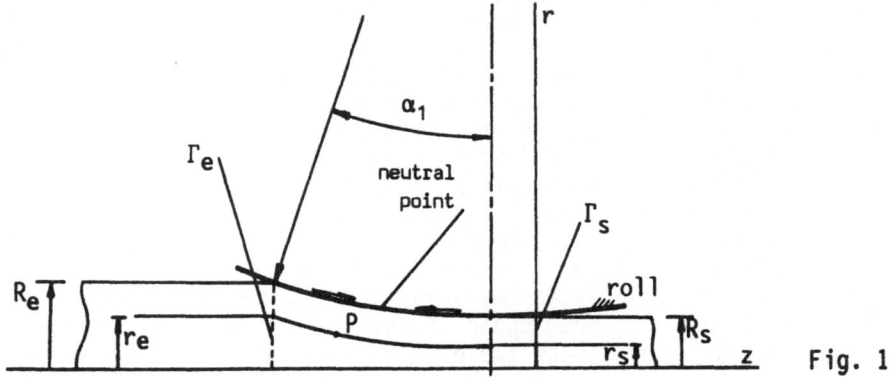

Fig. 1

From the incompressibility condition the components of velocity,

v_r and v_z , can be obtained by:

$$v_r = \frac{1}{b} \frac{\partial \psi}{\partial z} \qquad\qquad v_z = - \frac{1}{b} \frac{\partial \psi}{\partial r} \tag{3}$$

Differentiating the velocity components with respect to the variables r and z , the strain rate are calculated, and the total amount of plastic strain is evaluated by integrating the effective strain rate along the streamline.

The deformation takes place between two planes Γ_e and Γ_s normal to the z-axis (Fig. 1). The flow pattern in the plastic deformation region was assumed to be described by a family of ellipses (4) centered in the roll axis.

$$\frac{(z - C_z)^2}{a^2} + \frac{(r - C_r)^2}{b^2} = 1 \tag{4}$$

where a and b are parameters, and (C_r, C_z) is the center of the roll. The representation of the plastic flow by those curves seems to be adequate to the problem as equation (4), considers the z-axis and the plastic deformation region/roll interface as streamlines. All the flow lines are normal to the exit boundary.

The free parameters in equation (1) were determined by the least square method. The number of terms in the flow function was 36. This estimate provides for each streamline a good agreement with the initial flow conditions. The expressions for velocities are:

$$v_r = -(n+2) \sum_m \sum_n A_{mn} \; r^{2m+2} \; z^{n+1}$$
$$v_z = (2m+2) \sum_m \sum_n A_{mn} \; r^{2m+1} \; z^{n+2} \tag{5}$$

and for strain rates are:

$$\dot{\varepsilon}_r = -(n+2)(2m+2) \sum_m \sum_n A_{mn} \; r^{2m+1} \; z^{n+1}$$
$$\dot{\varepsilon}_z = -(n+2)(2m+2) \sum_m \sum_n A_{mn} \; r^{2m+1} \; z^{n+1} \tag{6}$$
$$\dot{\varepsilon}_{rz} = 0.5 \sum_m \sum_n A_{mn} \left| (-n^2-3n-2) \; r^{2m+2} \; z^n + (4m^2+6m+2) \; r^{2m} \; z^{n+2} \right|$$

3. Upper Bound Approach

The rolling torque can be calculated by the upper bound theorem:

$$\pi = \int_V \sigma \dot{\varepsilon} \; dV + \int_\Gamma \tau \; |\Delta v| \; dS - \int_{\Gamma_t} T_i v_i \; dS_{\Gamma_t} \tag{7}$$

where π is the power of the process and Δv and v_i are the tangential velocity discontinuities at the boundaries and the velocity of points at surface Γ_t where body tractions T_i are applied respectively. The surface Γ includes the roll/stock interface Γ_R and the entry boundary Γ_e . The velocity of each point of the plastic deformation region is calculated through the flow function. Besides the volume constancy a kinematically admissible velocity field should ensure that the normal components of velocity across discontinuity surfaces is continuous. In the present model, as the entry boundary is normal to the z-axis, the tangential velocity discontinuity is $\Delta v_{\Gamma e} = v_{\Gamma e}$. On this discontinuity surface the shear stress is assumed to be the shear yield stress of the material. Along the exit boundary there is no velocity discontinuity due to the shape of the streamlines.

For the materials used in this work the effective stress/strain curve was represented by the equation:

$$\bar{\sigma} = Y_0 \ (1 + \bar{\varepsilon}/b)^n \tag{8}$$

where Y_0 is the yield stress and n and b are constants (Table 2). Friction in the stock/roll interface was taken into account assuming a friction factor m : $\tau = m\bar{\sigma}/\sqrt{3}$.

4. Weighted Residuals Method

To obtain a complete solution for the rolling process the stress field has to be calculated, since the velocity, strain rate and strain fields were evaluated by the upper bound approach. To solve the equilibrium equations for plane strain rolling a stress function ϕ (r,z) is defined such that:

$$\sigma_r = \frac{\partial^2 \phi}{\partial z^2} \qquad \sigma_z = \frac{\partial^2 \phi}{\partial r^2} \qquad \tau_{rz} = -\frac{\partial^2 \phi}{\partial r \partial z} \tag{9}$$

The stresses should also satisfy the governing (Prandtl-Reuss) equations, and the following boundary conditions:

on the entry and exit boundaries, $\sigma_z = 0$

and on the roll stock interface, $(\sigma_r - \sigma_z) \dfrac{\sin(2\alpha)}{2} + \tau_{rz}\cos(2\alpha) = \pm m\bar{\sigma}/\sqrt{3}$ $\tag{10}$

The method of weighted residuals consists of proposing a solution in a form of a set of known trial functions as:

$$\bar{\phi} = \phi_0 + \sum_{j=1}^{n} C_j \phi_i \tag{11}$$

such that the n parameters C_j are obtained by minimizing the difference between the approximate $\bar{\phi}$ and exact ϕ solution. In the least square method used in this work, the integral of the square of the residual is minimized with respect to the free parameters. A full description of the application of this method to plasticity problems can be found in [5].

Taking into account the symmetry conditions of the problem the following stress function was selected, (with 25 terms):

$$\overline{\Phi} = \sum_m \sum_n C_{mn} \ r^{2m} \ z^{n+2} \tag{12}$$

Applying the expressions (9), the estimates for the stresses are:

$$\sigma_r = (n+2)(n+1) \sum_m \sum_n C_{mn} \ r^{2m} \ z^n$$

$$\sigma_z = (2m)(2m-1) \sum_m \sum_n C_{mn} \ r^{2m-2} \ z^{n+2} \tag{13}$$

$$\sigma_{rz} = -(2m)(n+2) \sum_m \sum_n C_{mn} \ r^{2m-1} \ z^{n+1} \qquad \sigma_\omega = 0.5 \ (\sigma_r + \sigma_z)$$

None of the boundary conditions is incorporated in the trial functions so that separate error functions have to be constructed to take them into account.

5. Computer Programme

The calculations were performed assuming rigid plastic materials, plane strain conditions and rigid rolls. The deforming zone was discretized in a mesh of 722 triangular elements, in order to evaluate the terms of equation (7), and to obtain the stresses by the weighted residuals. To compare the results with those determined experimentally and by the FEM the following cases were taken from the literature [1,3]:

Table 1

Case	w_0/h_0	h_0 (mm)	h_1 (mm)	reduc. R(%)	roll diam. (mm)	roll vel. (m/sec)	friction m
1	12.15	6.274	5.385	14.17	158.75	0.16	0.25
2	12.15	6.274	4.902	21.86	158.75	0.16	0.25
3	12.15	6.274	4.445	29.40	158.75	0.16	0.25
4	12.15	6.274	4.115	34.41	158.75	0.16	0.25
5	37.04	2.057	1.727	16.05	158.75	0.16	0.25
6	37.04	2.057	1.588	22.83	158.75	0.16	0.25
7	37.04	2.057	1.499	27.16	158.75	0.16	0.25
8	37.50	2.030	1.346	33.75	158.75	0.16	0.25
9	100.00	1.000	0.920	8.00	130.00	0.25	0.20
10	100.00	1.000	0.840	16.00	130.00	0.25	0.20
11	100.00	1.000	0.760	24.00	130.00	0.25	0.20
12	100.00	1.000	0.680	32.00	130.00	0.25	0.20
13	200.00	0.500	0.460	8.00	130.00	0.25	0.20
14	200.00	0.500	0.420	16.00	130.00	0.25	0.20
15	200.00	0.500	0.380	24.00	130.00	0.25	0.20
16	200.00	0.500	0.340	32.00	130.00	0.25	0.20

Table 2

Cases	material	Y_o (N /mm)	n	b
1 - 8	aluminium	50.3	0.260	0.050
9 - 12	steel	324.0	0.295	0.052
13 - 16	steel	358.0	0.300	0.044

For each case the programme provides the velocity, strain rate, strain and stress fields as well as the rolling torque, the normal pressure distribution along the roll surface and the roll separation load. The position of the neutral point is determined by the condition that the power required by the process π , should be equal to the power provided by the rolls, $\pi^* = T\Omega$ where T is the rolling torque per unit of width and Ω is the angular velocity of the roll.

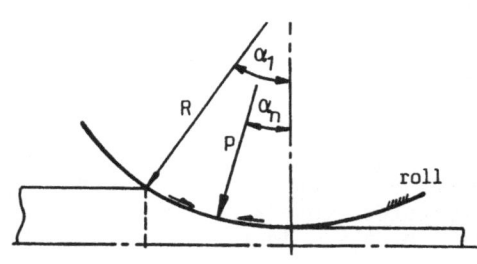

As the neutral point is placed somewhere in the contact arc (Fig. 2), the rolling torque is given by:

roll

$$T = \int_{\alpha_n}^{\alpha_1} \frac{-\alpha_n}{m} \frac{\overline{\sigma}}{\sqrt{3}} R^2 d\alpha - \int_0^{\alpha_n} \frac{\overline{\sigma}}{m} \frac{}{\sqrt{3}} R^2 d\alpha \quad (14)$$

Fig. 2

From the above conditions to determine the neutral point results an iterative process of calculation, since the evaluation of π^*, requires the previous determination of the torque T and π depends on the position of the neutral point.For the majority of the studied cases only five iterations were needed to obtain the convergence of the process (2 minutes of CPU time in a VAX 11-750 computer). The roll separation force per unit of width is calculated by:

$$P = R \int_0^{\alpha_1} \sigma_r d\alpha \qquad (15)$$

6. Results and Discussion

The results obtained with the approach described in the previous sections are compared with experimental data [3], and with those determined by FEM [3]. Figs. 2 and Figs. 3 show the variation of roll separation load and torque per unit of width respectively, for the cases summarized in Table 1. It is clear that the upper bound and the finite elements results are very similar. Figures also illustrate that the computed load is smaller than the experimental data, while the computed torque for aluminium is larger than the measured values. The upper bound values for steel are lower than the experimental ones.

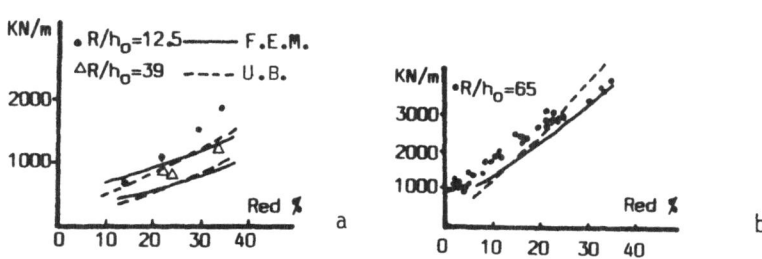

Fig. 2 - Roll Separation Load per unit of width.
    2a) Aluminium cases 1 - 8   2b) Steel cases 9 - 12

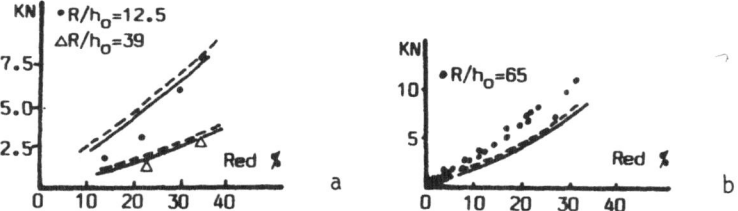

Fig. 3 - Torque per unit of width
    3a) Aluminium cases 1 - 8   3b) Steel cases 9 - 12

Fig. 4 illustrates the rolling pressure
variation for Aluminium case 1.
It is clear that the rolling pressure
distribution follows the pattern of the
"friction-hill". Both, the upper bound
and the finite element results do not
fit closely the experimental data.

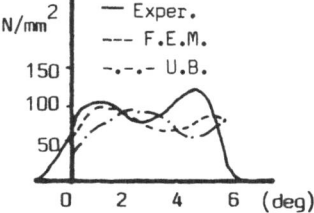

The discrepancies between the theoretical and experimental results are
explained on the grounds of the assumption of rigid rolls,as flattening
was not taken into account. Kobayashi and Li [1], pointed out that
"flattening contributes to an increase in the contact pressure and hence
the roll separating load over the values obtained under the rigid roll
condition". Lenard and Hwm [3],improved the FEM formulation introducing
the roll flattening and obtained closer results to the experimental measurements. Figs. 5, illustrate the distributions for Aluminium case 4.
The velocity field shows that the velocity increases from the entry to
the exit boundary.At the latter the vertical velocity component is zero
which agrees with the assumption of the exit boundary be normal to the
streamlines. From Fig. 5c,it can be seen that the strain rate is the
highest at the rolls gap entrance and diminishes towards the exit. The
stress distributions, Figs. 5e,f,make clear the "friction hill" effect.

320

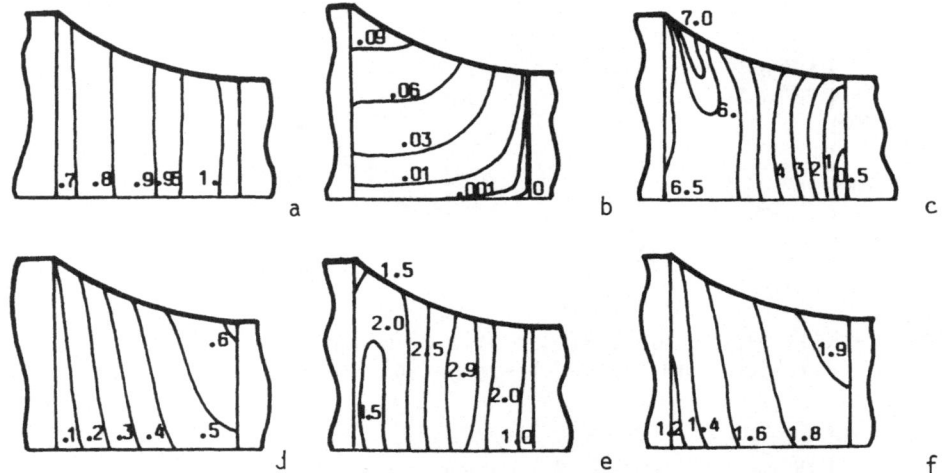

Fig 5. - Distributions produced by the model for Aluminium case 4.
5a) v_T/v_{exit} 5b) v_r/v_{exit} 5c) $\dot{\bar{\varepsilon}}$ 5d) $\bar{\varepsilon}$ 5e) σ_r/Y_0 5f) $\bar{\sigma}/Y_0$

7. Conclusions

An upper bound solution for plane strain rolling has been described. The velocity and strain rate fields are derived from a flow function based in prescribed streamlines in the plastically deformation region. The stress distribution is obtained by using these fields in conjunction with the weighted residuals method. The theoretical results follow closely those obtained by the finite element method, and simulates satisfactorily the rolling process. The major advantage, is that it provides to the user a complete information about all the process parameters, with small computing time (2 minutes in a VAX 11-750).

8. Acknowledgement

The authors wish to express thanks for the financial support provided by Junta Nacional de Investigaçăo Cientifica e Tecnologica,of Portugal

9. References

[1] - Kobayashi, S., and Li, Guo-Ji,"Rigid Plastic F. Element Analysis of Plane Strain Rolling", J. Engr. for Industry, 1982, p.55-64.
[2] - Lenard, J.G.,and Hwm,"The effect of Roll Deformation and Interfacial Friction on Roll Pressures in Cold Flat Rolling - a Finite Element Study",3th. Int. Conf. Struct. Anal. Systems "SAS", 1987.
[3] - Al Salehi, F.A. et al,"An Experimental Determination for the Roll Pressure Distribution in Cold Rolling", Int. J. Mech. Sci. 1973.
[4] - Marques, M.J.B., "Analysis of Tube Extrusion by a Flow Function", Int. Conf. Computational Plasticity, Barcelona 1987.
[5] - Marques, M.J.B.,"Det. of the Stress Field in the Hydros. Extr. of Tubes using the M.W.R",Int. Conf. Num. Met. For. Proc. Swansea 83

METALLOGRAPHIC VERIFICATION OF COMPUTER MODELLING OF HOT ROLLING

J H Beynon[1], A R S Ponter[1] and C M Sellars[2]
1   Department of Engineering, University of Leicester, UK
2   Department of Metallurgy, University of Sheffield, UK

ABSTRACT: Numerical modelling of the hot rolling of metals under industrial conditions has been conducted using a combination of techniques. Detailed experimental data have been used to predict microstructural changes in 316L stainless steel. These changes are shown to be highly sensitive to strain, strain rate and temperature and provide a valuable means of validating thermomechanical models.

1. Introduction

Traditional ways of validating numerical computations are by comparing computed and measured shape changes, or by comparing the forces necessary to effect those shape changes with the predictions. Computed temperature fields are also sometimes verified using isolated thermocouple or pyrometer readings from the laboratory or factory.

In hot working operations the microstructural changes within the deformed metal are also of vital importance. These changes are significant in two areas: firstly, having accurate flow stress data for the computation of metal flow and operating loads as work hardening and softening processes compete. Secondly, the microstructure which evolves largely determines subsequent room temperature properties; here the grain size is paramount. The metallurgical events of particular relevance to grain size are recrystallisation and grain growth. The kinetics and outcome of these microstructural events are highly sensitive to the thermomechanical treatment experienced during hot working.

2. Computational Techniques

The hot rolling process which has been modelled is the steady state rolling of flat wide strip. The initial gripping of the front end of the strip by the rolls, and the eventual departure of the end of the strip from the roll gap are not included. The geometrical constraint ensures plane strain conditions, and the considerable simplification to a two dimensional problem. The consideration of steady state rolling

J. L. Chenot and E. Oñate (eds.), Modelling of Metal Forming Processes, 321–328.
© 1988 by Kluwer Academic Publishers.

allows the use of the Eulerian approach whereby the metal flows through the stationary finite element mesh. Linear triangular elements are used together with a Petrov-Galerkin formulation to produce a directly coupled thermo-mechanical solution[1,2]. The computations described in this paper used a rigid, viscoplastic stress-strain model for the rolling stock. The present work assumes sticking friction between the stock and roll, though other friction conditions are available.

In multiple pass hot rolling, the periods between deformations are extremely important for the development of microstructure. During these periods, the temperature of the stock is falling due to cooling by radiation and convection, and the metal may recrystallise and even undergo grain growth. For the interstand computations, a finite difference program has been developed[3,4,5]. This program can also calculate surface oxidation and rapid cooling due to water sprays.

3. Materials Data

Thermomechanical modelling requires a detailed knowledge of microstructural and physical properties of the metal being rolled. Measurements of surface cooling rates in air and water have been carried out to calibrate the computations for interstand cooling. Laboratory rolling trials have been used to determine heat transfer coefficients for the cooling of the hot stock by the rolls.

Recovery, recrystallisation and grain growth kinetics have been determined as functions of hot working conditions using a simulator. This is a computer-controlled, servo-hydraulic, high-speed compression testing machine capable of accurately reproducing complex hot working sequences. Microstructural data have been obtained for a range of metals: aluminium and aluminium-1% magnesium[6], plain carbon and low alloy steels[7], and 304 and 316L stainless steels[8,9].

4. Example

The hot rolling of 316L stainless steel has been modelled for comparison with laboratory rolling results. Austenitic stainless steel was chosen for ease of metallographic observation. The alloy composition is (weight %): 0.024 C, 16.70 Cr, 12.20 Ni, 2.63 Mo, 1.50 Mn, 0.29 Si, 390 ppm N. It was reheated prior to rolling to 1160^0C to give an initial uniform grain size of 200 µm. It was rolled from a starting thickness of 25mm on a laboratory two-high mill in three passes, each of 25 % reduction with 15 s between removal from the furnace and the first pass, and between passes. The diameter of the rolls was 139 mm and the surface roll speed was 222 mm/s.

5. Microstructural Processes

Barbosa[9] determined the kinetics and products of static recrystall-isation and grain growth. The relevant equations for 316L stainless steel are those describing recrystallisation, since grain growth is slow under these conditions. Time for 50% recrystallisation ($t_{0.5}$, s):

$$t_{0.5} = 4 \times 10^{-15} \ Z^{-0.38} \ \varepsilon^{-3.6} \ d_0^{1.33} \ \exp(475000/RT) \qquad \text{for } \varepsilon \leq \varepsilon^*$$

where ε^* is the strain for the onset of dynamic recrystallisation - ie, recrystallisation concurrent with deformation - but is not attained for the conditions considered in this paper. ε is the von Mises effective strain, d_0 is the grain size of the metal before recrystallisation (μm), T is temperature after deformation (K), and R is the Gas constant (8.31 J/Kmol). The Zener Hollomon parameter is given by:

$$Z = \varepsilon' \ \exp(460000/RT')$$

where ε' is strain rate (/s) and T' is the deformation temperature (K). The fraction recrystallised (X) follows an Avrami type equation:

$$X = 1 - \exp[-0.693(t/t_{0.5})^k]$$

where k is a constant here found to be 0.6, and t is the time for the fraction X to recrystallise. The resulting recrystallised grain size (d_{rex}, μm) is given by:

$$d_{rex} = 470 \ d_0^{0.3} \ Z^{-0.1} \ \varepsilon^{-1}$$

This recrystallised grain size is that achieved when recrystallisation is complete. When recrystallisation is only partially complete, the current size of the recrystallised grains, d_r, is given by:

$$d_r = X^{1/3} \ d_{rex}$$

and the resulting mean grain size for the microstructure is

$$d = X^{4/3} \ d_{rex} + (1-X)^2 \ d_0$$

6. Computed Predictions

Figure 1 shows the outcome of the deformation imposed in the first pass as computed using the finite element program. The temperature on entry to the first pass showed a parabolic fall towards the surface to be expected from air cooling of the stock on leaving the reheating furnace. Most of the work of deformation goes into heating the metal, as is revealed at the centre of the stock. This temperature rise is obliterated nearer the surface by the chilling effect of the rolls. The strain imposed on the stock also varies with thickness. Around the centre of the stock, well away from the surface, the deformation is close to the overall von Mises effective strain for the deformation (0.33). Nearer the surface the strain increases as the rolls impose constraints on the flow of metal. This does not peak at the surface, however, since the rolls also chill the metal at the surface and this

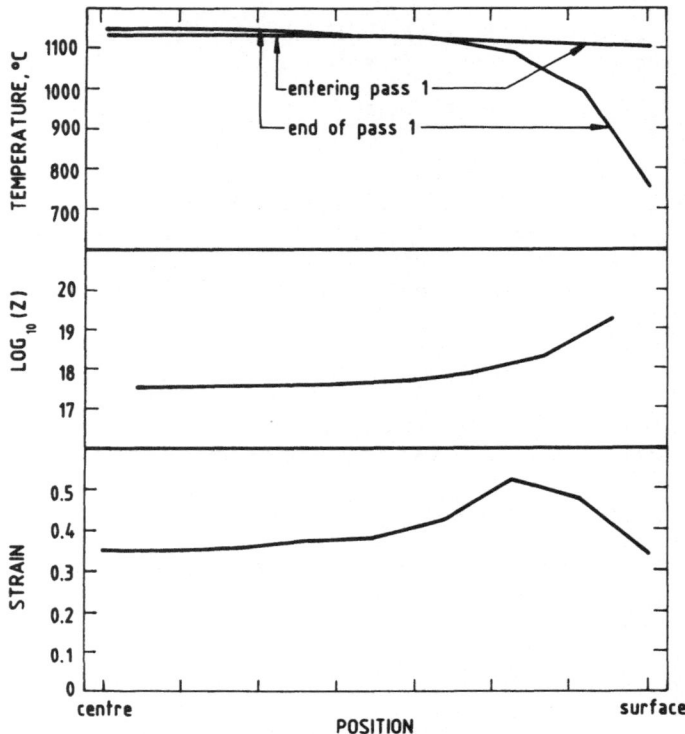

Figure 1: State of 316L stainless steel rolled stock on leaving the first pass of 25% reduction. "Z" is the Zener Hollomon parameter.

raises the flow stress, making deformation more difficult than just below the surface. The strain rate experienced by the stock during the rolling pass is also highly non-uniform and this is reflected in the non-uniform distribution of the Zener Hollomon parameter.

These locally varying values of temperature, strain and Zener Hollomon parameter all contribute to the development of microstructure during the 15 s interval before the next deformation. This development is computed with full allowance for the change in temperature as the stock surface regains some heat from the stock centre despite the normal surface cooling effects of convection and radiation.

Figure 2 shows the state of the metal just before entry to the second rolling pass. The temperature changes during the 15s interpass period are shown according to position through the thickness. Recrystallisation has occurred but is incomplete, as shown by the fraction recrystallised. The recrystallised grains have not yet reached their final size and the size of the unrecrystallised grains is changing as the new grains eat away the boundaries of the old. There is clear variation of these parameters through the thickness: a combination of the effects of varying temperature, strain and strain rate during the previous deformation.

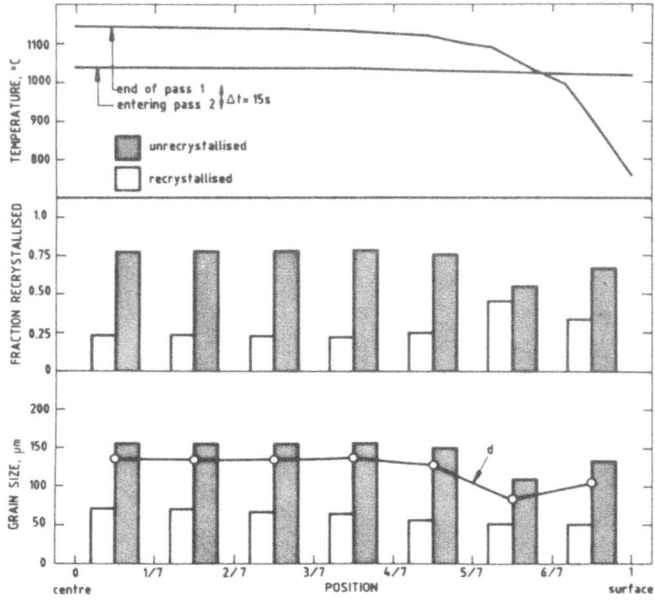

Figure 2:   Predicted state of rolled 316L stainless steel stock at start of pass 2 following a 25% reduction in pass 1 which ended 15s earlier. The original grain size was 200 µm.

The second deformation produces a similar pattern of inhomogeneity. Following 15 s of interpass cooling, recrystallisation is again incomplete with those parts which recrystallised before the second pass having behaved differently from those parts which still had residual strain from the first pass. The third deformation and subsequent 15 s interpass period resulted in no further recrystallisation.

7. Metallographic Observations

Several pieces of steel were rolled, each being taken a little further through the three-pass sequence before being quenched to "freeze" the microstructural state of the metal. Measurements of grain size have been extracted from earlier work[9].

Metallography revealed two populations of grain size just prior to the second pass, figure 3, as predicted in figure 2. The mean grain size measured half-way between the surface and centre just before the second pass was 135 ±15 µm, which compares well with the computed prediction of 137, figure 2. Just before the third pass, although four populations of grains were predicted, the two smallest should be indistinguishable as they are virtually the same size; furthermore, the amount of each is small. Three populations of grains were observed[9].

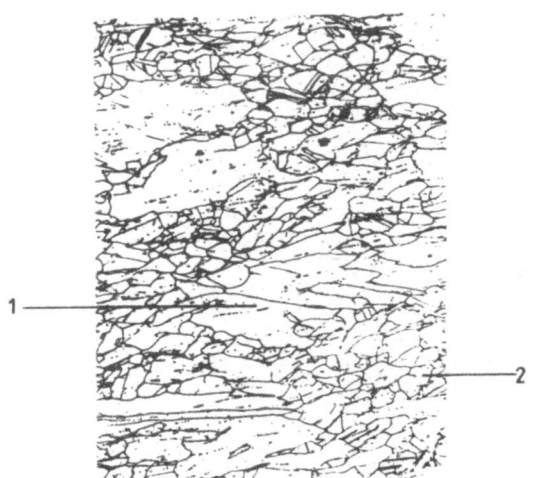

Figure 3: Observed micro-
structure in 316L stainless
steel half-way between the
surface and centre after a
25% reduction in pass 1
which ended 15s earlier.
"1" and "2" indicate
unrecrystallised and
recrystallised grains,
respectively. Mean grain
size: 135 ± 15 μm.

8. Discussion

The operation being modelled comprises a complex sequence of events,
each one of which demands careful treatment. The sensitivity of the
microstructural changes to the operating parameters can be gauged from
figure 4. Figure 4a shows the fraction recrystallised, X, 15s after
the end of the first pass as a function of strain in the pass. The
mean size, d, of the resulting mixture of recrystallised and unrecrystallised grains is also shown. The base values of temperature and Zener
Hollomon parameter, Z, used for this calculation were those from the
centre and were used together with an initial grain size of 200 μm.
Figure 4b shows X and d as functions of the temperature of the interpass period; (this is an equivalent value since account must be taken
of the non-steady temperature during this period). Figure 4c shows the
effect of variations in Z on the recrystallisation characteristics.
Each figure is marked with the conditions prevailing at the centre, C,
and at the peak strain, P. The latter was associated with the largest
fraction recrystallised and the smallest mean grain size, figure 2.
    Figure 4a shows the strong sensitivity of recrystallisation to
strain. The centre strain was computed to be 0.35. Maintaining the
same Z and interpass hold conditions, and just varying the strain to
0.52, the fraction recrystallised, X, would be expected to double from
0.22 to 0.44; and the mean grain size, d, to fall from 137 to 90 μm.
The values computed for the peak region, where the strain was 0.52, but
taking full account of local Z and interpass conditions, are very
close: 0.45 for X and 83 for d.
    X and d are also very sensitive to interpass temperature and to
Z, figures 4b and 4c. However, the higher Z at the peak, arising
partly from a lower deformation temperature, is counteracted by the
subsequent lower interpass temperature, and the net effect is small for
both X and d. Thus, in the example presented in this paper, it is the

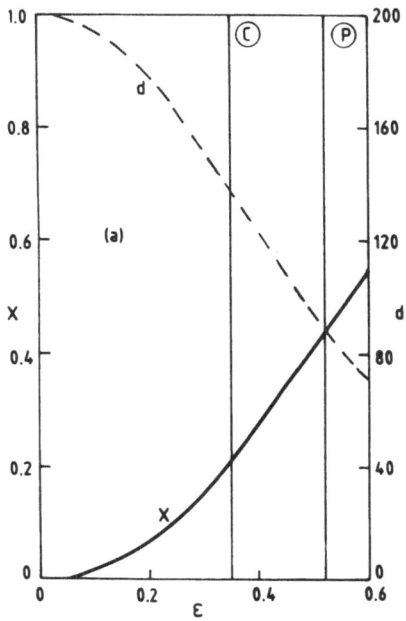

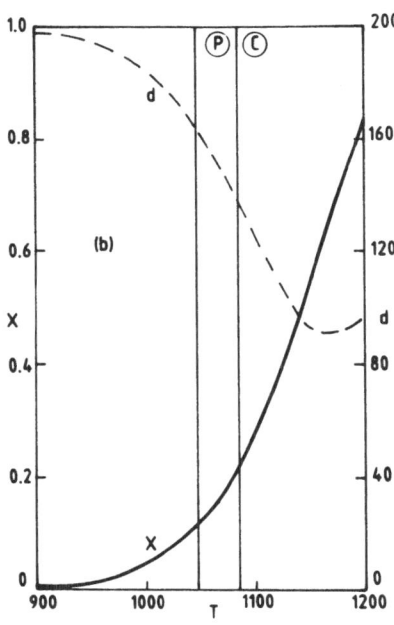

Figure 4: The sensitivity of
fraction recrystallised and
mean grain size following a
15s interpass period to:
(a) strain imposed in the
previous rolling deformation;
(b) equivalent temperature of
the 15s interpass period;
(c) Zener Hollomon parameter
for the previous deformation.
　　Values of strain,
interpass temperature, and
Zener Hollomon parameter from
the centre were used as the
basis for the calculations.
　　"C" and "P" indicate
conditions prevailing at the
centre and at the peak strain
(figure 1), respectively.

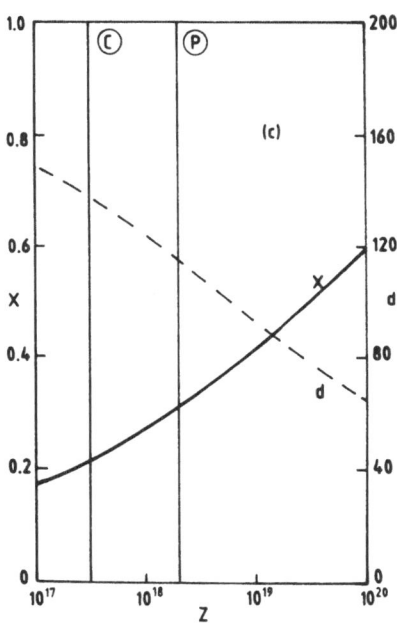

accuracy of the computed strain which is most critical.

The calculation of loads and torques in multiple pass rolling requires that allowance be made for the current state of the metal. The present model, with the assumption of rigid, zero work hardening, visoplastic behaviour, cannot take account of the mixture of recrystallised - and hence relatively "soft" - and unrecrystallised - "hard" - phases. Work is underway to introduce work hardening and work softening functions into the program to allow dynamic recovery and dynamic recrystallisation events to be modelled, together with full account being taken of the microstructural state of the metal as it enters each pass. This should not only improve the accuracy of the description of the state of the metal as it leaves each pass, but also allow more accurate computation of metal flow and hence of the rolling loads and torques required to perform the deformation.

Despite the simplifying assumptions of rigid viscoplastic flow behaviour and sticking friction, the results are most encouraging. Good data from relevant experiments have meant that relatively basic formulae for material behaviour have been used to good effect. The fraction recrystallised and particularly the resulting mean grain size can be measured with reasonable ease. Not only are these micro-structural events of major technological importance, but they also provide a highly sensitive means of validating thermomechanical models.

9. References

1. J H Beynon, P R Brown, S I Mizban, A R S Ponter and C M Sellars "Inclusion of metallurgical development in the modelling of indus-trial hot rolling of metals" from 'Numiform 86', Eds. K Mattiasson et al, A A Balkema (1986) 213-218.
2. J H Beynon, P R Brown, S I Mizban, A R S Ponter and C M Sellars "An Eulerian finite element method for the thermal and visco plastic deformation of metals in industrial hot rolling" from 'Computational methods for predicting material processing defects', Ed. M Predeleanu, Elsevier (1987) 19-28.
3. R A Harding, PhD Thesis, Sheffield University (1976).
4. C M Sellars and J A Whiteman, Metal Science, 13 (1979) 187-194.
5. C M Sellars, Materials Science and Technology, 1 (1985) 325-332.
6. E S Puchi, J H Beynon and C M Sellars "Simulation of hot rolling operations on commercial aluminium alloys" from 'THERMEC-88: Inter-national Conference on Physical Metallurgy of Thermomechanical Pro-cessing of Steels and Other Metals' Tokyo, Japan, 6-10 June 1988.
7. C M Sellars "The physical metallurgy of hot working" from 'Hot working and forming processes', Eds. C M Sellars and G J Davies, Metals Society, London (1980) 3-15.
8. D R Barraclough and C M Sellars, Metal Science, 13 (1979) 257-267.
9. R A N M Barbosa, PhD Thesis, Sheffield University (1983).

ACKNOWLEDGEMENT: The support of the Science and Engineering Research Council is gratefully acknowledged.

FINITE ELEMENT ANALYSIS OF TWO-ROLL HOT PIERCING

P. HRYCAJ[1], D. LOCHEGNIES[1], J. OUDIN[1], J.C. GELIN[2], Y. RAVALARD[1]
[1]Laboratoire de Génie Mécanique, GRECO Grandes Déformations et Endommagement, MECAMAT, Université de Valenciennes et du Hainaut Cambrésis, 59326 Valenciennes
[2]Laboratoire de Mécanique Appliquée, Université de Franche-Comté, 25030 Besançon

1.Summary

In two-roll piercing mill, the plastic velocity components are mainly located in the meridian planes of the billet and remain steady during the major duration of the process. To improve a more efficient control and achieve objective data banks, a two-dimension steady state finite element simulation has been performed

Considering an isotropic thermo-viscoplastic constitutive equation available for steel hot forging, in which consistency depends on temperature and strain, equivalent strain rate, stress, adiabatic temperature increase and heating rate distributions have been predicted by using an eulerian formulation implemented in ASTRID, the finite element software developed by the L.G.M. of Valenciennes.

The results show that the roll piercing process of seamless tubes is highly non-homogeneous, especially in the deformed zone located at the vicinity of the plug's nose. Strain rates, stresses and temperature increases are analysed in regard to different billet strain stress curves, initial heating temperatures and dimensions of the incipient central cavity which takes place in front of an ogival piercing plug.

2. Introduction

Roll piercing is the first forging step which is involved for the speedy transformation of a cylindrical billet into a seamless tube. In a two-roll (or three-roll) piercing mill, the billet is dragged by two (or three) rotating cone-shaped rolls which are inclined to the mill axis (Fig. 1). As soon as the billet rotates and moves forward into the rolls, a central cavity appears in the axis vicinity and an axial ogival plug is placed here to begin the tube forming. Most reports analyse the process parameters influence on roll force, roll torque and axial plug load [1, 2, 3] and some others the non-homogeneity of the flow by using experimental relations [4].

Successful competition in the seamless tubes business demands constant improvement of the process. For that objective, a finite element approach of the roll piercing process has been achieved in using a thermo-viscoplastic constitutive equation. The kinematic conditions at the billet interface have been expressed ; equivalent strain rate, stress distribution and temperature increase have been predicted. To illustrate the interest of the model, the influence of three specific parameters, grade of chomium in billet steel, piercing temperature, geometry of the central cavity, is discussed.

329

J. L. Chenot and E. Oñate (eds.), Modelling of Metal Forming Processes, 329–336.
© *1988 by Kluwer Academic Publishers.*

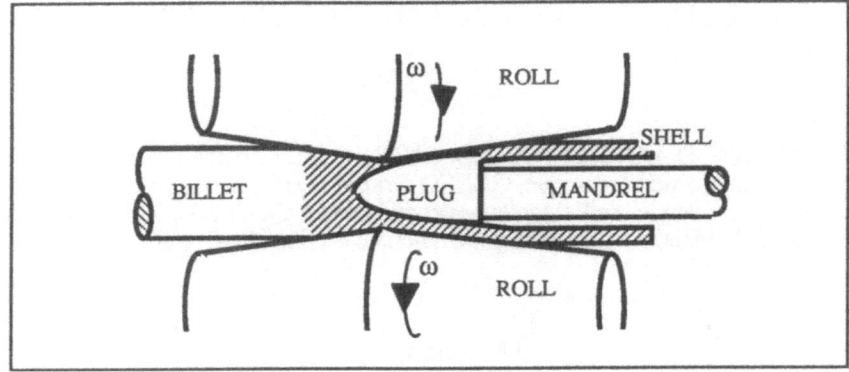

FIG. 1. Schematic view of an horizontal two-roll piercing mill.

3. MECHANICAL FORMULATION

3.1. MATERIAL BEHAVIOUR

In hot steel forging, an isotropic and viscoplastic law is well fitted. In that law, the deviatoric Cauchy stress tensor s is related to the viscoplastic potential W [5]

$$s = \frac{\partial W}{\partial D} \quad \text{with} \quad W = W \left(d + \frac{1}{3}\dot{e}I\right) = \frac{C\left(\theta, \bar{\varepsilon}^{p}\right)}{m+1}\left(\dot{\bar{\varepsilon}}^{p}\right)^{m+1}$$

in which D is the strain rate tensor, d its deviator , $\dot{e}$ the dilatancy rate, θ is the temperature, $\bar{\varepsilon}^{p}$ the equivalent strain, $\dot{\bar{\varepsilon}}^{p}$ the equivalent strain rate, $C\left(\theta, \bar{\varepsilon}^{p}\right)$ is the billet consistency which is temperature dependent and strain sensitive and m the isotropic strain rate coefficient of the billet material.

3.2. TEMPERATURE INCREASING

The adiabatic temperature increase is computed from :

$$\dot{\theta} = \frac{\kappa}{\rho C_{v}} C\left(\theta, \bar{\varepsilon}^{p}\right)\left(\dot{\bar{\varepsilon}}^{p}\right)^{m+1}$$

with κ dissipation coefficient which values are in the range 0.9 to 1.0, ρ billet density, C_{v} specific heat and $\dot{\theta}$ time derivative of the temperature.

The equivalent viscoplastic strain and the temperature are obtained by fully implicit formulations.

3.3. VARIATIONAL PROBLEM

Let Ω be the volume of the billet, Γ its boundary, Γ_t the billet-roll and billet-plug interface where the stress vectors are prescribed and Γ_v the interface part on which the velocities are prescribed. The finite element approach derives from the following functional :

$$G^o(v, p, \dot{e}) = \int_\Omega W\,(d + \frac{1}{3}\,\dot{e}\,I)\ d\Omega + \int_\Omega p\,(\,\text{tr}\,\nabla v - \dot{e}\,)\ d\Omega - \int_{\Gamma_t} t\ v\ d\Gamma$$

where p is the hydrostatic pressure.
The derivation of the functional G^o with respect to v, p and $\dot{e}$ leads to the finite element discretization. The non linear system is solved by Newton-Raphson method.

3.4. VELOCITY FIELD PRESCRIBED ON THE ROLL-BILLET INTERFACE

In hot piercing, slip usually occurs on the whole billet-rolls interface and the axial sliding velocity keeps the same direction from entry to exit of the mill.
    The axial component of the dragging velocity is expressed as

$$V_L(z) = \frac{\omega \sin \alpha}{2}\left(2\,R_m - |\,z\,|\frac{\text{tg}\,\beta}{\cos\alpha}\,(2 - \text{tg}^2\alpha\,)\right)$$

where ω is the rolls rotation speed, α the feed angle, β the rolls faces angle and R_m the maximum radius of the cone-shaped rolls. The mean axial velocity V_D of the billet is related to its axial entry velocity V_E

$$V_D(z)\,A(z) = V_E\,A_E.$$

A_E the initial cross section of the billet and A(z) the cross section at coordinate z. The apparent section $A_{ou}(z)$ model is given in Fig. 2 and is compared to experimental values.
    The axial component of the billet velocity at the rolls interface stands between V_L and V_D. When considering the longer piercing time occurence, a convenient model for the velocity boundary conditions is to linearize the prescribed axial velocity V_I, first for the increasing zone from V_E to V_M (which is the mean axial velocity in the z_m cross section) and second for the decreasing zone from V_M to V_S (Fig. 3).

4. FINITE ELEMENT MODEL

The finite element simulation of the steady state piercing of a D3 alloyed steel at 1200 °C has been achevied by using a 250 isoparametric element mesh with 306 nodes (Fig. 4) and the computing was realized with ASTRID software developed by the L.G.M of Valenciennes.
    The convenient strain stress relationship was the following [6] :

$$\sigma_0 = 10.26\ \dot{\varepsilon}^{\,0.121}\qquad \text{daN.mm}^{-2}$$

332

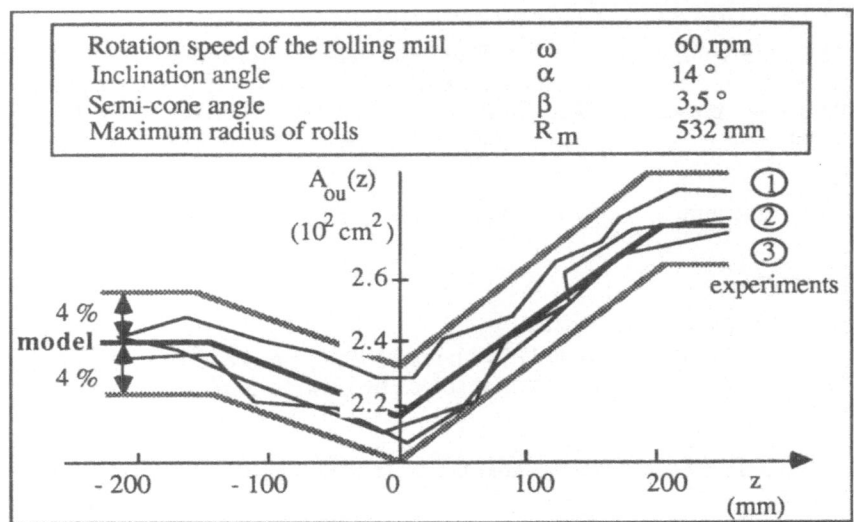

Rotation speed of the rolling mill	ω	60 rpm
Inclination angle	α	14 °
Semi-cone angle	β	3,5 °
Maximum radius of rolls	R_m	532 mm

FIG. 2. Apparent section evolution of the billet : model, measurements on pierced alloyed steels and including limits.

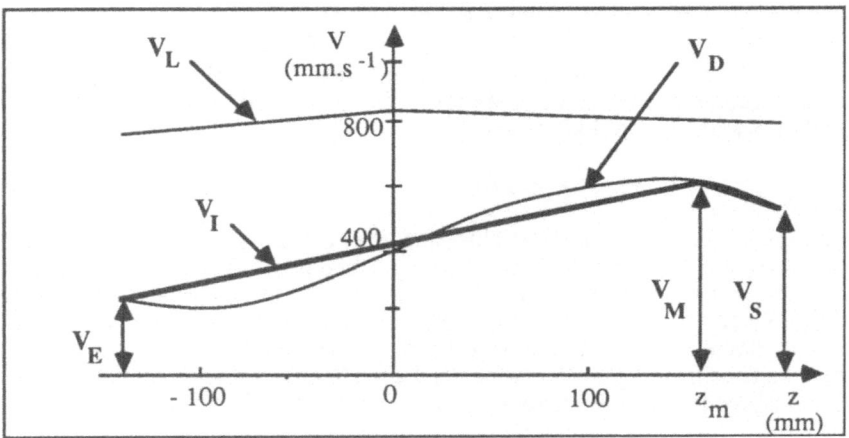

FIG. 3. Distribution of axial dragging velocity V_L, mean axial billet velocity V_D corresponding to parameters of Fig. 2 and V_I prescribed limiting velocity.

The incipient central cavity was limited by a conical surface which diameter at the plug interface equals to 10 mm and height equals to 27 mm.

Equivalent strain rates are found to be highest in the vicinity of the plug's nose-billet interface (Fig. 5), near the central cavity apex while axial stresses σ_{zz} are maximum at coordinate z_m which corresponds to the minimum cross section of the tube (Fig. 6). It is interesting to point out that around the nose of the piercing plug, σ_{zz} are compressive stresses and reach their maximum intensity.

The temperature increase map represented in Fig. 7 shows that the maximum is also observed near the plug's nose-billet interface.

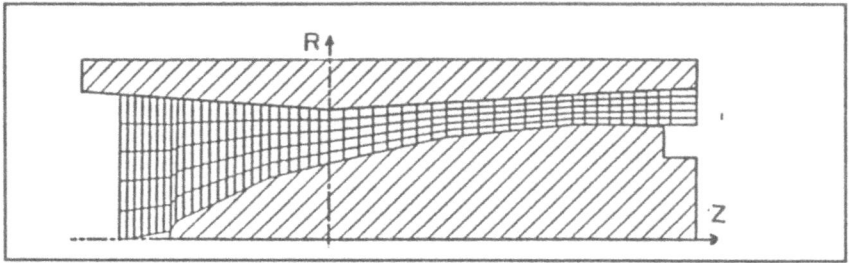

FIG. 4. Steady two-roll piercing simulation : representation of the mesh.

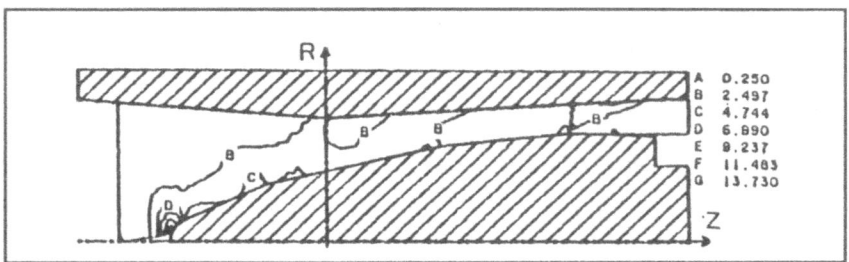

FIG. 5. Steady two-roll piercing simulation of D3 steel billet at 1200°C : equivalent strain rate distribution.

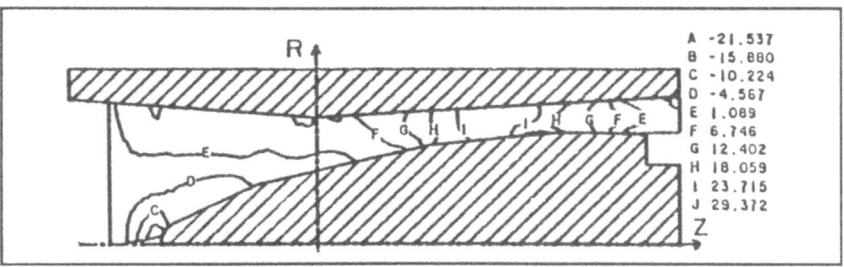

FIG. 6. Steady two-roll piercing simulation of D3 steel billet at 1200°C : axial stress distribution in daN.mm^{-2}.

These results are obviously sensitive to the rheology of the billet ; that is why it is interesting to assess the influence of rheological parameters. As Chromium is more and more commonly used in tube making, the influence of the percentage of chromium in steels has been analysed in terms of the rheological parameters C and m.

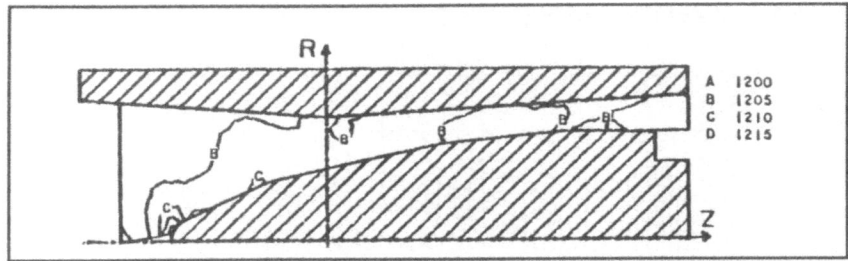

FIG. 7. Steady two-roll piercing simulation of D3 steel billet at 1200°C : temperature increase at time t equal to 3 s.

4.1. INFLUENCE OF A CHROMIUM GRADE IN STEELS

Figs. 8 show the evolution of heating rate and stress at the vicinity of the piercing plug's nose when using steels with percentage of chromium in the range from 1 % to 18 %. The heating rate increases from 200 to 650°C.s^{-1} with chromium percentage from 1% to 13% then it decreases a little bit. The compressive stresses σ_{rr} and σ_{zz} are all the higher as the percentage of chromium is high. These results may explain some plug wear occurences.

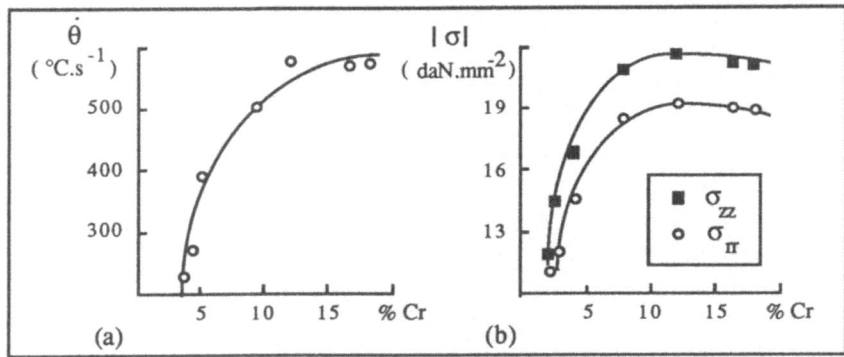

FIG. 8. Steady roll piercing simulation of chromium alloyed steels at 1200 °C :
(a) billet heating rate in the plug nose vicinity versus chromium percentage ;
(b) axial and radial stress in the plug nose vicinity versus chromium percentage.

4.2. INFLUENCE OF THE INITIAL PIERCING TEMPERATURE

Figs. 9 show the evolution of the heating rate and of the stresses when piercing 52100 steel billets with initial temperatures in the range from 950°C to 1150°C and with the same central cavity occurence in the form of a ø 10 mm x h 27 mm cone. 50 % decreases of heating rates and compressive stresses are predicted when the initial temperature is changing from 950 °C to 1150 °C .

4.3. INFLUENCE OF THE GEOMETRY OF THE CENTRAL CAVITY

Six typical dimensions are considered : (1) very little cavity, (2) ø 10 mm-h 27 mm, (3) ø 27 mm-h 27 mm, (4) ø 27 mm-h 9 mm, (5) ø 10 mm-h 9 mm and (6) ø 10 mm-h 3 mm (Fig. 10). The plug nose radius was equal to 15 mm. Compressive σ_{rr} stress increases when only the cavity diameter increases and compressive stress σ_{zz} increases when only the cavity length decreases. A very little cavity as considered in case (1) produces the highest values of heating rate and compression stress σ_{zz}.

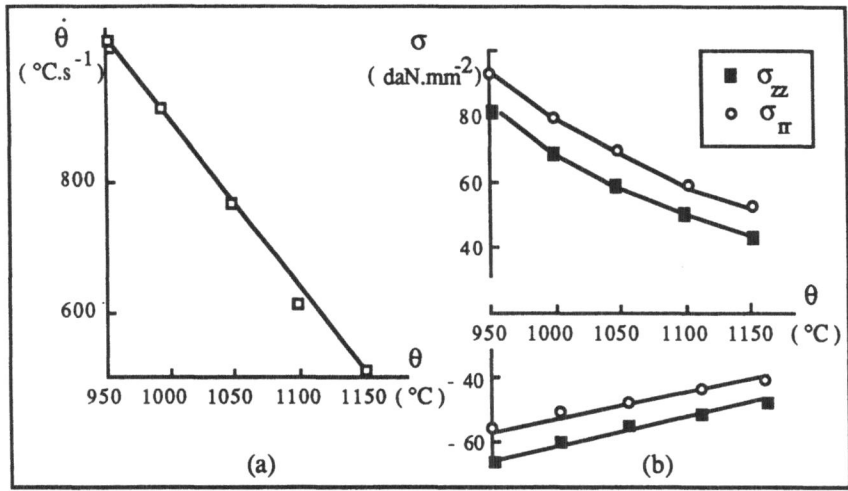

FIG. 9. Steady two-roll piercing of 52100 steel billets :
   (a) evolution of heating rate versus initial billet temperature ;
   (b) evolution of the extremum values of axial and radial stress versus billet temperature.

cone shaped central cavity	(1)	(2)	(3)	(4)	(5)	(6)
radius-length	0-0	5-27	13.5-27	13.5-9	5-9	3-5
$\dot{\theta}$ (°C.s^{-1})	562	250	390	386	380	380
σ_{rr} (daN.mm^{-2})	-11.6	-9.94	-12.2	-12.5	-12.7	-10.7
σ_{zz} (daN.mm^{-2})	-15.2	-12.25	-12.0	-12.8	-13.2	-14.4

FIG. 10. Steady two-roll piercing of 52100 steel billets at initial billet temperature of 1150 °C : billet heating rates and stresses at the vicinity of the plug nose versus typical cone shaped geometries of the central cavity.

336

5. Conclusions

Improvement, efficient control and analyses of seamless tubes manufacturing results could be obtained now by the extensive use of a new thermo-viscoplastic finite element model. The efficiency of the model is illustrated by predictions and analyses for specific aspects of the piercing :

(i) strain rate and stress distribution in the meridian planes of the billet to be pierced,

(ii) adiabatic temperature increase and related heating rate due to plastic deformation,

(iii) influence of the consistency and of the viscoplastic coefficient related to chromium percentage in the steel billet,

(iv) influence of the dimensions of the incipient cavity located in front of the plug's nose.

It appears that the finite element model developed in ASTRID software would be able to confirm and improve validation of a processing scheme obtained by other speedy approaches, such as the kinematic element method and in this way, it appears to be a powerful mean for secure management of piercing mills.

Acknowledgements

The authors would like to thank Nord Pas de Calais Region, C.N.R.S., Ministry of Research and Higher Education and Vallourec Industries for supports.

References

[1] MIZUNUMA S., KAWAHARADA M., YANAGIMOTO S., Analysis for predicting process parameters in cross rolling mills, Ro.M.P., 1985, pp. 329-340.

[2] ZOROWSKI C.F. , HOLBROOK R.L. , Influence of the mill set up on hollow geometry produced by rotary piercing, 8th Int . Mach . Tool Design and Research Conf., 1968, pp. 1041-1058.

[3] ERMAN E., The influence of the processing parameters on the performance of the two-roll piercing operation, J. of Mech. Working Technology, 1987, 198, pp. 167-179.

[4] BLAZINSKI T.Z., Geometry factors and inhomogeneity of flow in rotary piercing and cold drawing of seamless tubing, Int. J. Mech. Sci., 1979, Vol. 21, pp. 527-536.

[5] GELIN J.C., LOCHEGNIES D., Simulation numérique en grandes déformations plastiques de la mise en forme de tubes, Ecole d'Eté Matériaux Mise en Forme Pièces Formées, GRECO C.N.R.S. Grandes Déformations et Endommagement, G.I.S. Mise en forme, 1987, F-Oléron.

[6] ALTAN T., OH S., GEGEL H., Metal Forming, Fundamentals and Applications, ASM Series in Metal Processing, 1986.

Modelling the Mechanics of the Longitudinal Tube Rolling Process

Dr. K.Baines/Dr. I.M.Cole

Dept. of Mech. and Prod. Eng.

University of Aston

Aston Triangle

Birmingham

West Midlands, B4 7ET

England

Professor D.H.Sansome

Technoform-Sonics Ltd.

9 Enterprise Trading Estate

Pedmore Road

Brierley Hill

West Midlands, DY5 1TX

England

J. L. Chenot and E. Oñate (eds.), Modelling of Metal Forming Processes, 337–345.
© *1988 by Kluwer Academic Publishers.*

1. Introduction

The longitudinal rolling of tube through two or more grooved rolls is carried out in a series of roll stands. These mills generally form the last and an important sequence in the complex process of hot seamless tube manufacture.

A tube-reducing mill consists of a number of longitudinal stands, each containing a number of equi-oriented grooved rolls, presenting to the on going tube, the required pass shape. Successive sets of grooved rolls are arranged to have gradually decreasing pass size such that the initial diameter of the tube is reduced to the size finally required as it passes longitudinally down the mill. Two grooved rolls comprise the more common stand arrangement, but the inherently more efficient three-roll configuration is becoming more prevalent in modern tube reducing mills.

When a tube, which is unsupported internally, is rolled in a single stand without front or back tension or compression, the tube thickness and velocity often increase with the diametral reduction. Under such conditions, the thickening which takes place due to the purely compressive action of the grooved rolls, is said to take place under no-pull or sinking conditions. This condition arises in a multi-stand mill when increasing roll speeds at each successive stand are adjusted to produce neither tension nor compression in the lengths of tube between each stand. Furthermore, the whole tube length will thicken by an amount dependent upon the overall reduction that takes place and the diameter of the rolls.

If it is arranged for each stand to operate at a higher speed relative to the preceding stand, i.e. in excess of that which is required for the previous sinking condition, then tension in the tube is generated between the stands and the process becomes one of stretch reducing. Such an arrangement not only reduces the tube diameter but lessens tube thickening between the stands and the roll loads are reduced as a consequence of tension.

Another variant of the process is the mandrel mill. Here a mandrel is inserted in the tube to provide internal support to the bore whilst the tube wall is being deformed between the roll grooves and the mandrel at each roll stand. As a consequence of this internal support, a much greater reduction of the wall thickness is possible at each stand, resulting in a shorter mill, i.e. fewer stands. However, as a result of the use of a mandrel and a smaller number of stands, the final tube is larger in diameter and must be passed for further processing i.e. stretch reducing. The complete mandrel mill, often referred to as the continuous mill, is highly productive but the range of tube sizes it can produce is limited. Furthermore, for economic reasons, these mills are unsuitable for the manufacture of large tube sizes.

The three-roll arrangement of the process is widely used and its use is increasing since three-roll mills possess most of the advantages associated with the four-roll design. Thus, slip between the roll and tube is minimal, there is little marking of the tube and a greater draft per stand is obtained than with the conventional two-roll mill; also it is not as expensive as a four-roll mill.

Researches by Cole[1](1969), Haleem[2](1978) and Labib[3](1982) included detailed commentries on the paucity of sound analytical and experimental techniques relating to the tube rolling process. Thus work programmes were planned to place the mechanics of longitudinal tube rolling on a reliable foundation to assist mill operators to optimise the process. Optimisation of the process is defined as the production of the maximum reduction of area per stand consistent with acceptable tube quality. Thus the work programmes attempted to predict the performance of two- and three-roll longitudinal tube rolling mills with a minimum of further ad hoc experimental work. This approach also provided a more unified theoretical analysis to more general tube rolling conditions by establishing the minimum total work done per unit volume to deform the tube.

For the two-roll work Cole[1] employed a converted milling machine as a single stand two-roll mill, with pin loadcells incorporated in the groove of the upper roll. To simulate hot steel, lead tube was rolled in a circular groove which represented the finishing stand of a production mill. Cole conducted experiments to investigate the roll groove pressure distribution for sinking, stretch-reducing and mandrel rolling. This work also included a very comprehensive literature survey on tube rolling and of the methods of measurement of the roll groove contact pressures.

Haleem [2] continued Cole's [1] work, improving the performance of the pin loadcells and assessing the roll groove pressure for specific tube outside diameter to thickness ratios (D/t). Sinking and stretch-reducing conditions were investigated for the two-grooved roll arrangement. He formulated a new theoretical approach for the mean roll pressure based on the energy principle, noting a better correlation with experiment than had previously been displayed by the equilibrium analyses. Haleem also studied the neutral zone and the pre-contact or "free" plastic deformation zone relevant to the process parameters.

Cole's [1] and Haleem's [2] single stand mill was employed by Labib [3] for an investigation into the mandrel rolling of thin lead tube through two-grooved rolls. Rolling trials assessed the behaviour of the roll separating force (R.S.F.), roll torque and roll groove pressure distribution for specific tube D/t ratios and various conditions of the mandrel surface. Again pin loadcells were used to measure the pressure distribution around and along the roll groove, the latter result enabling the arc of contact to be established.

His results indicated a non-uniform pressure distribution and the existence of a free deformation zone i.e. tube deformation prior to the contact zone. Labib also noted that the roll loads and torques were greatly affected by changes in tube thickness and friction condition at the mandrel-tube interface. Furthermore, he noted the application of front and back tension to the tube decreased the roll loading; increasing the mandrel speed produced a decrease in the roll torque.

A new theoretical approach based on the energy principle was proposed and compared with existing equilibrium analyses. The total work done was then calculated and the tube and mandrel velocities assessed for application to a multi-stand mill. In general, good agreement was obtained between the theoretical predictions and experimental results.

In 1984 Baines[4] examined the longitudinal tube rolling process utilising the three-roll configuration. He designed and instrumented a complete experimental three-roll mill for the rolling of lead tube as an analogue material for hot steel. A novel type of roll loadcell was incorporated and its design and testing discussed. Employing three roll sizes of 170 mm, 255 mm and 340 mm shroud diameter, precise tube specimens of various tube diameter to thickness ratios were rolled under sinking and mandrel rolling conditions. Theoretical studies utilising the equilibrium and energy methods were applied to both conditions. Baines noted that. in general, the energy approach gave better comparison with experiment, especially for mandrel rolling. The influence of the tube deformation zones on the two processes and on the subsequent modification of the tube-roll contact length was observed.

2. Theoretical Analyses of the Sinking and Mandrel Rolling Processes

A comprehensive assessment of the mechanics of the longitudinal rolling of the tube through two-and three-grooved rolls was made by Cole [1]. He detailed a critical analysis of the available theoretical treatments for the sinking, stretch reducing and mandrel rolling processes. All the theories, primarily for the two-roll configuration, were based on the classical slab equilibrium approach, the majority of them being Russian in origin. Cole commented on the limitations of these approaches and suggested the energy or apparent strain method as an alternative technique for a more realistic analysis of the processes.

The main criticism of the equilibrium approach is the distortion of the assumed principal planes by the inevitable shear stresses. A friction stress will naturally produce a rotation of the longitudinal and circumferential principal planes. The thinner the tube wall thickness the greater this effect will have on the assumed principal planes in the main tube deformation zone away from the roll-tube interface.

The energy method was applied to the sinking and stretch reducing processes by Haleem [2] and to the mandrel rolling process by Labib [3]; in both cases the two-roll configuration was considered.

Haleem directed his attention toward the strain energy or apparent strain method of analysis for determining the mean roll pressure.

He stated that the work done per unit volume of rolled tube comprises three components:

1. Work of homogenous deformation W_h, which represents the minimum possible work required to change the tube shape.
2. Work done against friction W_f at the tube-roll interface.
3. Redundant work W_r required to shear the tube as it passes into and out of the deformation zone.

Labib also applied the energy method to mandrel rolling, noting that the work done per unit volume of the rolled tube was the sum of the following components:

1. Work of homogeneous deformation W_h.
2. Frictional work, which is the work done to overcome friction between tube and rolls and between tube and mandrel.
3. Redundant work, which is the work lost due to unnecessary internal shearing of the tube material produced by the constraints imposed on the material flow.
4. External work $W_{a,b}$ supplied by the applied front and back tensions.

Both Haleem and Labib continued by noting that the total external work done by the rolls per unit volume, as shown in Fig.1, W_{te} could be written as:

$$W_{te} = W_h + W_r + W_f$$

In the case of rolling with applied front and back tensions, the total work done per unit volume, W_T is:

$$W_T = W_{te} + W_{a,b}$$

where $W_{a,b}$ is the work done by the applied front and back tensions.

The apparent strain concept then yields:

$$\bar{\varepsilon}_a = \bar{\varepsilon}_h + \bar{\varepsilon}_f + \bar{\varepsilon}_r$$

and the total internal work of deformation W_{ti} is the integral of the true stress/strain curve between the limits of zero strain and $\bar{\varepsilon}_a$.

Thus:

$$W_{ti} = \bar{\sigma}_y \, \bar{\varepsilon}_a \quad \text{where } \bar{\sigma}_y \text{ is the mean yield stress of the material at strain } \bar{\varepsilon}_a.$$

Conservation of work gives: $W_{te} = W_{ti}$

Thus each apparent strain component of W_{te} is assessed and summed to give the total mean roll pressure p_m.

Work of homogeneous deformation W_h:

By definition and from measurement of the principal strains $\varepsilon_l, \varepsilon_r, \varepsilon_c$:

$$W_h = \bar{\sigma}_y \, \bar{\varepsilon}_h = \sqrt{2/3 \left\{ \varepsilon_l^2 + \varepsilon_r^2 + \varepsilon_c^2 \right\}}$$

Work done against friction W_f :

The mechanics of friction are complex and determination of the exact functional relationship between the frictional shear stress τ and the other variables is difficult. For metal forming without lubrication there are two types of friction:

1) Coulomb sliding friction, where $\tau = \mu p$

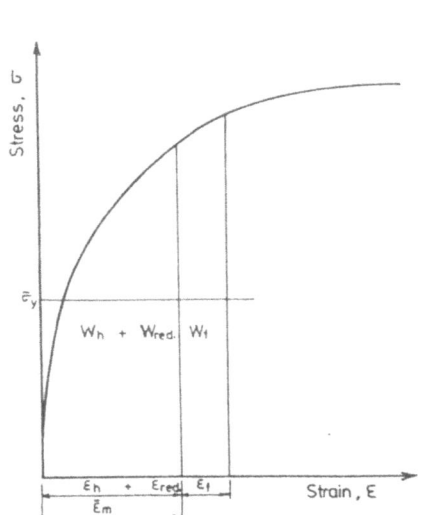

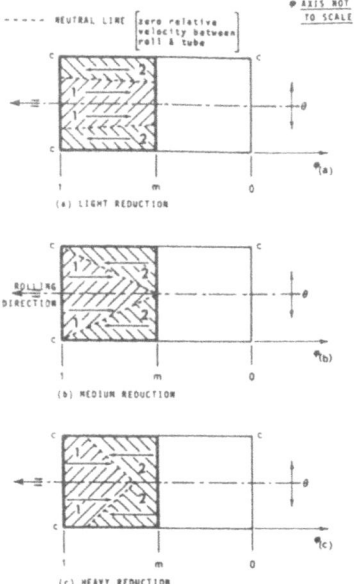

Figure 1.   Total work of Deformation

Figure 2.   Proposed deformation zones
and relative velocity modes (sinking)

2) constant shear yield stress friction, where $\tau = m \ \sigma_y/\sqrt{3}$ with

m for a given set of conditions is in the range $0 \leqslant m \leqslant 1$ i.e. from frictionless to full sticking conditions.

It has been shown by many workers that in hot rolling a combination of sliding and sticking friction is prevalent. Consequently both Haleem and Labib adopted the constant shear stress concept since only average roll pressures are required and friction loss calculations are considerably simplified.

The work done against friction per unit time is given by:

$$\dot{W}_f = \int \tau \, v_r \, ds = A_0 \, U_0 \, W_f$$

ds is the elemental surface of the contact.
v_r is the relative velocity between the two surfaces in contact i.e.

   roll/tube in tube sinking; roll/tube and tube/mandrel in mandrel rolling.
A_0 is the initial cross sectional area of the tube.
U_0 is the initial tube velocity.

Experimental evidence by Haleem and Labib suggested that the surface of contact, i.e. the friction interface between tube and roll groove can be divided into two equal zones.

However Haleem's assumed model for the relative roll-tube velocity direction over the contact surface as shown in Fig.2.(a) is only valid for small tube reductions, since it is implicit in his model that only a small change of roll radius along the neutral line (zero relative roll - tube velocity) is being considered. Baines [4] stated that, for constancy of volume, this percentage change in roll radius must approximate to the percentage change in tube cross-sectional area. Thus he proposed that better models for the behaviour of the neutral line for increasing tube reductions are those illustrated in Fig.2.(b) and (c).

Redundant work W_r:

Both Haleem and Labib evolved analyses which indicated minimum contribution to the total work done by the internal shearing of the tube material. However Haleem drew attention to the presence of a free zone, thus acknowledging the existence of circumferential bending and unbending of the tube as it passed from the free to the controlled deformation zone. Baines noted that this contribution to the redundant work is significant for thicker tubes and, by considering full plastic bending and unbending of the tube in these deformation zones as shown in Fig.3., estimated that W_r was proportional to the ratio: tube thickness/roll radius.

Total Mean Roll Pressure (p_m):

By assuming that the pressure distribution round the groove and along the arc of contact is uniform, the total mean roll pressure can be calculated by equating W_t to the work done by the externally applied roll pressure.

The total work done by the external roll pressure per unit volume is:

$$W_{te} = \int p_m \; s \; dt/V$$

where p_m = total mean roll uniform pressure,

  s = surface area of contact,
  V = rolled material volume = $A_o v_o$,
  dt = change in tube wall thickness.

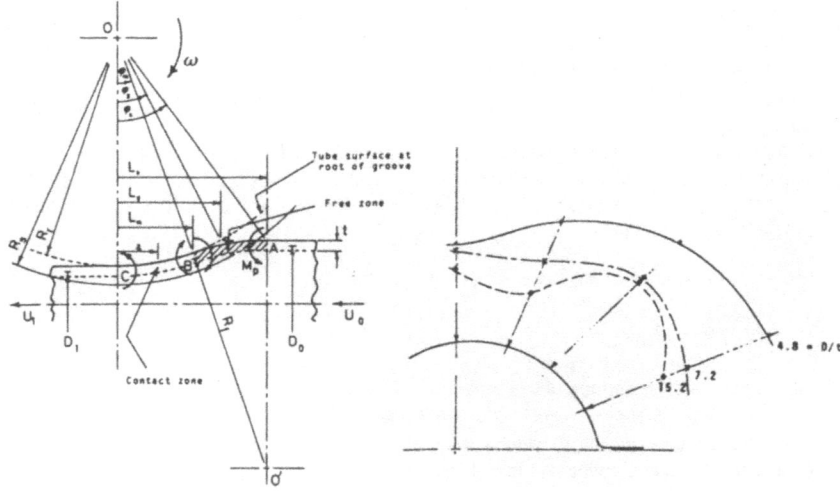

Figure 3. Effect of the free deformation   Figure 4. Distribution of the roll
zone on the tube-roll arc of contact   pressure round the groove (Haleem)

Equating the external work to the internal work i.e. $W_{te} = W_{ti}$, Haleem and Labib applied the apparent strain concept to the two-roll configuration to give the mean roll pressure(p_m) for tube sinking and mandrel rolling.

The apparent strain or energy method does not provide information on the distribution of the pressure round the roll groove, or along the arc of contact, only the mean roll pressure is evaluated; however this is quite sufficient for most practical applications.

Roll Separating Force (R.S.F.):
Both Haleem and Labib acknowledged that the product of the total mean roll pressure and the horizontal projection of the area of contact on each roll gave the roll separating force.

Rolling Torque (T):
To evaluate the rolling torque Haleem introduced the concept of the lever arm as used in flat rolling. Here the roll separating force is assumed to act at some distance from the roll axis. This distance, the lever arm, is generally expressed as a ratio of the length of the arc of contact. For hot flat rolling the ratio is usually assumed to be 0.5. Haleem then considered various definitions of the arc of contact length and for round to oval passes employed a mean value for the ratio of 0.66. A similar approach was utilised by Labib.

3. EXPERIMENTAL TUBE SINKING AND MANDREL ROLLING MILLS

To formulate the mechanics of longitudinal tube rolling two- and three-roll experimental mills were designed and instrumented [1],[2],[3] and [4], and prime consideration was given to the determination of the:
(i)   roll groove pressure distribution and roll separating force,
(ii)   roll torque,
(iii)   front and back tube tensions.

The above factors were then be related to the following variables:
(i)   ratio : roll diameter/tube diameter
(ii)   ratio : tube outside diameter/tube wall thickness (D/t),
(iii)   percentage reduction of the tube cross-sectional area,
(iv)   reduced tube strain distribution.

4. EXPERIMENTAL RESULTS

4.1. Haleem
   1. The deformation zone consists of two axial zones; a free zone where plastic
      deformation takes place prior to the tube contacting the rolls, and a contact zone.
   2. The free zone decreases the length of the arc of contact. The measured arc of contact
      was found to be about $2^{-0.5}$ of the calculated value in the absence of the free zone.
   3. Pin loadcells noted the occurrence of a pressure peak at or near the entry plane of
      the tube in the roll gap. Comprehensive roll pressure distribution curves were also
      discussed; Fig.4 is a typical distribution.
   4. Test results were presented for the rolling of oval tubes through oval grooves, a
      condition relevant to tube rolling practice.

4.2. Labib
   1. The R.S.F., rolling torque and mean roll pressure (Fig.5) increased with both
      reduction of area and D/t ratio. Thus higher reductions should be confined to the first
      stands in a production mill where the tube thickness is greatest.
   2. The pressure distribution round the roll groove perimeter was extremely irregular;
      the highest peak occurred at the root of the groove.

344

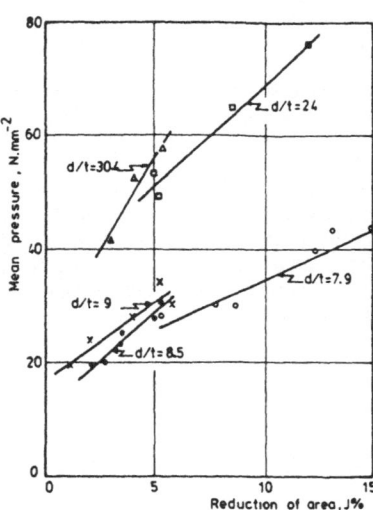

Figure 5. Effect of changing the reduction of area and D/t ratio on the mean roll pressure

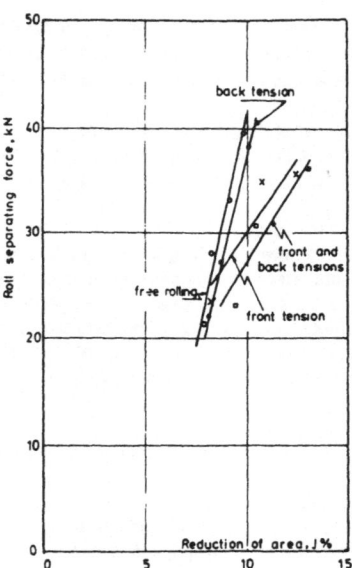

Figure 6. Effect of applying front and back tube tensions on the R.S.F.

3. Friction on the mandrel surfaces greatly influences the rolling loads and torques. The effect is even greater as the tube gets thinner. Increasing friction between tube and mandrel hinders the flow of the tube material in the longitudinal direction and results in lower reductions of area and higher rolling loads and torques.

4. In the absence of front and back tube tensions, the rolling torque varies linearly with the rolling load.

5. As shown in Figs.6 and 7 front and back tube tensions reduce the rolling load; also, as anticipated front tension reduces the rolling torque; back tension increases it. Thus higher reductions of area can be accommodated with tension without increasing the rolling loads.

6. A free zone, where deformation takes place prior to the tube contacting the rolls can occur in mandrel rolling. In this case the length of the arc of contact was higher than the theoretical length calculated from the geometry of the tube and the roll groove.

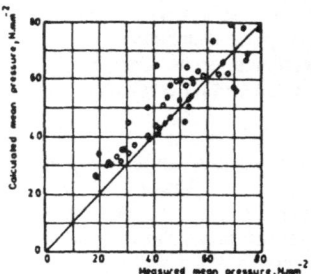

Figure 8a.   Measured and calculated mean roll pressure

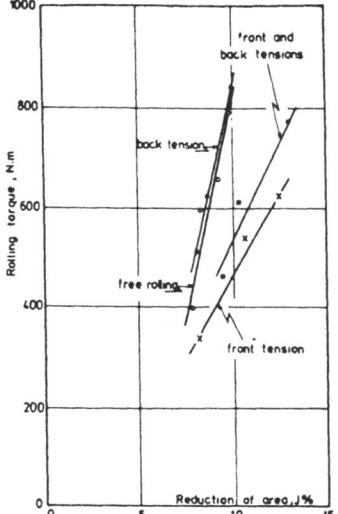

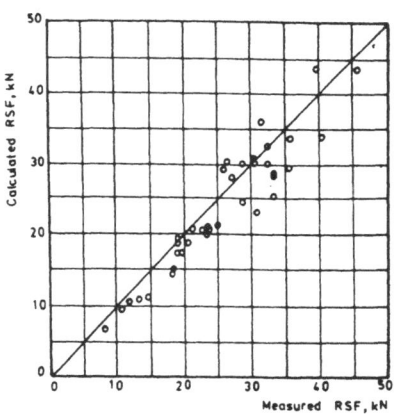

Figure 7.   Effect of applying front and back
tube tensions on the rolling torque

Figure 8b.   Measured and calculated R.S.F.

5. CONCLUSIONS

The analysis of the tube rolling processes is complex, especially mandrel rolling. Until recently all the theories involved the classical approach by considering the equilibrium of forces acting on an element of the tube in the deformation zone. They also included a yield criterion.   The resulting differential equations were then solved for the appropriate boundary conditions.   As commented upon by Cole [1], these theories, many semi-empirical, appear to be only of partial validation and alternative theoretical analyses utilising the strain energy concept were attempted, with some success, by Haleem [2] for sinking and by Labib [3], see Fig. 8, for mandrel rolling.

These new theoretical analyses, developed by Haleem and Labib and based on the strain energy of deformation, compare favourably with their test results and was an improvement on the equilibrium technique employed by many other workers. Labib also developed a computer programme to select the combination of rolling parameters to yield the minimum total work done per unit volume.

6. REFERENCES

1. Cole, I.M.
    "An investigation of the rolling of cylindrical tube by grooved rolls"
    Ph.D. thesis, 1969, Aston University, Birmingham, England.
2. Haleem, A.S.
    "An investigation into the longitudinal rolling of tubes through
    two-grooved rolls"
    Ph.D. thesis, 1978, Aston University, Birmingham, England.
3. Labib, O.M.
    "Mechanics of rolling tube on a mandrel through two-grooved rolls"
    Ph.D. thesis, 1982, Aston University, Birmingham, England.
4. Baines, K.
    "The mechanics of longitudinal tube rolling through three-grooved rolls"
    Ph.D. thesis, 1984, Aston University, Birmingham, England.